Rebels, Scholars, Explorers

REBELS, SCHOLARS, EXPLORERS

WOMEN IN VERTEBRATE PALEONTOLOGY

Annalisa Berta and Susan Turner

Johns Hopkins University Press

Baltimore

Johns Hopkins University Press
2715 North Charles Street
Baltimore, Maryland 21218-4363
www.press.jhu.edu

Library of Congress Cataloging-in-Publication Data

Names: Berta, Annalisa, author. | Turner, S. (Susan), 1946–, author.
Title: Rebels, scholars, explorers : women in vertebrate paleontology /
    Annalisa Berta and Susan Turner.
Description: Baltimore : Johns Hopkins University Press, [2020] |
    Includes bibliographical references and index.
Identifiers: LCCN 2020009420 | ISBN 9781421439709 (hardcover) |
    ISBN 9781421439716 (ebook)
Subjects: LCSH: Women paleontologists—History. | Women in science—
    History. | Paleontology—History. | Vertebrates, Fossil—Research—History.
Classification: LCC QE705.A1 B47 2020 | DDC 560.92/52 [B]—dc23
LC record available at https://lccn.loc.gov/2020009420

A catalog record for this book is available from the British Library.

*Special discounts are available for bulk purchases of this book.
For more information, please contact Special Sales at
specialsales@press.jhu.edu.*

Johns Hopkins University Press uses environmentally friendly book
materials, including recycled text paper that is composed of at least
30 percent post-consumer waste, whenever possible.

To all those women who have gone hunting bones,
both in the field and in museum collections

CONTENTS

SEVEN
Challenges and Opportunities 207
Women in STEM, geosciences, and paleontology

Women have long been pioneers in vertebrate paleontology from the well-known Mary Anning in the early 19th century to the present day, with an increasing presence over the past 50 years. This book celebrates the history of women "rebels," "scholars," and "explorers" who have contributed to vertebrate paleontology worldwide. We delve into their lives and work, examining how the lure of our discipline has spread and shaped understanding of the history of life, especially that of our backboned ancestors on Earth. Gaining professional employment came slowly (and in some countries, this is still the case), with few women recognized seriously before the late 19th to early 20th century (e.g., Falk, 2000). But over time the fossil discoveries and scientific contributions of women scientists have paved the way for later generations to pursue diverse careers in vertebrate paleontology (VP).

Although there are histories of different aspects of VP, in most the work undertaken by women is mentioned only in passing, if at all. Until now there has been no complete history of women in vertebrate paleontology, although there are biographies of a few (e.g., Anning, Dorothea Bate, and Margaret Matthew Colbert); some are also mentioned in histories and compilations (e.g., Aldrich, 2001; Sarjeant, 1978–1987; Spalding, 1993; see also Bibliographic Sources and Further Reading). We aim to redress the imbalance. In addition to examining the lives and contributions of women vertebrate paleontologists, we also provide highlights of the important challenges that women have overcome to gain degrees and employment, as well as to attain leadership positions in professional societies and on international research projects.

As an important component of our book, we launched a GoFundMe project that enabled us to videotape interviews of a cross section of women members at the 2018 Albuquerque, New Mexico, meeting of the Society of Vertebrate Paleontology, providing diverse perspectives of their VP experiences.

Many other women and male mentors have responded to written interview questions or provided CVs (see Appendixes and online resources at https://jhupbooks.press.jhu.edu/title/rebels-scholars-explorers). Our online database, Women in Vertebrate Paleontology (see ResearchGate project, www.researchgate.net), provides a definitive list of important women in VP, both past and present, including researchers, educators, artists and illustrators, collectors, preparators, curators, and collection managers, and we hope to add to Wikipedia and other online sites. To date we have uncovered more than 1,200 women vertebrate paleontologists (VPs) (and counting!) who have made significant contributions to the field. Unfortunately, the constraints of space render it impossible for us to fully explore each woman's history here. Names of women VPs are bolded on first mention.

Our book is organized into seven chapters. The first provides a review of the emergence of vertebrate paleontology as a science. This is followed by a summary of the vertebrate fossil record, highlighting the contributions of women in specific research areas and techniques employed, and a brief history of paleontological professional organizations worldwide. The second through fifth chapters chronologically examine the histories of women, profiling major contributors to the field against a historical backdrop of political, social, and cultural changes during the 19th and 20th centuries and into the 21st century. The history reflects our chosen title, for we see women "rebels," "scholars," and "explorers" contributing to vertebrate paleontology in these and many other ways. The sixth chapter chronicles the important contributions of women artists, preparators, collection managers, and outreach educators to vertebrate paleontology. The book's final chapter considers both challenges and opportunities faced by women VPs. In part, the gender gap in VP is due to historical bias, with fewer women scientists in earlier decades, but there are other barriers to participation by women (e.g., implicit bias, sexual harassment, family issues, and the pay gap). We review long-term gender trends in STEM, the geosciences, and vertebrate paleontology, identify strategies that are working, and highlight additional actions needed to ensure diversity and inclusion, thus broadening the reach of our science.

This book was written to serve two audiences—students and professional paleontologists—as well as women scientists and interested people in other fields. By highlighting our history and sharing some of our stories, we hope this book brings recognition to women in vertebrate paleontology and will encourage girls and young women to follow their passions for hunting bones.

We first want to thank all the women VPs past and present for their hard work, commitment, and passion; they are creating a better future for vertebrate paleontology and science in general. We have benefited from the advice and assistance of many colleagues. For photographs, anecdotes, and data, we thank Mike Archer, Nancy and Jim Askin (Castleberry family), Meghan Balk, Nathalie Bardet, Chris Bell, Zoya Bessudnova, Alain Blieck, Niels Bonde, Arnis Brangulis, Dan Brinkman, Don Brinkman, Angela Buscalioni, Sandy Carlson, Karen Chin, Rich Cifelli, George Corner, Liping Dong, Margarita Erbajeva, Robert Finch, Richard Forrest, Zhicun Gai, Lilita Ilga Gailite, Mireille Gayet, Michal Ginter, Tom Goodhue, Mark Goodwin, Gavin Hanke, Pat Holroyd, Jeanette Hope, Barry Hughes, Frankie Jackson, Philippe Janvier, Eva Koppelhus, Sandrine Ladeveze, Oleg Lebedev, Kath Lehtola, Margaret Lewis, Elisabeth Locatelli, John Long, Jing Lu, Ervins Lukseviks, Irina Malakhova, Elga Mark-Kurik and her son Simo, Joseph McKee, Priscilla McKenna, Sam McLeod, Roger Miles, Angela Milner, Dick Moody, Joe Nielsen, Leslie Noè, Jingmai O'Connor, Rob van Oudjehans, Michelle Pinsdorf, Tuo Qiao, Emily Rayfield, Pat and Tom Rich, Julia Sankey, Tamaki Sato, Hans-Peter Schultze, Claudia Serrano, Misha Shishkin, David Spalding, Joy Stacey, Hans-Dieter Sues, Darren Tanke, Tony Thulborn, Suyin Ting, Kelsey Vance, Dave Weishampel, Lars Werdelin, David Whistler, Hongyu Yi, Robert Scott Young (Museum of Comparative Zoology Archives, Harvard University), Darla Zelenitsky, and Meemann Zhang. Agnese Lanzetti is gratefully acknowledged for her assistance organizing data from the Women in Vertebrate Paleontology database and providing graphic line drawings based on it.

Special thanks to GoFundMe donors for their generous support of our videotaped interviews of women in VP conducted at the 2018 Annual Meeting of Vertebrate Paleontology in Albuquerque, New Mexico (see videotape highlights posted to the Society of Vertebrate Paleontology website and excerpts in this book): J. David Archibald, Gloria Bader, Catherine Badgley, Brian Beatty, Anna Kay Behrensmeyer, Chris Bell, Virginia Benson, Annalisa Berta, Theresa Berta, Albert Berta Jr., Robert and Sara Boessenecker, Emily Buchholtz, Julia Clarke, William A. Clemens, Margery Coombs, Philip Currie, Larisa DeSantis, Jaelyn Eberle, Eric Ekdale, John Flynn, Scott Foss, Mark Goodwin, Greg Harris, Patricia Holroyd, Bonnie and Lois Jacobs, Christine Janis, Karen Kinney, Margaret Lewis, Judy Massare, Julie Meachen, Ashley Morhardt, P. David Polly, Rachel Racicot, Meredith Rivlin, Jessica Theodor, Susan Turner, Blaire Van Valkenburgh, and an anonymous donor.

Thanks to all videotaped interviewees, including Catherine Badgley, Anna Kay Behrensmeyer, Emily Buchholtz, Julia Clarke, Larisa DeSantis, Catherine Early, Jaelyn Eberle, Anjali Goswami, ReBecca Hunt-Foster, Christine Janis, Zerina Johanson, Bolortsetseg Minjin, Jingmai O'Connor, Emily Rayfield, Kristi Curry Rogers, Ana Valenzuela-Toro, and Blaire Van Valkenburgh, plus those who responded to our written questionnaires (Chris Bell, Anusuya Chinsamy-Turan, Jenny Clack, William Clemens, Margery Coombs, Philip Currie, John Flynn, Romala Govender, Frankie Jackson, Louis Jacobs, Kathy Lehtola, Carolina Loch, Bruce MacFadden, Marisol Montellano, Don Prothero, Ken Rose, Natalia Rybnczinski, Tamaki Sato, Judy Schiebout, Pat Vickers-Rich, Anne Warren, Lisa White, Joanne Wilkinson, and Darla Zelinitsky) and all other women VPs who have answered our emails and questions about their lives (see Appendixes). We gratefully acknowledge the considerable time, effort, and expertise of our videographer and director/producer, Robin Rupe, and her production company, Volti Subito, in Albuquerque, New Mexico.

In addition, a Facebook fund-raiser assisted us to continue the research and to present at the History of Geology Group's "100 Years of Women's Fellowship of the Geological Society Conference" in May 2019. We would like to thank the following for their help: Michael Archer, Steve Avery, Mike Boyd, Rob Clack, Neil Clark, Joana Damani, Götz Dreismann, Merv Fildes, Allan Grey, Annie and John Henry, Tori Herridge, Andrew Hewson, Peter Hubbard, Wayne Itano, Christine Janis, Susan Jenkins-Temple, Juan Liu, Guy Loosemore, Simone Meakin, Ashley Morhardt, Anne Musser, Alex Nearney, P. David Polly, Claudette Prince, Martha Richter, Ian Rolfe, Marg Rutka, Thomas Teoh, Jessica Theodor, Tony Thulborn, Pat Vickers-Rich, Veronica Watson, Belinda Whitwell, Barney Wolfram, Don Wotton, and anonymous donors.

Archivists Ellen Alers, Ted Benioff, Deborah Shapiro, and Ginger Yowell were especially helpful in locating Society of Vertebrate Paleontology records at the Smithsonian Institution Archives. Archivist Timothy H. Horning at the University of Pennsylvania, Yolanda Bustos at the Los Angeles Natural History Museum, Nicola Woods at the Royal Ontario Museum, and Christel Schollaardt and Karien Lahaise at the Naturalis collection archives and images, Leiden, helped with photographs. Librarian Wendy Cawthorne and archivist Caroline Lam gave great help at the Geological Society of London, as did former librarian Meg Lloyd at the Queensland Museum, and Philippe Janvier provided assistance at the Gaudry Library and Palaeontological Library, Muséum national d'Histoire naturelle, Paris. Suzanne Richie with the American Association of Science is acknowledged for providing fellows records.

Finally, we are grateful to the friends and family who supported us; to our colleagues Margery Coombs, Kath Lehtola, David Spalding, and Blaire Van Valkenburgh, who read the entire book manuscript or various chapters and provided valuable comments; and to our reviewers, David Archibald and Catherine Badgely, who did a sterling job and offered us great encouragement. A special thanks to our editor, Tiffany Gasbarrini, and the talented Johns Hopkins University Press production team including Esther P. Rodriguez for their patience, advice, and expertise. Kimberly Giambattisto, Ashley Moore, and Juliana McCarthy are acknowledged for their thorough and meticulous edits of the manuscript.

Rebels, Scholars, Explorers

CHAPTER 1

Introduction

> To travel millions of years back into the past, which is what paleonto-
> logical study amounts to, is much more fascinating than the most exotic
> geographical travel we are able to undertake today.
>
> Zofia Kielan-Jaworowska, 1965

We cannot evaluate the contributions of women vertebrate paleontologists (VPs) today without knowing what happened in the past. So we begin by considering the science of vertebrate paleontology, starting with its origin and early development. We look at the participation of women within the larger context of emerging professional VP organizations in various countries. Historical trends in the participation in VP by women, from collectors to researchers, are outlined and serve as a framework for exploring the lives and contributions of these women.

By the late 17th century in Europe, fossils were understood to be the remains of once-living plants and animals (e.g., Adams, 1938). One of the first European naturalists to recognize fossil vertebrates was the Danish anatomist Nicholas Steno (1638–1686), who observed that triangular objects known as "tongue stones" were in fact fossil teeth from lamniform sharks (e.g., Rudwick, 1972; Hoch, 1985). Vertebrate paleontology as a scientific endeavor has its true beginnings in the late 18th century, especially with the work of French naturalist Georges Cuvier (1769–1832; e.g., Taquet, 2006; Liston, 2013). To interpret fossil remains, Cuvier compared them with the anatomy of living animals, showing, for instance, that fossil elephants (i.e., mammoths and mastodons, the latter from early collecting in America; Rudwick, 1972) were distinct from living elephants. Thus, he laid down the principles of comparative anatomy that we use to this day. Cuvier is also partly responsible for the concept of extinction—that is, that some species no longer exist. Supporting this concept was recognition of the significance of fossils in aging different rocks by the engineer and self-taught geologist William Smith (1769–1839). Later researchers, such as pioneer geologists Charles Lyell (1797–1875) and Charles Darwin (1809–1882), also discerned that the geological record revealed a succession of different species that evolved over time on Earth. It is against this backdrop of fossils, faunal succession, evolution, and extinction that the discipline of VP has developed. By the late 1800s women were becoming involved in the burgeoning science, especially in collecting and illustrating fossil vertebrates (see Chapters 2, 6).

Although vertebrate paleontology had strong geological roots, it is recognized today as a diverse discipline that also encompasses aspects of zoology, biology, anthropology, ecology, physics, chemistry, and computer science. We next provide a brief review of the major research areas, subdisciplines, and techniques of paleontology. VP focuses on the fossils of backboned animals, including bones and teeth, scales, and related structures—horns, beaks, skin impressions, footprints, and so on—from the earliest fish and their relatives to humans. There are women fossil hunters involved in all aspects of VP, although many have gone unrecognized until now.

The research interests of women VPs are varied, including phylogeny and taxonomy, the study of the evolutionary relationships of organismal groups using techniques such as morphological (e.g., photography, computed tomography [CT], and surface scanning) and, more recently, molecular phylogenetics and genomics. Traditionally, a major subdivision of vertebrate paleontology based on its geologic roots has been dating rocks and the faunas contained within them using the fossils, as well as, more recently, various isotopic and paleomagnetic methods (Chapters 2, 3, 4, and 5). Related to the stratigraphic framework of fossils (biostratigraphy) are their past distributions. Paleogeography involves comparing the spatial distributions of fossils and analyzing changing geography and climatic conditions through time (Chapters 4 and 5). Taphonomy, the science of burial, unravels what happens to fossils after the death of an organism and before its fossilization (Chapters 4 and 5). Derived from this is the study of traces of organisms (i.e., ichnology; e.g., tracks, burrows, coprolites) (Chapters 2, 4, 5). Fossils are also used to understand evolution, both patterns and processes (i.e., macroevolutionary studies) such as extinction, diversity, and speciation (Chapter 5). Paleontologists also study communities and ecosystems and how they respond to environmental change, and more recently they have integrated fossils with developmental data (Chapter 5).

Women vertebrate paleontologists have been at the forefront in all research fields from the earliest times (e.g., Turner et al., 2010a; Switek, 2010). In addition to working in these research areas and on techniques, women have been collectors, curators and collection managers, educators and outreach specialists, preparators, and artists, all making major contributions to our understanding of fossils as animals (Chapter 6).

The thrill of discovering the remains of past life is what captures the interest of many women vertebrate paleontologists. As Judith Schiebout of Louisiana State University pointed out in our interview with her, "VP is a way to see much more of Earth's story, beyond human-recorded history." For Carolina Loch, who specializes in unlocking the secrets hidden in long-buried marine mammal teeth at the University of Otago, New Zealand, "being able to handle and study specimens that are millions of years old has always felt like the ultimate humbling experience to me [and it] gives us so much appreciation for the vast history of our planet."

The appeal of vertebrate paleontology for others, as discussed in our interviews with women VPs, is the recognition that it is a multifaceted, integrated discipline that combines many fields of study. As Anjali Goswami from the Natural History Museum of London succinctly put it, "Paleontology is everything I love about biology, but with the added component of time that just made it so much more interesting to me." According to Larisa DeSantis at Vanderbilt University, she was "debating between the fields of conservation and paleontology. It was either/or, but the really great thing is that the fields have come together where we can now use the fossil record to ask questions of relevance to conservation." The fact that VP can provide insights into environmental stewardship is another aspect that attracts women scientists with interdisciplinary interests. Australian Suzanne Hand, director of the Palaeontology, Geobiology and Earth Archives Research Centre in Sydney, observed, "Often, VP involves a focus on both extinct and extant species, and there is increasingly recognized relevance of the fossil record to local and global issues in modern conservation biology and policy." Noted rebel, scholar, and explorer Catherine Badgley at the University of Michigan said, "It is not so much my individual discoveries that are my proudest accomplishments, but more the fact that I feel as though I have been learning how to integrate . . . all those different perspectives and partly to gain a greater understanding of where humans fit in."

History of Paleontological Professional Organizations

The histories of various paleontological organizations provide background for consideration of the role of women in VP; in the chapters that follow, we also consider the relationship of various women VPs to such professional societies. Britain can claim the first geological and paleontological society in its Geological Society of London (GSL), founded in 1807 (Lewis and Knell, 2009). Its first 100 years were dominated by pioneering paleontologists (including Lyell and Darwin), and toward the end of the century a few of them were advocating that women should become members. This would not happen, though, until after World War I (1914–1918), a time during which women took on many nontraditional roles. By that time many women, at least in the English-speaking world, had entered universities, gained degrees, and begun professional careers in science. In 1919 women were finally allowed to become fellows of the GSL (Herries Davies, 2007), and 100 years later that event was celebrated, including for those women vertebrate paleontologists who became fellows (Burek and Higgs, 2020; Chapters 3, 4). Although a formal scientific organization for geology began in Britain, a separate vertebrate paleontology society was never formed, with VP members firmly favoring the GSL or the Palaeontological Association, formed in 1957. However, annual informal meetings (the Symposium of Vertebrate Palaeontology and Comparative Anatomy [SVPCA], begun in 1952; Plate 1) brought together vertebrate paleontologists and comparative anatomists and, later, preparators (Symposium of Vertebrate Palaeontology Preparators, beginning in 1991), finally joining forces on occasion with the Society of Vertebrate Paleontology (SVP; see later in this chapter), as in 2009 in Bristol. Often, these meetings had associated field trips. Wider Europe gained an annual vertebrate paleontology meeting, the European Association of Vertebrate Paleontologists (EAVP) in 2003.

The Paleontological Society of Russia (later the All-Russia and All-Union Paleontological Society), which was founded in 1916, focused on VP, although the older Imperial Society of Naturalists of Moscow (later the Moscow Society of Naturalists; e.g., Khain and Malakhova, 2009) founded in 1805, encompassed early work in the discipline. Two pioneer Russian women in VP, Maria V. Pavlova and Ana B. Missuna, were founding members of the Paleontological Society; Pavlova also served on the council (Zhamoida and Rosanov, 2016). Since 1992, with the loss of the SSR countries, the Russian Paleontological Society has seen a major plunge (50%) in membership. Additionally, social change from Soviet times to the Russian Federation has affected jobs and work conditions for women VPs (see Chapters 3–5).

The first exclusively paleontological organization in the United States was the Society of American Vertebrate Paleontology (SAVP), formed in 1902 (Williston and Hay, 1903). In 1908, a group that was not officially a part of the SAVP began the Paleontological Society (PS), which was affiliated with the Geological Society of America (GSA). The group of PS founders included two future charter members of the SVP, S.W. Williston and F.B. Loomis (see more on charter members in Chapter 4). The first PS meeting was held in 1909, and the SAVP membership was invited to merge with this society. Most SAVP members did join the PS, after which the budding SAVP was disbanded (Romer and Simpson, 1940; Cain, 1990).

Approximately 35% of PS members were former members of the SAVP. The 1909 merger had the effect of expanding the old SAVP to include invertebrate and plant paleontologists (Cain, 1990). For the next two decades, the arrangement remained the same, though with some important changes within the organization. By 1934 the PS had grown to 428 members, but less than 15% of its membership had direct interest in vertebrate paleontology. This lack of interest in VP among most PS members, substantiated by the results of a poll of vertebrate members, prompted the 1934 PS Organization Committee to recommend that a Section of Vertebrate Paleontology be established. Members of this vertebrate division of the PS presented 12 research papers on fossil vertebrates in a separate room at the 1934 PS meeting held in Rochester, New York (Cain, 1990, table 3). The Section of Vertebrate Paleontology lacked a constitution or any other defined basis of organization of its own. In 1934 and 1935, respectively, two of the three vice presidents of the PS (also known as chairmen) were vertebrate paleontologists: Chester Stock (1892–1950) and Elmer Riggs (1869–1963). In 1938 the president of the PS, Charles W. Gilmore (1874–1945), was a vertebrate paleontologist, and in 1939 the vice president, Alfred Sherwood Romer (1884–1973), was also a vertebrate paleontologist.

As the Section of Vertebrate Paleontology's autonomy expanded during the 1930s, its status within the PS grew problematic. In their 1940 "Information Bulletin," Romer and G.G. Simpson (1902–1984) put forth their case for abandoning the PS and forming a separate society. Both Romer and Simpson were interested in attracting zoologists, biologists, and comparative anatomists and in encouraging biological studies of fossils. The SVP was formally constituted on December 28, 1940, immediately following dissolution of the PS Section of Vertebrate Paleontology (Simpson, 1939–1940). Romer was elected provisional president and Simpson provisional secretary-treasurer; the latter "was in Venezuela at the time and couldn't protest" (Romer, 1944:22). As further discussed in Chapter 4, three women signed the constitution, Tilly Edinger, Anne Roe Simpson, and Nelda Wright.

The newly formed SVP continued its affiliation with the PS and the GSA; the latter was still recognized as the principal organization for all paleontologists (Simpson, 1941, 1942). The strong relationship of the SVP to both of these societies continues to this day. For example, the GSA provided financial support for early volumes of the Bibliography of Fossil Vertebrates (BFV), an index to the world's literature of vertebrate paleontology covering published references from 1500 to 1993 (e.g., Romer, 1958). As discussed in Chapter 4, women played a key role in compiling and editing the BFV. In 1948, the SVP joined the American Geosciences Institute (AGI) as one of the founding societies, and the AGI printed the BFV from 1973 to 1980. The SVP and the University of California Museum of Paleontology (UCMP) supported the BFV from 1980 until it ceased publication in 1996, owing to the accessibility of VP literature in computerized databases and the increasing costs of indexing. The BFV Online was created by John Damuth in 1994 with support from the SVP and a grant from the Dinosaur Society, which resulted in its expansion to include references from the older printed volumes, but it is not yet complete (see SVP website). We have used the BFV to obtain information about publications by women VPs (see Bibliographic Sources and Further Reading).

Since its beginning, the SVP has often held meetings either before or after the GSA at the same location. Among collaborative projects with the PS is the joint SVP/PS Distinguished Lecturer Program, initiated in 2018 to promote presentations by paleontologists on research, evolution, and the nature of science (see SVP website). Other SVP society affiliations followed, including with the International Palaeontological Association in 1974, the Association of Systematic Collections in 1976, and the Coalition for American Heritage in 2019 (Wilson, 1990; D. Polly, pers. comm., 2019).

At the 1940 organizational SVP meeting, members voted to hold annual meetings to conduct business and facilitate research presentations from members. Members also voted to provide the *SVP News Bulletin* for the information of its members, with women as the earliest editors (see Chapter 4). In addition to the United States, Mexico and Canada were initially considered domestic constituents of the SVP (see, e.g., *SVP News Bulletin* 1–10). In 2013 the Canadian Society of Vertebrate Paleontology began to hold its own annual meetings in conjunction with an annual conference. Before the 1950s, there were few women vertebrate paleontologists in Latin America (Herbst and Anzótegui, 2016; Chapters 3 and 4), but today in South America there are several geological and paleontological societies (e.g., the Asociación Paleontológica Argentina, the Asociación Chilena de Paleontología, the Sociedad Geológica de Chile, and the Sociedade Brasileira de Paleontologia), although none are exclusively devoted to vertebrate paleontology.

Despite early get-togethers of vertebrate paleontologists in Australia in the 1970s at large geological conferences (e.g., the 1973 Third Gondwana Symposium; Plate 2), there were not enough practicing vertebrate paleontologists to form a formal society. A failed attempt was made in Australasia to form a vertebrate paleontology section within the Association of Australasian Palaeontologists in 1986. This resulted in a break-off group that set up a biennial Conference on Australasian Vertebrate Evolution (Turner, 1990; Tedford, 1991b), which continues with meetings in Australia and New Zealand, and in conjunction with the SVP in Brisbane, in 2019 (Plate 3).

Vertebrate paleontologists in Asian countries belong to various professional paleontological organizations. The Paleontological Society of China had 1,500 active members a decade ago according to Xiao et al. (2010). The Society of Vertebrate Paleontology of China was established in 1984 and affiliated with the Institute of Vertebrate Paleontology and Paleoanthropology of the Chinese Academy of Sciences in Beijing (IVPP), which was originally founded in 1929 as the Cenozoic Research Laboratory, part of the Geological Survey of China. Pioneer vertebrate paleontologist Chung Chien "C.C." Young (Yang Zhong-Jian) (1897–1979) created an institute for VP in 1953 (named the IVPP in 1957; Sun and Zhou, 1991; Young also produced the first Chinese VP journal, *Vertebrata PalAsiatica*, in 1957. Conferences on all aspects of vertebrate paleontology are held in China, including annual meetings of the SVPC (Lucas, 2001). The Paleontological Society of Japan, founded in 1935, has VP members, holds biannual meetings, and publishes an international journal, *Paleontological Research*. Founded in 1984, the Paleontological Society of Korea meets annually and publishes a semimonthly journal, the *Journal of the Paleontological Society of Korea*.

Several paleontological organizations exist in Africa with the active participation of women VPs. The Paleontological Society of South Africa, founded in 1979, holds biennial meetings and publishes the electronic journal *Palaeontologia Africana*. Since 1994 the Johannesburg-based Paleontological Scientific Trust supports research, education, and public outreach in all areas of paleontology.

Women in Vertebrate Paleontology Database

In order to understand historical trends in the representation of women in VP, we need to examine the available data. We have built a database to track women paleontologists (information on nonbinary/other individuals was not available until recently (e.g., for SVP beginning in 2017) but they warrant inclusion in the future) and their research interests in broad taxonomic groups (Box 1.1): early vertebrates and jawed fish, amphibians, reptiles (with dinosaurs and birds as separate subcategories), mammals, and mammal-like groups (i.e., mammal-like reptiles, stem mammals), with primates/hominoids as a further, separate subcategory (see, e.g., online appendix S1). We also track various subdisciplines, including taphonomy (including trackways and trace fossils), evolution (including macroevolution and taxonomy), and paleobiology (including pathology, paleohistology, conservation paleontology, paleoecology, biostratigraphy, and isotope geochemistry). In online appendix S1, we categorize the data using five time periods, the same as those used in Chapters 2–5: (1) 18th to mid-19th century (1700–1850), (2) late 19th to early 20th century (1851–1939), (3) mid-20th century (1940–1975), (4) late 20th to 21st century (1976–2000), and (5) 21st century (2001 to the present). We used known and estimated birth and death dates as a guide to the time of research or employment activity of women and found that a

Box 1.1 **The Vertebrate Fossil Record**

We highlight here the contributions of women vertebrate paleontologists to the understanding of the evolution and radiation of vertebrates. *Fish* is an informal name for animals that live in the sea and generally have fins and scales, but scientifically, the term encompasses many interrelated groups. A number of scholars concentrate on and specialize in each, such as agnathans, the jawless fish. These first evolved 500 million years ago or more, making use of calcium phosphate to develop the first external armor of plates and scales made up of dentine and forms of bone; we are still exploring their early history.

Women scientists with research expertise on jawless and other early vertebrates have increased in number in recent decades (e.g., Valentina Karatajūtė-Talimaa, Elga Mark-Kurik, Susan Turner, Tiiu Märss, and Živilė Zigaitė; Chapters 4 and 5). From the jawless fish came the gnathostomes, the jawed fish, in which the first jaws and teeth evolved. Examples include different forms of "sharks" (chondrichthyans—cartilaginous skeletons) and "bony" fish (osteichthyans—bony skeletons), which are currently a source of much discussion (e.g., Carole Burrow, Moya Meredith Smith, and Katherine Trinajstic; Chapter 5). At least one group, the rhizodonts (e.g., Zerina Johanson, Yu-zhi Hu, and Tuo Qiao; Chapter 5), gave rise to all other vertebrates. Among the jawed fish, some took the next step, including some sharks, and had live young; we now have evidence of internal fertilization and live birth in fossil placoderms (an extinct armor-plated group) going back some 370 million years (e.g., Trinajstic; Chapter 5).

The fish that made it onto land are generally called amphibians, the earliest tetrapods (four-footed beasts). They are still with us today as salamanders, frogs, toads, and newts. These animals still had to lay their eggs in water or moist places, even if they had the capacity to live on land (e.g., Jennifer Clack and Anne Warren; Chapters 4 and 5).

The next step tetrapods took was to protect their eggs with a hard shell and three membranes (amniotic egg) to develop a wholly land-based existence as reptiles. A diversity of amniotes evolved, and some eventually stood upright on two feet, becoming bipedal. At this stage, there evolved groups such as lizards, snakes, crocodiles, and turtles, along with extinct lineages such as ichthyosaurs, plesiosaurs, and pliosaurs (e.g., Betsy Nicholls, Judy Massare, Zulma Gasparini, and Sue Evans; Chapter 5)—some of which went back to the water—and dinosaurs (e.g., Catherine Forster, Susannah Maidment, Emily Rayfield, and ReBecca Hunt-Foster; Chapter 5). Some of these groups took the next step and had live young (e.g., Nathalie Bardet, Mary Schweitzer, and Kristi Curry Rogers; Chapter 5).

Within amniotes, some evolved feathers and some hair or fur coverings. The next steps in evolution led to conquering the air, for instance, by pterosaurs (e.g., Cherrie Bramwell; Chapter 4) and birds (e.g., Hildegarde Howard, Patricia Vickers-Rich, and Julia Clarke; Chapters 4–5). Although mammal ancestors evolved more than 300 million years ago, with early synapsid forms such as cynodonts and dicynodonts (e.g., Erika von Huene, Anusuya Chinsamy-Turan, and Gillian King; Chapters 3–5), they did not reach their peak until the early Cenozoic (about 64 million years ago), after the demise of the nonavian dinosaurs. Mammals include the monotremes (which retain egg laying), marsupials, and placentals, all of which suckle their young (e.g., Zofia Kielan-Jaworowska, Karen Sears, Suzanne Hand, and Kat Piper; Chapters 4 and 5) and have reached the coldest poles and hottest deserts. Some would say placental mammals made their greatest leap by reentering the water and producing some of the largest animals that have ever lived on Earth, the whales (e.g., Maureen O'Leary, Ana Valenzuela-Toro, Romala Govender, and Annalisa Berta; Chapter 5). Evolution of the primates, the apes, and, later, the human lineage has been the focus of a number of women vertebrate paleontologists (e.g., Mary and Maeve Leakey, Pat Shipman, and Mary Silcox; Chapter 5).

number were active during two time periods and that several were active over three time periods, resulting in a total of 1,219 women VP records. This is a selective and conservative list; the basis for inclusion was a measurable contribution to VP, which for researchers, educators, and curators required a meeting presentation or publication. Many other women (e.g., solely members or delegates of various international societies) were not included on this list but are recorded in our computer database (see the ResearchGate Women in Vertebrate Paleontology project page).

Taken together, our data reveal that the total number of women VPs recorded in the selected time periods shows the largest increase during the 21st century (961 records), representing a 70% increase since 1940 (Figure 1.1; online appendix S1). On a subject level, our data show that the earliest women VPs (1700s–1850s) studied fish and reptiles in near-equal numbers. Few worked on dinosaurs during the late 19th and early 20th centuries, and women in this field have only reached their highest numbers in recent decades. Women paleontologists who favored work on fossil mammals began to appear in the late 19th century and by the middle of the 20th century dominated the field, continuing on to the present. Women VPs studying primates reached their highest number, nearly 10% of the total, during the late 20th and early 21st centuries (Figure 1.2; online appendix S1).

In terms of employment, our data show that the earliest women VPs began primarily as collectors and artists, with a few, such as bone-hunting icon and rebel for her time Mary Anning, contributing academic insights (Chapter 2). Bibliographers and editors are the next largest employment category for women VPs, increasing during the 1940s and 1950s, as discussed in Chapter 4. Women VPs occupying academic and research positions increased in number through the 20th century, forming 80% of the total from the 1970s to the present. Since the 1970s to the present, the number of curators has held steady at 4%–5%, with preparators between 4% and 6% of the total, and outreach scientists and educators at 1%–2% (Figure 1.3; online appendix S1).

Our geographic data show that women VPs were mainly European until the 1940s, when a larger proportion came from North America (coinciding with the establishment of the SVP), coinciding with a considerable increase in members from the former Soviet Union (Chapter 4). Beginning in the late 19th century and continuing to the mid-20th century, women VPs from Asia (mostly the former Soviet Union) made up about 20% of the total. Since 2000 there has been a 6% increase in women VPs from South America and a smaller increase in those from Africa (Figure 1.4; online appendix S1).

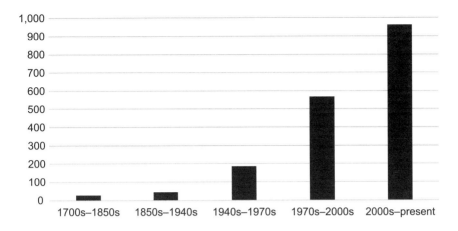

Figure 1.1. A selection of women VPs through time.
Sources: SVP News Bulletins; BFV; and SVP, EAVP, CAVEPS, and SVPCA annual meeting programs and abstracts.

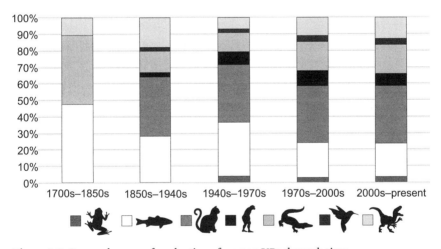

Figure 1.2. Research areas of a selection of women VPs through time.
Sources: SVP News Bulletins; BFV; and SVP, EAVP, CAVEPS, and SVPCA annual meeting programs and abstracts.

As is true for other fields in the science, technology, engineering, and math (STEM) disciplines, women are underrepresented in VP (Hill et al., 2010; Stigall, 2013b; Holmes et al., 2015; Johnson, 2018). This underrepresentation of women in VP is due, in part, to historical bias, with fewer women scientists in earlier decades, as discussed in Chapters 2–6. Other barriers to

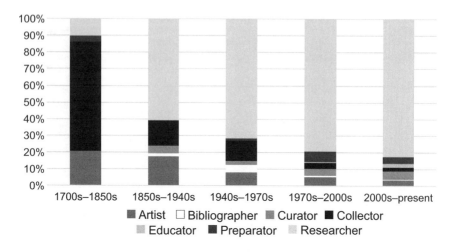

Figure 1.3. Positions occupied by a selection of women VPs through time. *Sources: SVP News Bulletins*; BFV; and SVP, EAVP, CAVEPS, and SVPCA annual meeting programs and abstracts.

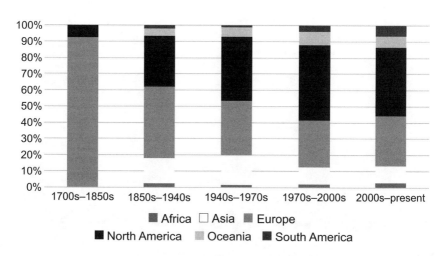

Figure 1.4. A selection of women VPs by country through time. *Sources: SVP News Bulletins*; BFV; and SVP, EAVP, CAVEPS, and SVPCA annual meeting programs and abstracts.

participation of women (e.g., implicit bias, sexual harassment, and family issues) are reviewed in Chapter 7, where we investigate strategies for achieving gender parity and inclusiveness. There is considerable evidence that gender parity matters (Page, 2007). We live in a diverse world, whether in scientific discipline, work experience, gender, ethnicity, or nationality.

Paleontology can provide a prism through which scientists from across the world can share their own cultural heritages and better understand those of others. Bolorsetseg Minjin, founder of the Institute for the Study of Mongolian Dinosaurs, commented in our interview with her, "I really wanted to learn about these animals that lived millions of years ago and at the same time I didn't realize that my country had such a rich history."

For paleontology we need a variety of minds asking a range of questions that involve incorporating data from the fossil record. After all, pushing boundaries is an inherent part of what women in this field do. Jaelyn Eberle, at the University of Colorado, Boulder, encapsulated the feelings shared by many women in paleontology in our interview with her: "I cannot think of a time when I did not want to be a paleontologist. . . . I think it's exciting to go to places that few people, few paleontologists have ever experienced before. It's the last frontier in some ways in our science."

In this book we present histories of the lives and work of the diverse women who have hunted fossil bones. We also consider various ideas and multidisciplinary approaches to achieving gender parity within the VP discipline and beyond. By exploring ways to recruit, support, and retain women in vertebrate paleontology, we aim to better position our science for the future.

CHAPTER 2

Early Discoveries and Collection of Fossil Vertebrates, 18th to Mid-19th Century

My dear Mr Owen, my whole collection is always, à vos orders, & available generally for science & scientific purposes, & I think it very desirable the Hordle bones should be kept together, in my keeping, as one of these days, at my death, if not before, my collection [will] become public property.

Barbara Hastings to Richard Owen, undated letter
(see Kölbl-Ebert, 2004:83)

There has been a historical lack of acknowledgment of women in vertebrate paleontology. In his survey of the history of geoscience in the Age of Revolution (1776–1848), Martin Rudwick mentions Mary Anning but summarily dismisses the contributions of women "of all social classes," noting that "their significance should not be forgotten, but neither should it be exaggerated in the interests of an anachronistic egalitarianism or in the name of political correctness" (Rudwick, 2005:37). When you read histories of vertebrate paleontology, such as Robert Howard's *The Dawnseekers* (1975), Url Lanham's *The Bone Hunters* (1973), or even the books of noted vertebrate paleontologist Edwin Colbert, such as *Men and Dinosaurs* (1968) and *The Great Dinosaur Hunters and Their Discoveries* (1984), there are no or very few women mentioned. Historians Michele Aldrich (1982), Mary Creese and Thomas Creese (e.g., 1994), Margaret Rossiter (e.g., 1995), Martina Kölbl-Ebert (e.g., 1997), Susan Turner et al. (2010a), and others cited here have begun to address this concern.

We know almost nothing of such women outside Europe. Was Lady Xia, grandmother of the first Chinese emperor, Qin Shihuang, the first fossil bone hunter? Her 2,200- to 2,300-year-old tomb, when found, was like a zoo or museum, with bird and mammal bones and an unusual gibbon skull. As reported by scientists from the Zoological Society of London and University College London (UCL), the gibbon was likely a pet, which appears to have been common among Chinese royals, and it was described as a new extinct gibbon, *Junzi imperialis* (Turvey et al., 2018).

Vertebrate fossil collecting took off in earnest after the discovery of large bones—mammoth, mastodon, and giant reptilians (including dinosaurs)—in both Europe and the Americas, aided by the understanding from Georges Cuvier and others in France (e.g., Servais et al., 2012; see Chapter 1) and the work of William Buckland (1784–1856) in Britain (e.g., Gordon, 1894; Rupke, 1983; Cadbury, 2000) that these were extinct animals. Discovery of fossil crocodiles in the Jurassic of Yorkshire in 1758 and a *Mosasaurus* skull in Maestricht in 1780 contributed to the late 18th- to early 19th-century enthusiasm for fossil collecting among the rising middle classes of Britain, Europe, America, and farther afield (e.g., Barber, 1980; Spalding, 1993). Women wanted to learn, and they gained information, at least about the "bones," from the new public libraries and museums, and even a few local societies welcomed them (e.g., Sheets-Pyenson, 1988; Pyenson and Sheets-Pyenson, 1999; Rolfe et al., 1988).

In this early era, there were a few women collecting fossil vertebrates, but by the 18th century in Britain, women were making their mark in our science. One was Anna **Gurney** (1795–1857), a Norfolk Quaker scholar who acquired fossil mammal bones in the Cromer Forest bonebed near Norwich, England, in the early 18th century (Cleevely, 1983). Late in life she became an associate of the British Archaeological Association, having been the first woman to join the society (Lane, 2001). We know even less about others, such as Miss **Cowderoy** (unknown–1852) or Miss **Orless** (dates unknown), except that they collected reptile or other vertebrate specimens from England and France for museum collections (Cleevely, 1983). In France, Mme **de Christol** (179?/18??–18??; unmarried name unknown) possibly was related or married to Jules de Christol (1802–1861), but she is regarded as a pioneer archaeologist and prehistorian who found evidence of the association of stone tools and animal bones in the caves of southern France in the 1820s. There are few references to her (e.g., an 1828 paper is cited by Osborn, 1915), but she is credited with contributing to the early understanding of the antiquity of man (Boule, 1923; Kass-Simon and Farnes, 1990).

It is hard to find out much about early paleontological women because they are mostly missing from biographic compilations (e.g., about five in Lambrecht et al., 1938, contra Cleevely, 1983). William Sarjeant (1978–1987), Marilyn Ogilvie and Joyce Harvey (2000), and Catharine Haines with Helen Stevens (2001) did much better listing several women in vertebrate paleontology. Although women during this time may not have actually been scientists—the profession had hardly taken off, and Mary Somerville (1780–1872) was cited

as the first "scientist" by William Whewell in 1833—they did begin collections, and they thought and wrote about their fossil discoveries. There may
be no women counterparts of Cuvier or Charles Lyell, but here we com-
memorate the able and sometimes ambitious women whose bone-hunting
lives began during the late 18th to mid-19th century (Table 2.1). They suc-
ceeded in producing important contributions to vertebrate paleontology,
notably influencing and assisting various areas of research (Gunther, 1904;
Cleevely, 1983; Turner et al., 2010a; Kölbl-Ebert and Turner, 2017). Recent
research is helping to correct some of the myths that have accumulated
about these women (e.g., Aldrich, 1990; Falk, 2000; Burek, 2004; Taylor and
Torrens, 2014a, b).

Nineteenth-century women who initially were mostly in Britain collected
fossils as a hobby (e.g., Etheldred Benett, Eliza Gordon Cumming, Barbara
Hastings); some turned them over or sold them to men for description, but as
the century unfolded, a few ventured into print, either in collaboration with
male family members or colleagues or as sole authors (e.g., Mary Anning,
Mary Buckland, Elizabeth Philpot). As Kölbl-Ebert (2002:4) writes: "This [fo-
cus on fossils] fitted nicely into the traditional female role as 'caretakers of
life.' It was socially acceptable for them to have an affinity to fossils as former
life-forms." British women including Anning, Gordon Cumming, Hastings,
and the Philpot sisters (Edmonds, 1978, discussed in more detail later) were
especially important as fossil collectors. Those with enough money and time,
such as Etheldred Benett and Barbara Hastings, were also able to hire collectors, purchase specimens, buy books, and correspond with the leading scientists of the time to identify their fossil discoveries (e.g., Torrens, 1995;
Goodhue, 2002, 2004; Davis, 2009).

The contributions of women "amateurs" (lovers of paleontology) of the
later 18th to early 19th century were generally accepted by their male counterparts (e.g., Torrens, 1995). For the most part, these women, like men at this
time, were not university trained or members of scientific societies. The earliest geological "club," the GSL, founded in 1807, was primarily a dining and
discussion forum for men and did not admit women to meetings until 1904,
and not to fellowship until 1919 (Herries Davies, 2007; Kölbl-Ebert, in press).
Likewise, the Linnean Society in London, though founded in 1788, did not
admit its first women fellows until 1905 (Gould, 1993). It was not until 1831
that women were allowed to attend geological lectures by Lyell at King's
College, London. A number of women had asked to be admitted to the talks,
but Lyell refused because he thought that women in a classroom would be

TABLE 2.1

Notable women VP pioneers of the 18th to 19th centuries

Name	Dates	Country	Major contribution
Mary **Anning**	1799–1847	England	Collected/sold reptiles, research
Molly **Anning**	1764–1842	England	Collected?/sold fossils; mother of Mary
Carrie Adeline **Barbour**	1861–1942	United States	Curator; collected mammals; charter member of SVP
Etheldred Anna Maria **Benett**	1776–1845	England	Collected reptiles; wrote book
Pliss Boggart	fl. 1890s	United States	Museum work? University of Wyoming Geological Museum, Old Main, 1899
Mary **Buckland**	1797–1857	England	Fieldwork; collected/illustrated bones for Cuvier, Buckland
Carolina von **Budberg**	17/18??–1847	Latvia	Collected fish
Sarah and **Mary Congreve (Congrieve)**	1737–1836; 1745–1823	England	Collected reptiles
Mme **de Christol**	179?–18??	France	Stone tools and animal remains, prehistory
Eliza Maria **Gordon Cumming**	179?–1842	Scotland	Collected fish
Barbara Yelverton **Hastings**	1810–1858	England	Collected reptiles
Harriet Sophia **Holland Hutton**	ca. 1835–1908	England	Fieldwork; collected fossil vertebrates
Harriett Mary **Hutton**	1873–1937	England	Collected reptiles
Mary Leeds	1858–1922	England	Collected and prepared reptiles
Anna **Leidy**	183?–1913	United States	Fieldwork; collected fossil vertebrates
Mary Ann **Mantell**	1795–1869	England	Collected dinosaur teeth
Charlotte **Murchison**	1788–1869	England	Collected fossil vertebrates
Rachel Brewer Peale	1744–1790	United States	Assisted in Peale's museum
Elizabeth **Philpot**	1780–1857	England	Collected fish, reptiles
Margaret **Philpot**	1786–1845	England	Collected reptiles
Mary Louise Philpot	1777–1838	England	Collected reptiles
Ann Marie **Redfield**	1800–1888	United States	Tree of life; fossil vertebrates
Mary Hone **Smith**	1784–1866	England	Collected reptiles
Hester **Stanhope**	1776–1839	England/Lebanon	Collected fish

Sources: Lambrecht et al., 1938; Sarjeant, 1978–1987; Cleevely, 1983; Burek and Higgs, 2007; modified from Turner et al., 2010a; Kölbl-Ebert and Turner, 2017; *SVP News Bulletins;* and internet.
Notes: Includes early collectors and geologists/archaeologists significant to VP. Where names are in bold, see more information in the text.

"unacademical." Several days later the mathematician, astronomer, and scientific writer Mary Somerville commented, "It is decided by the Council of the University that ladies are to be admitted to the whole course, so you can see what in[va]sions we are making on the laws of learned societies, reform is nothing to it" (Kölbl-Ebert, 1997, quoted in Patterson, 1983:93).

United Kingdom

The best-known pioneer fossil collector, often considered the first woman vertebrate paleontologist, is Mary Ann **Anning** (1799–1847), a carpenter's daughter with little formal education. As a child in 1810, she began collecting fossils in the Jurassic strata with her father and brother near her home at Lyme Regis on the English Channel coast (Plates 4). High cliffs near Lyme comprise medium- to thinly bedded shales, limestones, and sandstones that continuously yield fossils such as ammonoids, belemnites, nautiloids, gastropods, and bivalves, as well as vertebrates. All can be collected and are still sold in local shops (Davis, 2009; https://jurassiccoast.org). After her father's death, to survive, 11-year-old Mary and her mother, Molly (1764–1842), continued the business of selling fossils that her father had begun (e.g., Goodhue, 2004; Evans, 2010). Anning developed a remarkable skill not only in locating fossils but also in extracting them (e.g., Spalding, 1993; Creese and Creese, 1994; Torrens, 1995; Goodhue, 2002, 2004). She also made careful scientific observations and drawings, from which notable paleontologists and geologists including Cuvier, Buckland, the Swiss American geologist Louis Agassiz (1807–1873), and others benefited in their published scientific research (Torrens, 1995; Kölbl-Ebert, 2002; Turner et al., 2010a). She became an acknowledged expert in many aspects of paleontology, so that George Roberts (1834:284) saw her as the "worthy, Miss Mary Anning, whose name is already well-known in every part of the world where persons read of the discoveries and progress of science"—no small feat for a lower-class woman in the early nineteenth century (Plate 5).

Only a few highlights of her significant contributions to vertebrate paleontology will be provided here. Among her important firsts were specimens of Jurassic marine reptiles, a flying reptile, and fish. In 1811, at the age of 12, after her brother Joseph actually made the find, she excavated the skull and later skeleton of a marine reptile, an ichthyosaur. Anning sold the specimen for a much-needed £23. The ichthyosaur was described, without acknowledgment of her, by Everard Home (1814). The "Anning" reptile was later identified as *Ichthyosaurus* in 1821 by English geologists Henry De La Beche

(1796–1855), Anning's childhood friend (Morgan, 2014), and William Conybeare (1787–1857). Later in 1836 this specimen was illustrated by Buckland, another of Anning's friends and one of the most renowned early 19th-century British geologists. Anning, however, is not mentioned in these early scientific papers as the discoverer. The Philpot sisters (see later in this chapter) had earlier recovered fossil fragments of this taxon, but the identity of the fossils were as yet unrecognized.

Mary Anning went on to make many other important discoveries, including several additional well-preserved ichthyosaur skeletons. From a scientific standpoint, however, perhaps her most important find was a largely intact plesiosaur, which she located in 1824. This grand Lyme Regis 3-meter fossil skeleton, with its incredible snakelike neck, was an immediate international sensation. Up to that time, plesiosaurs had been poorly known, and Cuvier had rejected as fanciful Conybeare's and De La Beche's initial (1821) assertions about the structure of this animal. But when he read Conybeare's (1824a, b) further descriptions of *Plesiosaurus dolichodeirus* based on the intact specimen, Cuvier admitted he had been wrong and pronounced Anning's fossil a major discovery—the first plesiosaur. Again, nowhere in the description of this discovery is Anning's name mentioned. A second specimen, for which Anning had paid a mere £3 to sailors, was then bought on behalf of Cuvier in Paris for the meager sum of £10. Though at first he doubted Anning's honesty and the authenticity of the specimens, he and Anning entered into correspondence and she sold him specimens on a regular basis (Taquet, 2006). Once her fossils' authenticity was confirmed, the scientific community began to recognize their paleontological value. As her fame grew, Anning sold other specimens for considerably more, as much as £100–£200 in the 1830s, which equates to as much as £11,500–£23,000 today ($US14,000–$US28.000) (Torrens, 1995; Davis, 2009).

Flying reptiles, pterosaurs, had been known since the late 1700s but only from the Jurassic of Germany. In 1828 Anning discovered the first pterosaur skeleton outside Germany. This time she was credited for her discovery at Lyme Regis when Buckland (1829b) announced *Pterodactylus macronyx* at a GSL meeting. The taxon was later renamed *Dimorphodon* by English anatomist and paleontologist Richard Owen (1804–1892) (Owen, 1859).

Anning was also the discoverer of Jurassic fish, notably the ray-finned fish *Dapedius* sp., which she called her "turbot"; the shark *Hybodus*; and the primitive chimaerid *Squaloraja polyspondyla* (Taylor and Torrens, 1986; Goodhue, 2002; Davis, 2009). Louis Agassiz named a fossil hybodont shark for her as *Acrodus anningiae* (Agassiz, 1837). Her observations played a key role in the discovery that coprolites were fossilized feces (Häntzschel et al., 1968; Torrens, 1995; Davis, 2009). "Bezoar" stones (for the gallstones of bezoar goats that they resembled), as they were called at the time, were common in the abdominal cavities of the reptiles she found, and when broken open, they sometimes contained fossilized fish bones, scales, and the bones of smaller ichthyosaurs. Buckland (1829a, 1835) acknowledged Anning's insights about them.

Although Anning's remarkable fossil discoveries were important for generating public interest in vertebrate paleontology, nonetheless, as a woman, she was not eligible to join the GSL and, as we and others have noted, she rarely received credit for her scientific contributions. Indeed, she wrote in a letter, "The world has used me so unkindly, I fear it has made me suspicious of everyone" (Anon in Dickens, 1865; see Taylor and Torrens, 2014b). Clearly, she felt slighted. Anning complained to another fossil-hunting friend, Anna Maria Pinney (1812–1861) of Somerton Erleigh, Somerset, who confided to her diary that "the whole world has used her ill and she does not care for it, according to her account these men of learning have sucked her brains, and made a great deal by publishing works, of which she furnished the contents, whilst she derived none of the advantages" (Lang, 1955:55). Nevertheless, at the behest of her friends De La Beche and Buckland, an annuity of £25 (about US$2,875 in today's money) was acquired from the British Association for the Advancement of Science for her and her mother, Molly. Before the end of her life, Mary Anning did gain honorary status at both the Dorset Museum (Torrens, 1995) and the GSL.

As paleontologist Stephen Jay Gould (1941–2002) writes: "The beginnings of British vertebrate paleontology in the early nineteenth century owe more to the premier collector of this or any other age, Mary Anning, than to Buckland or Conybeare or Hawkins or Owen or any of the men who then wrote about her ichthyosaurs and plesiosaurs" (Gould, 1993:16). Anning's fossils were especially important in helping dispel religious beliefs of the time, which supported the world's creation in seven days and opposed the ideas of extinction and evolution. Apart from notes and drawings in her letters, Anning left few written records of her activities; she was, however, familiar with scientific literature, having transcribed papers and illustrations for her personal library. Reportedly, she taught herself French so that she could read the works of Cuvier (Davis, 2009). Her only publication is a short note to Edward Charlesworth, editor of the *Magazine of Natural History*, concerning details of the fossil shark *Hybodus* (Anning, 1839), in which she stated that a conversation with Agassiz had confirmed that the supposed frontal spine was in fact a tooth (Creese and Creese, 1994). Anning died of breast cancer on March 9, 1847, just short of 48, gaining an affectionate obituary from her friend (De La Beche, 1848).

Although she grew up in poverty and lacked much formal education, Anning's influence was worldwide and has continued. She taught herself so well that she was consulted by many eminent scientists of the era—her specimens were discussed, published, and obtained for collections, and in return, influential people came to visit her in Lyme—from Georges-Louis Leclerc, count de Buffon; Cuvier; Etienne Saint-Hilaire; and the king of Saxony in Europe (e.g., Torrens 1995; Turner et al., 2010a; Massare and Lomax, 2014; Lomax, 2019). Numerous books have been written about Anning, reportedly more biographical works than any British biologist other than Charles Darwin—many, especially the fiction and recent films, adding to her iconic status (Torrens, 1995; Taylor and Torrens 2014a; Table 2.2).

Anning's legacy is her role as the first true professional in vertebrate paleontology. Her contribution added such value to the contemporary Lyme Regis Jurassic Coast area that it gained UNESCO World Heritage status. In recent years and particularly important for STEM, Anning has become a popular female role model for young girls and her story has been incorporated into Dorset and other elementary school curricula (Turner et al., 2010a).

Alongside Anning, Etheldred Anna Maria **Benett** (1776–1845; Plate 6) is recognized as the first woman geologist (Nash, 1990; Burek, 2004). Benett was born into a wealthy family in Wiltshire, England, and was self-taught in geology and paleontology to some extent, because, like many of the middle-class women of that time, she could not receive a higher education (Torrens et al., 2000; Burek, 2001; Turner et al., 2010a). Her relative Aylmer Bourke Lambert (1761–1842), a founding member of all the important scientific societies in London, helped Benett develop a love of fossils and gain connections. Being unmarried, financially independent, and free from major family commitments, she spent much of her time and energy fossil collecting (Creese and Creese, 1994), becoming both a researcher in paleontology and an accomplished artist. At least by 1809, Benett had amassed an impressive fossil collection, mostly of invertebrates but also vertebrates, based on material she collected and purchased from others. Despite not being able to join scientific societies, she developed professional relationships and corresponded with a

TABLE 2.2
Selection of Mary Anning books and notable memorabilia

Date	Author	Title	Type	Notes
1828	Anon.	"Another discovery by Mary Anning of Lyme: an unrivalled specimen of *Dapedium politum* an antediluvian fish"	Newspaper article	*Salisbury and Winchester Journal*
1829	Anon.	*Edinburgh Philosophical Journal*	First list of British geological collections/collectors	Anning et al.
1857	Anon. (Frank Buckland, William Buckland, George Roberts)	*Chambers's Journal*	Article	Taylor and Torrens, 2014a
1858	Anon.	*Eclectic magazine* (New York)	Article	Copy of 1857 article
1859	Henry Rowland Brown	*Beauties of Lyme Regis*	Guidebook	Taylor and Torrens, 2014a
Post-1861	Sculptor?	Pair of German porcelain nodding figures	Lyme Regis Museum	Anning and brother or De La Beche: (M. Taylor, pers. comm., 1997)
1865	Anon.	In Charles Dickens, *All the Year Round*	Article	18 years after death
1908	Terry Sullivan	"She sells seashells on the seashore"	Rhyme	
1925	Charles Edmund Brock	*The World in the Jurassic Age*	Children's encyclopedia	Cites Forde (see next entry) as a man
1925	Harriott Anne Forde*	*The Heroine of Lyme Regis: The Story of Mary Anning, the Celebrated Geologist*	Biography	
1935–1936	William Dickson Lang	*Mary Anning, of Lyme, Collector and Vendor of Fossils* et al.	Biographical research	
1950, 1954	William Dickson Lang	"Mary Anning and Anna Maria Pinney"	Biography, article	
1963	Richard Curle	*Mary Anning, 1799–1847*	Biography	
1969, 1981	John Fowles	*The French Lieutenant's Woman*	"Victorian-era romance with postmodern twist" and 1981 film	Wikipedia
1980	Lynn Barber*	*The Heyday of Natural History*	General history	
1981	S.R. Howe, Tom Sharpe, and Hugh Torrens	*Ichthyosaurs: A History of Fossil "Sea-Dragons"*	Book	
1982	John Fowles	*A Short History of Lyme Regis*	Book	Reprint 1991
1991	Sheila Cole*	*The Dragon in the Cliff*	Novel	Reprint 2005

Year	Author	Title	Type	Notes
1995	Marie Day*	Dragon in the Rocks: A Story Based on the Childhood of the Early Paleontologist, Mary Anning	Children's book	
1995	Hugh Torrens	"Mary Anning (1799–1847) of Lyme; 'the greatest fossilist the world ever knew'"	Scholarly biography	British Journal for the History of Science
1995/1996	Crispin Tickell	Mary Anning of Lyme Regis	Book	
1997	Dennis B. Fradin	Mary Anning: The Fossil Hunter	Children's book	
1998	Nigel Clarke	Mary Anning 1799–1847: A Brief History		
1998	Tom Sharpe and P.J. McCartney	The Papers of H.T. De La Beche (1796–1855)		
1999	Jeannine Atkins*	Mary Anning and the Sea Dragons	Children's book	
1999	Catherine Brighton*	The Fossil Girl	Children's book	Michael Dooling, illustrator
1999	Don Brown	Rare Treasure: Mary Anning and Her Remarkable Discoveries	Children's book	Reprinted 2003
1999	Dennis Dean	Gideon Mantell and the Discovery of Dinosaurs		
1999	John J. Socha	"Anning, Mary: English, 1799–1847"	Scholarly biography	In R. Singer (ed.), Encyclopedia of Paleontology
2000	Deborah Cadbury*	The Dinosaur Hunters: A True Story of Scientific Rivalry and the Discovery of the Prehistoric World	Nonfiction book	
2000	Sally M. Walker*	Mary Anning: Fossil Hunter	Children's book	Afterword about the scientist's legacy
2002	Tom W. Goodhue	Curious Bones: Mary Anning and the Birth of Paleontology	Children's book	
2004	Tom W. Goodhue	Fossil Hunter: The Life and Times of Mary Anning (1799–1847)	Scholarly biography	
2006	Lawrence Anholt	Stone Girl Bone Girl	Children's book	
2006	Patricia Pierce*	Jurassic Mary: Mary Anning and the Primeval Monsters	Children's book	
2007	Caroline Arnold*	Giant Sea Reptiles of the Dinosaur Age	Children's book	Laurie A. Caple, illustrator
2008	Hugh Torrens	Anning, Mary, 1799–1847	Biography	Oxford Dictionary of National Biography
2009	Tracy Chevalier*	Remarkable Creatures	Fiction	E-book

(continued)

TABLE 2.2 (continued)

Date	Author	Title	Type	Notes
2009	Shelley Emling*	*The Fossil Hunter: Dinosaurs, Evolution, and the Woman Whose Discoveries Changed the World*	Biography	
2010	R.T.J. Moody, E. Buffetaut, D. Naish, and D.M. Martill (eds.)	*Dinosaurs and Other Extinct Saurians: A Historical Perspective*	Scholarly essays	Includes S. Turner*, C.V. Burek*, and R.T.J. Moody, 2010a
2010	Joan Thomas*	*Curiosity*	Novel	
2017	Maria Eugenia Leone Gold* and Abagael Rosemary West*	*She Found Fossils*	Children's book	
2017	Beth Strickler*	*Daring to Dig: Adventures of Women in American Paleontology*	Children's book	Alana McGillis, illustrator
2018	Christopher Powell	*Lyme Regis Monographs*	Nine historical essays	General history including Buckland, Conybeare Herridge*, 2019
2019	Francis Lee	*Ammonite*	Dramatic biopic film	

Note: Asterisk indicates a woman author.

number of 19th-century English geologists and paleontologists, including Buckland, Gideon Mantell (1790–1852), and James Sowerby (1757–1822). She was a correspondent and acquaintance of both Mantell and his wife, Mary Ann (see later in this chapter), and in 1819 Mary Ann Mantell accompanied Benett "in her carriage to see the collections of my [Mantell's] geological friends" (Curwen, 1940:6).

Benett was a contemporary of William Smith, considered a founder of British stratigraphy, and undoubtedly knew of his ideas. She started study-ing and collecting fossils in earnest in about 1800, when she was 23, and she herself gauged the value of using fossils to correlate the rock formations that she studied and proffered a stratigraphic section of Wiltshire's Upper Chicks-grove Quarry to the GSL in 1815. Benett's catalogs of Wiltshire fossils in-clude more than 1,900 specimens representing 696 species, and, as such, she made significant contributions to the region's biostratigraphy. After her death, much of her collection was purchased by the English expatriate and physician Thomas Bellerby Wilson (1807–1865), who donated it to the ANSP. Other, smaller portions of her collection reside in London, in Saint Peters-burg, and at the Leeds Museum, while other specimens were given to vari-ous researchers and later dispersed (Cleevely, 1983). Among notable fossil vertebrates that she collected was a vertebral centrum, the holotype of *Ich-thyosaurus trigonus*, transferred to a new genus (*Macropterygius*; Owen, 1840). Although Owen did not figure the specimen, the holotype bearing Benett's handwritten label was found in the ANSP's Vertebrate Paleontology Collection (Spamer et al., 1989). A systematic list of Benett's fossil vertebrate remains (their appendix D) including shark, bony fish (*Leptolepis*), crocodile, turtle, plesiosaur, ichthyosaur, and a mammal is provided by Torrens et al. (2000). Interestingly, when the Imperial Moscow Society of Naturalists (Khain and Malakhova, 2009) offered her membership in 1836, they had mis-takenly thought that Benett was a man.

Mary **Congreve** (or Congrieve) (1745–1823) of Warwick was another En-glish fossil reptilian hunter of which little is known, and there might even have been two collecting sisters, Mary and Sarah (1737–1836) (Turner et al., 2010a). Hugh Torrens (pers. comm. to T. Goodhue, April 18, 1996) believes that, unlike the Philpots, Congreve and perhaps her sister primar-ily collected fossils found by others rather than excavating the fossils them-selves. Conybeare was well acquainted with at least one Congreve, for he de-scribes examining an ichthyosaur skull in her possession (Goodhue, 2004).

In Scotland, Eliza Maria **Gordon Cumming** (née Campbell) (1797?–1842) was an aristocrat living on the shores of the Moray Firth, on beds of the Middle Devonian Old Red Sandstone. Her 1839 discovery of the placoderm fish *Coccosteus* at Lethen Bar lime quarry was major. She was a skilled painter who, along with her eldest daughter, Anne Seymour Conway Baker Cresswell (1818–1858), drew and painted their finds. Gordon Cumming corresponded with Agassiz, Buckland, and Roderick Murchison, who all visited and, notably Agassiz (1833–1844) used her collection. She sent illustrations, letters, and specimens around Europe and intended to publish her drawings and theories on how these fish would have appeared in life. Some of these illustrations survive in the archives of the GSL. Her highly respected work was praised both by Agassiz and by the Scottish paleontologist Hugh Miller (1802–1856), who wrote that she had collected the remains and distributed them among geologists with the greatest liberality. Gordon Cumming had studied the remains with great care, and prepared drawings of all her most perfect specimens with a precision of detail and artistic talent, which few naturalists could hope to match (Agassiz, 1844–1845:vii; Miller, 1852; Norquay, 2012). Agassiz's (1833–1844) masterwork on fossil fish benefited yet again by his use of Gordon Cumming's collection and expertise, allowing him to write up and disseminate *her* findings (Cleevely, 1983; Aldrich, 1990; Creese, 2007; Orr, in press). During her 13th pregnancy, Gordon Cumming was impatient to get back to her studies, writing to Roderick Murchison, "I am breathless to be at work again" (Geological Society, London, Murchison Papers GSL M/G11/1a-b April 21, 1842) in vain because she died that day from complications following the birth (Kölbl-Ebert, 2007; quoted by Norquay, 2012:51). Agassiz named a Devonian ray-finned fish in her honor (Appendix 4). Many of the fossils in Gordon Cumming's collection were personally identified by Agassiz (Andrews, 1982) and are now held by the National Museum of Scotland, Edinburgh, the Natural History Museum, London (NMH), and the University of Neuchatel, Switzerland.

Barbara Rawdon **Hastings** (née Yelverton) (1810–1858; Plate 7), Marchioness of Hastings, was born into nobility, the only child of Henry Edward Yelverton, the Baron Grey of Ruthin (Lambrecht et al., 1938; Dadley, 2004; Kölbl-Ebert, in press). Her father died when she was an infant. Lady Grey de Ruthyn, as she was known, married George Augustus Francis Rawdon Hastings, the 2nd Marquis of Hastings, in 1831 and had six children. After that husband's early death, she married for a second time, in 1845, to Captain (later Admiral) Hastings Reginald Henry, who later took the name Yelverton, and had another child. She was a prodigious collector of fossil vertebrates, initially near the Isle of Wight, off the south coast of England, in 1840, but she also purchased fossils from around the world. Following her second marriage and after her children had grown, she transferred her collecting efforts to coastal Hampshire at Hordle and Beacon Cliffs, west of Milford-on-Sea. She made detailed stratigraphic observations of the cliffs and published three papers on her work (Hastings, 1848, 1852, 1853). During this time she maintained a productive, enthusiastic correspondence with Richard Owen, then at the Royal College of Surgeons Hunterian Museum in London, and relied on him to identify the fossils she discovered. She wrote to Owen, whom she referred to as "the British Cuvier," near the birth of her 7th child about one such fossil discovery: "Of the Hordle *Tri[onyx]*—I have a large series of bones and carapaces for you. Pray call one Barbarae I shall be charmed and just now I move about very Tortoise like" (Hastings, undated a, quoted in Kölbl-Ebert, 2004). Later, in fact, Owen did name a new turtle species in her honor, *Trionyx barbarae* (Owen and Bell, 1849) (Appendix 4).

Hastings was also considered a skilled preparator, and her knowledge of anatomy helped her repair broken fossils. Owen commented on the extremely fragile and crumbly character of some of the most valuable of her finds—turtle shells and crocodile skulls when they were first extracted—and the care with which she had "readjusted, cemented and restored them to their original condition" (Owen, 1848; quoted in Creese and Creese, 1994:31). In a letter to Owen, Hastings revealed her personal connection with collecting: "I am enchanted my dear Mr. Owen at my good fortune—You know I have had these remains floating abt in my head ever since 39 when you named that Phalanx found by me in the same spot, and figured in yr mammalia but I have never been able to return there till now—I brought home after a most arduous miry wet walk the other day two more *Iguanodon* like teeth" (Hastings, undated b, quoted in Kölbl-Ebert, 2004:81). Given her social position and her commitment to fossils, Hastings was able to present a paper and exhibit her turtle shell and crocodile skull fossils at the British Association for the Advancement of Science meeting at Oxford in 1847; after this meeting, in 1848, Owen renamed the skull *Crocodilus hastingsiae*, which had earlier been named in her honor (Wood, 1846; Creese and Creese, 1994). In addition to Owen, Hastings also was visited by Lyell and Buckland, as well as other noted geologists and paleontologists. By 1855 her fossil-collecting days were apparently over (increasing ill health) and she sold a large number of Tertiary mammals and reptiles to the British Museum in London (now the Natural History Museum [NHM]) for £300 (approximately US$40,000 in today's money) (Cleevely, 1983).

Hastings's contributions to vertebrate paleontology are mainly those fossils she and local women and children collected from the Eocene Lower Headon Beds of Hordle Cliff. The "Hastings Collection" consisted of several thousand fossil specimens from England and Europe, including "a series of choice cabinet Lias specimens from the collection of Mary Anning" (Cleevely, 1983:146). Her interest and excitement in her fossils and desire to learn more about them are evident, as she wrote to Owen, "Can you tell me [a] book [that] is likely to be of service to me in my Paleontological researches? I have Cuvier—I should think that your notes on the Com[parative] Anat[omy]. of the vertebrata wh I have ordered wd be a great assistance—I am dying to see the Chelonian book" (Hastings in litt.; see Kölbl-Ebert, 2004). Hastings died in Rome at only 48 years old while traveling to visit her husband in the Royal Navy.

Harriet Sophia **Holland** (later **Hutton**) (ca. 1835–1908) and Harriet Mary **Hutton** (1873–1937) were English mother-and-daughter fossil collectors. Sophia's father was a well-known agriculturalist, and their home was surrounded by the Jurassic Inferior Oolite, which made ploughing difficult but which produced excellent fossils, such as the ammonites used by William Smith to work out English stratigraphy. When living at Dumbleton in Gloucestershire, Harriet also collected Liassic reptiles, mentioned by Thomas Wright in the local field club (Cleevely, 1983). Her interest in fossil collecting was a hobby passed on to her daughter Mary, whose collection was given to the Cheltenham Museum on her death. Later, some of Mary's reptilian fossils were passed on to other museum collections, including that of the Gloucester Museum, which houses the crocodile *Steneosaurus* and other saurian fossils. Before her death, Mary became one of the rare women who was elected a fellow of the GSL (Turner et al., 2010a; Turner, in press).

Several other British women worked with their husbands as geological assistants, making observations, creating illustrations, and assisting in the collection of specimens. Charlotte Murchison (née Hugonin) (1788–1869), wife of geologist Roderick Impey Murchison (1792–1871), influenced him to leave his first career in the army and introduced him to minerals, rocks, and fossils, accompanying him on excursions and making her own collection. In Lyme Regis she "went fossilizing" and enriched her collection; later she corresponded with Anning, even promoting her in society (Kölbl-Ebert,

1997a:40). Anning was grateful for Charlotte's kindness in mentioning her skeletons to different collectors (Kölbl-Ebert, 1997a).

Two more important women, Mary Mantell and Mary Buckland, both made noteworthy contributions to VP. Mary Ann **Mantell** (née Woodhouse) (1795–1869), wife of Gideon Mantell, has been credited with the discovery (Mantell, 1822; Colbert, 1984:16, writes "unwittingly"; we might say seren-dipitously) of the first fossil tooth of the dinosaur *Iguanodon* (see Chapter 6), which led to her husband's publication announcing the discovery (Creese and Creese, 1994; Dean, 1999; Torrens, 2008; Turner et al., 2010a). Gideon him-self says that Mary found the first specimens of the teeth (Mantell, 1822; dis-cussed in Weishampel and White, 2003:90). Mary **Buckland** (née Morland) (1797–1857; Plate 8) was not only William Buckland's wife but also a 19th-century pioneer English paleontologist, marine biologist, and scientific il-lustrator (Torrens, 2004). In terms of vertebrate paleontology, she is perhaps best known for her fossil illustrations (e.g., see also Chapter 6). She had made a name for herself before marriage as "a scientific draughtswoman, a collec-tor of fossils and a woman with many scientific accomplishments" (Kölbl-Ebert, 1997b:33; Turner et al., 2010a). Her interests in natural science were encouraged at a young age by her father and then by Oxford physician Chris-topher Pegge and his wife, who took on her upbringing, a not uncommon practice in the late 18th to early 19th century, when children were often sent to another home (e.g., McGowan, 2001; Worsley, 2017). She accompanied Buckland on various field trips; made observations, models, and sketches; and assisted in the preparation of papers and books (Kölbl-Ebert, 1997b; Turner et al., 2010a). Sadly, their honeymoon coach trip traveling the continent led to a terrible accident where both were injured, and this may have contributed to their ill health later in life (Morgan, 2019).

Her husband disapproved of women appearing as official participants in scientific gatherings, stating, "If the meeting [of the British Association for the Advancement of Science] is to be of scientific utility, ladies ought not to attend the reading of the papers—especially in a place like Oxford" (Gordon, 1894:123). This did not hinder him from relying on the scientific abilities of his wife, as was written on them, "His prose, corrected by his wife Mary Morland, was very fine indeed, and probably superior to that of any of his col-leagues" (Rupke, 1983:7). When William awoke with an idea about fossil tracks at two o'clock one morning, Mary helped him decipher footmarks found in a slab of sandstone by covering the kitchen table with pie crust dough while he fetched their pet tortoise and together they confirmed his intuition, that tortoise footprints matched the fossil prints of *Chirotherium* (Bowden et al., 2010), a relative of crocodiles. In addition to having her own fossil collection, Mary was actively involved in the Oxford University Museum, preparing and repairing fossils. Her time to pursue science, however, was limited given her responsibilities raising five of the couple's nine children that survived childhood and performing other household tasks, but her support and aid throughout allowed William to achieve prominence (e.g., Kölbl-Ebert, 1997b; Morgan, 2019).

Elizabeth **Philpot** (1780–1857), the better known of three sisters (Margaret, 1786–1845, and Mary, 1777–1838), was an English amateur paleontologist and artist who collected fossils from Lyme Regis beginning in 1805. Although all three sisters contributed to fossil collecting, it was Mary who corresponded with leading geologists of the time William Buckland, Conybeare, and De La Beche about the collection, which was well known for its fossil fish. It also contained the fossil teeth that Buckland combined with a partial skeleton found by Mary Anning when he described in 1829 the pterosaur *Pterodactylus macronyx*. Despite a nearly 20-year age difference, Elizabeth was a close friend of Anning and they were often seen fossil collecting together. Later, Agassiz visited Lyme Regis to collect and study fossil fish with Elizabeth and Anning. He was impressed by their knowledge and wrote in his journal, "Miss Philpot and Mary Anning have been able to show me with utter certainty the dorsal fins of sharks that correspond to different types" (Emling, 2009:166–170). In appreciation for their work, he named a fossil fish species *Eugnathus philpotae* after Elizabeth and another two fish species after Anning (Plate 9; Appendix 4). In a letter dated December 9, 1833, Elizabeth wrote to Mary Buckland and enclosed a sketch of an ichthyosaur head that she painted using ink from a fossil squid of the same 200-million-year-old rocks as the ichthyosaur (More Than a Dodo, 2016, accessed June 11, 2019). In addition to letters from Elizabeth Philpot, the Philpot collection of nearly 400 fossils is held in the Oxford Museum of Natural History and contains a number of type specimens (Powell and Edmonds, 1976; Davis, 2009).

Mary Hone **Smith** (1784–1866) was an amateur collector in Tunbridge Wells, England. She spent considerable time in the acquisition of Cretaceous fossils, collecting her own specimens from local chalk quarries. Like Anning and others, she also purchased material from quarrymen and exchanged with other collectors. Gideon Mantell was well acquainted with her; he visited her in Tunbridge and inspected her fossil collection from the Chalk, including a beautiful specimen of 30–40 vertebrae with ribs, and the jaws and teeth of "a small lizard, allied to the Agama; such a beauty! and from Kent" (Curwen, 1940:158, 198). Some of her fossil reptiles were used by Mantell and Owen and figured in natural history museum catalogs; other fossils are listed in the Brighton and Nottingham Natural History Museums and in the Institute of Geological Sciences, Keyworth (Cleevely, 1983; see also Creese and Creese, 1994; Kölbl-Ebert, 2004; Turner et al., 2010a).

Middle East

Hester Lucy **Stanhope** (1776–1839; Plate 10), noted eccentric British aristocrat and traveler through the Middle East, dabbled in fossil bone hunting near her adopted home of Sidon, Lebanon. This daughter of the great Chatham family and niece of and private secretary to British prime minister William Pitt the Younger abandoned her European life and became a legend in the Middle East (Blanche, 1959; Gibb, 2005). In 1810, at 33 years old, she left behind her British life and went on to make pioneering archeological finds in Askelon (Israel). She gained great respect from Arab tribal elders. They and visitors alike found a statuesque, commanding woman, nearly six feet tall (two meters), dressed in voluminous Turkish-style trousers, not veiled but with her head shaved and covered by a turban. The poet Lord Byron thought of her as "that dangerous thing, a female wit" (Byron, 1973; Nicholas, 1982). Stanhope got hooked on fossils along with other curios, as was typical of the time in any "Grand Tour," when the wealthy made collections during their overseas journeys (e.g., Jardine et al., 1996; Burek and Higgs, 2007). She collected and acquired from near her adopted home of Sidon, at a site near Sahel Alma, Hajoula (or Haqel), and amassed a collection from the long-known Middle Cretaceous fossil-fish-bearing beds of Lebanon. Her fish collection was given (in part) to the NHM (Nicholas, 1982; Cleevely, 1983). Stanhope died alone in her home, the former monastery at Djoun, in June 1839, at the age of 63.

United States

American women participated in paleontology in the late 18th and early 19th centuries mainly as assistants or by illustrating scientific papers. Women were learning about vertebrate paleontology from discoveries of mammoth and mastodon bones and visiting museums such as those founded by Charles Willson Peale (1741–1827) in 1786. Women and children were welcome at this first major US natural history museum in Philadelphia (see the painting that depicts a Quaker woman gazing in wonder at the bones in the museum: Sellers, 1980:ii). Peale's wife, Rachel (1742?–1790), and the wives of vertebrate paleontologists, such as Anna Harden **Leidy** (183??–1913), assisted their husbands with fieldwork, microscopy, and writing and illustrating their papers (Sellers, 1980; Warren, 1998; see Chapter 6). Joseph Leidy (1823–1891) did not marry until late, and he may have chosen Anna because of her interest in biology. She went out west with him to Wyoming in 1872 and collected fossil vertebrates, especially mammals (Warren, 1998:146). The Leidys both attended American Association for the Advancement of Science (AAAS) meetings in Philadelphia (Ewing, 2017).

Colleges were founded in the United States in the 17th century—Harvard, Yale, and Dartmouth—but it wasn't until 1833, when Oberlin College first admitted women, that coeducational colleges were started (Bailey, 1994; Rossiter, 1982). The first institution to establish a graduate program for women was Bryn Mawr College in 1885, although female students were still segregated from their male colleagues (Bailey, 1994; Ogilvie, 1986; Rossiter, 1982). Paleontology was taught as part of geology courses at the early women's colleges, notably Mount Holyoke (founded in 1837), although it was not until the 20th century that graduate programs in paleontology were established (Aldrich, 1990). Florence Bascom (1862–1945), considered by many the first American female geologist, influenced the field significantly (Sarjeant, 1978–1987; Burek and Higgs, 2007). She was born in Williamstown, Massachusetts. Her mother was a suffragette, her father a professor and then president of the University of Wisconsin. Her interest in geology was aroused after a visit to Mammoth Cave in Kentucky. She received a PhD from Johns Hopkins University in 1893 by special dispensation, as women were not admitted officially until 1907. She is important because she taught the next generation of young women, one of whom, Mignon Talbot, was to make a significant contribution to American vertebrate paleontology early in the next century (see Chapter 3).

One early American college-educated woman who added to the debate over Cuvier's and Agassiz's ideas was Ann Marie **Redfield** (née Treadwell) (1800–1888). She is perhaps best known for having produced an early tree-like although nonevolutionary wall chart showing animal relationships (Redfield, 1857). She also wrote about fossil vertebrates. Redfield's educational wall chart yielded great influence at the time (Butts, 2010). Redfield was born in Canada to wealthy parents and was well educated. She moved to Clinton, New York, for postgraduate classes at the precursor of Hamilton College, gaining a master's degree from the now defunct Ingham University, which has the distinction of being the first institution of higher learning for women in the United States (Archibald, 2014).

Compared with their British female colleagues, American women were slower to enter the field of vertebrate paleontology even as "amateur" collectors. In George Gaylord Simpson's (1942) paper on the beginnings of vertebrate paleontology in North America covering the time period up to 1842, not a single woman is mentioned even as a collector (but see Aldrich, 1982, for coverage of women in VP during this time). The first woman scientist was elected to the AAAS in 1850, but women neither presented papers nor held elected office until much later. Beginning in the 1870s, women were elected fellows of the AAAS in recognition of their scientific achievements (Rossiter, 1982). It wasn't until the late 1800s that women with formal university training in geology and paleontology, such as Talbot and others, appeared in increasing numbers in Britain, the United States, and Russia (see Chapter 3). Sketching was taught by schools for girls (e.g., Troy Female Seminary in New York), and as in Europe, drawing was considered an appropriate genteel skill that women could display without being considered unfeminine (Aldrich, 1990; Kölbl-Ebert, 2012; Turner et al., 2010a; see more in Chapter 6).

Russian Empire

Russian women were collecting fossil bones early in the 19th century. Carolina von **Budberg** (17/18??–1847), a young German Baltic aristocrat, was one of the first to find fossil remains in the Livonian Governorate of the Russian Empire (modern Latvia), collecting placoderm fish bones at Lake Burtnieki. Budberg showed her collection and graciously presented several interesting fossils to the German scientist Georg Friedrich Parrot (O. Lebedev, pers. comm.). Parrot (1836) then described these fossil fish and gave them to the Saint Petersburg Academy of Sciences; he noted that she had already been collecting 10 years earlier, probably from 1820 to 1826, but we do not know her age at the time.

There was also interest in the sciences among women in the Russian imperial family that can be traced back to the 1850s. Maria Nikolaevna **Romanova** (1819–1876), daughter of the emperor Nicholas I, sent the Moscow Society of Naturalists the lower jaw of a fossilized Ice Age rhinoceros in 1855. According to Olga Valkova (2008:140), the society was particularly honored by this gesture, so it asked for permission "to beautify the list of its honorable members by including the name of Her Highness." Nikolaevna is thus recorded in the imperial society, and this provides testament to her interest, albeit indirectly, in paleontology.

Developing Interest in Fossil Bones

The work of the women discussed here became part of the growing understanding of fossil bones from around the world. Even early in the 19th century, women writers took on the role of educators and contributed literature to teach the following generations about vertebrate paleontology. Maria Hack (1777–1844) was a science educator who wrote a popular manual specifically for children about geology—her 1832 *Geological Sketches and Glimpses of the Ancient Earth* made use of Agassiz's and earlier vertebrate finds and espoused Lyell's gradualism framework. Her book was intended for use by mothers educating their sons about the new fossil finds, including Conybeare's (actually Mary Anning's) plesiosaur. Mary Roberts (1788–1864), a rather idiosyncratic writer, was—like Hack—a lapsed Quaker, but she retained a strong religious moralistic world view (Gould, 1993). Roberts particularly disagreed with Cuvier on the "carnivorous" elephant, or Mastodon, of Ohio, which he had described as herbivorous; she did not accept this, based on its enormous grinding teeth, which she thought typical of a carnivore. In contesting his authority, Roberts wished to demonstrate the compatibility of scripture and geology. For her, the days of creation, for example, were clearly to be understood as literal.

A final 19th-century woman writer who brought vertebrate fossils to children was Arabella Burton Buckley (1840–1929), who had worked as Lyell's secretary and for many years was within the inner circle of rising geologists. Her *Winners in Life's Race; or The Great Back-Boned Family*, with a preface written in 1882, reflects the new Darwinian agenda, which dealt with reptiles and saurians but strangely no dinosaurs per se. Her work has influenced many up to the present day (Barber, 1980; Turner et al., 2010a).

By the middle of the 19th century, the discovery and identification of so many extinct vertebrates by both men and women had led to much discussion in Europe, the Americas, and Australia about the significance of these bones. Finally, after two decades of thought, and with the encouragement of his friends and particularly his wife, Emma, Charles Darwin published *On the Origin of Species by Means of Natural Selection* in 1859 (Healey, 2001). It was an instant hit. It was to change thinking on all aspects of life. In America, Joseph Leidy was the first to write to congratulate Darwin and admit that he had been thinking along the same lines but under a fog until he read the book (Warren, 1998). Leidy was to influence at least one young woman (Chapter 3) before his star waned. Darwin's (1859) *Origin* cleared the way for new ideas, including those about the nature and role of women. Young women growing up in the late 19th century were becoming captivated by natural history and fossils (Barber, 1980), spurred on by Darwin to think of becoming scientists and even paleontologists. Young Beatrix Potter, Sophia Holland's granddaughter, best known for her children's books, was one (Gardiner, 2000). Her interests in geology and natural history grew, and she developed her well-known first love of collecting, drawing, and painting, with some fossil fish teeth among her specimens (Lear, 2007). Potter also photographed fossils and identified them by reference to NHM collections (Gardiner, 2000).

Summary

Several common themes regarding early female scientists (Burek and Higgs, 2007), including those in VP, are described in this chapter. First, they were mostly British or Western European, and often born into influential families (i.e., Benett, Budberg, Gordon Cumming, and Hastings), which allowed them to work voluntarily and usually for no pay. Second, women scientists of the 18th to mid-19th century had a pioneering, even rebellious, spirit, since society as a whole did not support their ambitions. Most women did not have access to universities, and scientific societies were slow to accept them. Although science was largely seen as a masculine pursuit, several prominent male scientists of the day (i.e., De La Beche, Buckland, Lyell, Mantell, Murchison, and Owen) had wives or women assistants and even encouraged them to write and illustrate their fossil discoveries; this allowed some women to advance and achieve recognition in the field. Despite a legacy of tradition, prejudice, and adversity, women contributed significantly to the early development and history of vertebrate paleontology. Among the notable early contributions to VP by women are specimens of marine reptiles, including the first plesiosaur and ichthyosaur, as well as the first pterosaur skeleton discovered outside Germany, collected by Anning; extensive fossil fish collections (Anning, the Philpots), together with precise, detailed specimen drawings (made by Gordon Cumming and her daughter), to be described by Agassiz; and the first fossil dinosaur tooth of *Iguanodon*, discovered by Mary Mantell.

CHAPTER 3

Women in Vertebrate Paleontology, Late 19th to Early 20th Century

> Nothing in science has any value to society if it is not communicated.
> Anne Roe (Simpson), *The Making of a Scientist* (1953:17)

The latter half of the 19th century brought increasing wealth to the middle classes in the developed nations, such as Britain and the United States. The opening of colleges and universities to women in the 19th century in the United States, Europe, and elsewhere marked the participation of university-educated women in geological and paleontological research. Women began to gain work within the discipline; Michele Aldrich (1982), Martina Kölbl-Ebert (e.g., 2012), and Kölbl-Ebert and Susan Turner (2017) have explored their growing emancipation. By late century, women such as Harriet Huntsman and Cecilia Beaux (see Chapter 6) were often employed by early American male paleontologists, including Edward Drinker Cope (1840–1897), to illustrate reports for state geological surveys. The number of libraries and museums also grew (e.g., Sheets-Pyenson, 1988; Pyenson and Sheets-Pyenson, 1999), providing new opportunities. We focus on the careers of women vertebrate paleontologists and zooarchaeologists during this time who made important fossil discoveries and significant, lasting contributions (Table 3.1).

Most women VPs were still in the United States and Britain, but young women also began to train in France, Germany, and Russia. The major shift in Russia occurred in 1917 with the Russian Revolution, after which more women gained university educations across the Soviet bloc, and geology, biostratigraphy, and comparative zoology flourished. This and the events of World War I led to a surge in employment for women as they took jobs vacated by men in universities and other institutions. The post–World War I period was particularly important in opening up employment on a more egalitarian basis. Nevertheless, in this era, if women married, they usually lost their jobs.

Africa

African-born Marjorie Eileen Doris **Courtenay-Latimer** (1907–2004), although not a vertebrate paleontologist, is discussed because she is well known as the discoverer in 1938 of the coelacanth, a "living fossil" thought

TABLE 3.1

A selection of women VPs active in late 19th to-mid 20th centuries

Name	Dates	Country	Research	Major contribution
Karoline "Lotte" **Adametz**	1879–1966	Austria	Mammals	Research
Annie Montague **Alexander**	1867–1950	United States	Vertebrate paleontology	Collector, museum
Dorothea Minola Alice **Bate**	1878–1951	Wales, England, Mediterranean	Zooarchaeology of mammals, birds	Collector; research, museum
Elizaveta I. **Beliaeva**	1894–1983	Russia/USSR	Mammals	Research; SVP member
Marjorie **Courtenay-Latimer**	1907–2004	South Africa	Fish	Collector first coelacanth
Agnes **Crane**	1852–1911	England	Fish, mammals	Research
Mathilde **Dolgopol de Sáez**	1901–1957	Argentina	Fish, reptiles	Research
Tilly **Edinger**	1897–1967	Germany, United States	Reptiles, mammals	Paleoneurology; SVP charter member
Madeleine **Friant**	1897–197/8?	France	Mammals	Research
Dorothy Garrod	1892–1968	England, Mediterranean	Zooarchaeology Curator of mammals, humans	Collector, fieldwork
Zinaida **Gorizdro-Kulczycka**	1884–1949	Russia/USSR/Poland	Fish	Research
Vera Isaacovna **Gromova**	1891–1973	Russia/USSR	Mammals	Research; SVP member
Ina von **Grumbkow**	1872–1942	Germany	Dinosaurs	Site management; drawings
Eileen Mary Lind Hendriks	1887–1978	England	Geology	Collector; fish named for her
Fanny **Hitchcock**	1851–1936	United States	Fish	Research (Paleozoic sharks)
Erika von Hoyningen-**Huene**	1905–1969	Germany	Reptiles, mammals	Research
Anna Boleslavovna **Missuna**	1868–1922	Russia/USSR	Fish	Curator; research
Maria Vasil'evna **Pavlova**	1854–1938	Russia, France	Mammals	Research; teacher
Helga **Pearson**	1899–1975	England	Reptiles, mammals	Research
Antje **Schreuder**	1887–1952	Netherlands	Mammals	Research
Igerna Brunhilda Johnson **Sollas**	1877–1965	England	Fish, mammals	Research
Mignon **Talbot**	1869–1950	United States	Geology, reptile	Research

Sources: Wikipedia and references herein.

Notes: Still primarily from English-speaking and other European countries. Where names are in bold, see more information in the text.

to have been extinct at least 65 million years and important in the story of the transition of vertebrates from water to land (e.g., Thomson, 1991, 1992; Weinburg, 1999). Courtenay-Latimer was born in East London, South Africa. She was encouraged by her parents to follow a variety of nontraditional interests, including ornithology, natural history, and cultural history. Although she trained as a nurse, given her local natural history expertise, she was hired at the age of 24 as the first curator at the East London Museum. Courtenay-Latimer's desire to see and collect unusual specimens was known to fishermen. In December 1938, she received a telephone call that such a fish had been brought in. She hauled the fish to her museum and tried to find it in her books, without success, but realized that it was different, having thick, hard scales, four limb-like fins, and a strange "puppy dog" tail. Eager to preserve the fish but having no facilities at the museum, Courtenay-Latimer reluctantly sent it to a taxidermist to skin and gut. Several weeks later her friend, noted South African ichthyologist J. L. B. Smith (1897–1968), came and recognized that the fish was a coelacanth; in 1939 he named it for her, *Latimeria chalumnae* (see Smith, 1956).

It would be fourteen more years before another coelacanth was caught, in 1952, from the Indian Ocean after an international effort by Smith and colleagues to find more specimens (Thomson, 1991). In 2000 a further surprising coelacanth was found, identified again by a woman, this time in a local Sulawesi fish market, and recently more specimens have been recovered from Indonesian waters (e.g., Erdman, 1999). Recent analysis of the coelacanth genome has confirmed a close link between lungfish and tetrapods, with coelacanths occupying a side branch of the vertebrate lineage. Nevertheless, the discovery of the coelacanth is considered one of the most important zoological finds of the 20th century. Although best known for this discovery, Courtenay-Latimer continued her broad interest in the natural history of her part of the world, investigating dodo eggs. In 1935 she excavated the skeleton of the Triassic dicynodont *Kannemeyeria simocephalus* from a site near Tarkastad in the Eastern Cape (Courtenay-Latimer, 1948). Courtenay-Latimer died on May 17, 2004, at the age of 97 (e.g., Williams, 2004).

German Viktorine Helene Natalie "Ina" von **Grumbkow** (1872–1942) accompanied her husband, geologist and volcanologist Hans Reck (1886–1937), to Africa, where she contributed to the success of the East African fossil expeditions of the Museum für Naturkunde Berlin. In 1912 she was given the responsibility of supervising paleontological excavations of Jurassic fossils (including dinosaurs) at Tendagaru in the southeastern part of today's Tanzania. The excavations were a major undertaking involving up to 400 indigenous workers. When her husband went on long expeditions to other areas, including the important site of Olduvai (see Chapter 4), which he discovered, she assumed full responsibility for supervising and maintaining the excavations. During this time she kept records, wrote about her work, and made sketches of the workers, landscape, and wildlife that were subsequently used as a basis for her oil paintings (Maier, 2003; Mohr, 2010; Turner et al., 2010a).

Europe

United Kingdom. Dorothea "Dorothy" Minola Alice **Bate** (1878–1951) was a British paleontologist and zooarchaeologist (Swinton, 1951; Sarjeant, 1978–1987). She was born in Carmarthen, Wales, and came from a military background, her father rising to chief of police when she was young. Despite little formal education and her claim that her education was only briefly interrupted by school, she grew up in Wales with a love of nature and then moved to the ancient coal-mining district of the Forest of Dean in the west of England (Shindler, 2005). Bate became one of the most formidable and well-respected vertebrate paleontologists of the 20th century, with a record of collecting and identifying material from over 60 sites in Europe and the Middle East. She may have been the first (in the Old World, at least) to make use of four-wheel-drive vehicles, using a Ford pickup to do fieldwork in Jordan before World War II (Shindler, 2005).

Bate's scientific pursuits began in the Forest of Dean, where she excavated the bones of small Pleistocene mammals from a cave into which she scrambled, which led to her first scientific publication (Bate, 1901). At the age of 19 she had gained a position at the British Museum (Natural History) (BMNH, now the NHM), reportedly by "demanding a job" (Shindler, 2005, 2007; Kölbl-Ebert and Turner, 2017). She became the first woman employed (actually a paid unofficial officer), sorting bird skins and later preparing fossils, and eventually she became curator of fossil mammals. Bate maintained her position at the BMNH for 50 years and pursued vertebrate paleontology, geology, ornithology, and anatomy. Her work was mostly unpaid; she was given small amounts of money depending on the number of fossils curated plus occasional travel funds. Her paleontological contributions were varied, with research on topics ranging from Tertiary and Holocene bonebeds to cave faunas. She worked in Europe and farther afield in the Mediterranean in Cyprus

(1903–1911), Gibraltar, Corsica, and Sardinia (see Shindler, 2005). She also worked on a collection of fossil ostriches from China made by British surgeon and ornithologist Percy Lowe (Shindler, 2005).

She worked mainly with other women, such as the English archaeologist Dorothy Garrod (1892–1968) at Mount Carmel and geologist Elinor Wight Gardner (1892–1980) in Bethlehem (Plate 11), and mentored Gertrude Caton-Thompson (1888–1985) (Irwin-Williams, 1990). Bate pioneered zooarchaeology, and she was the first to study Palestinian faunas, beginning in 1928. She described several new Pleistocene mammal species not formerly known to have been living in these areas and constructed one of the first quantitative curves of faunal succession (Bate in Garrod et al., 1928; Bate, 1932; Garrod and Bate, 1937).

Bate's fieldwork was not without intrigue and foul play. She was asked by the Palestinian Antiquities Department to assist in the recovery of fossil bone fragments found in Bethlehem. After climbing 12 feet (3 meters) down a partly dug well and being covered in mud from head to foot, she identified a piece of tusk from Palestine's first elephant, *Elephas planifrons*. Although her work in Bethlehem seemed guaranteed, this was not to be because James Starkey, an archaeologist working on excavations nearby, persuaded the authorities to give the site to him. For Bate (being British and far too polite to make a scene), her only choice was collaboration. For the next three field seasons (1935–1937), she excavated at Bethlehem alongside her colleague Gardner. And although these two women were the force behind the project, Starkey was the official head. However, he was murdered in 1938 by an angry mob that confronted him at the opening of the Palestine Museum of Archaeology. His death marked the end of the Bethlehem excavations, and given increasing political unrest in the Middle East and the onset of World War II that precluded further fieldwork, Gardner's stratigraphic work and Bate's discoveries of fossil elephants, giant tortoises, and early horses presented the first real understanding of the faunal history of Bethlehem (Shindler, 2005; TrowelBlazers, undated).

Finally, in March 1940 Bate received acclaim for her judicious and careful work on Pleistocene faunas, climate, and paleoenvironment from the president of GSL, who awarded her the Wollaston Fund. Then it was realized that she was not a member, so she was hastily elected a fellow of the GSL several months later. Sadly, Bate's own end was marred with ill health after a diagnosis of cancer in 1941, but as she had become a Christian Scientist, she received no real treatment and died isolated. After her death in 1951 and glowing obituaries (e.g., Howard, 1951; Swinton, 1951), her work seems to have been forgotten until revived by historian Karen Shindler's comprehensive biography, recently revised (Shindler, 2017).

Apart from self-taught Mary Anning, another 19th-century woman in Britain who wrote scientifically was Agnes **Crane** (1852–1911) of the Brighton Natural History Society, who was clearly interested in Darwin's ideas on evolution. From a prominent family, she was a self-taught paleontologist who concerned herself with invertebrate as well as vertebrate natural history. Toward the turn of the century, Crane spoke at the society on the origin of birds and mammalian horns and hoofs (Creese, 1998). She offered a discussion regarding descent with modification and provided a stratigraphic distribution chart on the affinities of certain living and fossil fish, particularly lungfish (Crane, 1877). Crane had written to the authorities at the British Museum (NH) and traveled to London to examine specimens before giving her paper at the society.

Another early British vertebrate paleontologist, Helga Sharpe **Pearson** (1899–1975), was born into an academic family; her father was influential English mathematician and biostatistician Karl Pearson of UCL, who was a prominent social Darwinist and supporter of women's education. UCL had gained a supplemental charter in 1878, making it the first British university to award degrees to women, including in science. Helga entered that university just after World War I and after gaining her BSc, she became assistant to David Meredith Seares Watson (1886–1973) in the Zoology Department. During a short visit to the United States in 1921–1922, she was encouraged by American Museum of Natural History (AMNH) paleontologist William Diller Matthew (1871–1930) to study and describe several fossil peccary skulls of *Thinohyus* (now recognized as junior synonym of *Perchoerus*) from the Oligocene White River badlands (Pearson, 1923; Prothero, 2009). She continued to postgraduate studies under Watson, "D.M.S." as he was affectionately known, who encouraged women to become trained scientists, including his own daughter, Janet Vida Watson (1923–1985), who became the first female president of the GSL in 1982. In 1912, Watson was appointed as a lecturer in vertebrate paleontology at UCL (Parrington and Westoll, 1974). Pearson went on to further research on the anatomy and evolution of fossil suids—her early paper on Oligocene peccaries was cited by young Edwin Colbert in 1933, and her last was on Chinese suids (Pearson, 1928). Her other main interest was dicynodont therapsids (Pearson, 1924a, b). Work on the skull of *Kannemeyeria* allowed Pearson to identify the presence of a vomeronasal organ, not known before. She interpreted the lifestyle of dicynodonts as fossorial, which Gillian King (1990b) later disputed. A comparative anatomist, Pearson worked on various small mammals, including the platypus, and after gaining her MS, she returned, as Helga Hacker, to studies at UCL in the mid-1950s (Barry Hughes, pers. comm., 2012).

Like Bate, our next woman benefited from a wealthy upbringing and the increase in women's freedom to gain education and then jobs toward the end of the 19th and beginning of the 20th centuries. Igerna Brunhilda Johnson **Sollas** (1877–1965), born in Dawlish, Devon, was the second daughter of paleontologist William Johnson Sollas (1849–1936) and Helen Coryn (Creese and Creese, 2006). Although not strictly a VP, she did train in geology and zoology and was educated at Alexandra School and College, Dublin, and then at Newnham College, Cambridge, where she became a fellow (1904–1906) and lecturer in zoology (1903–1913), working in evolutionary genetics alongside William Bateson. With her father, Sollas carried out a "ground-breaking" study on the enigmatic Devonian fish *Palaeospondylus gunni* Traquair (Sollas and Sollas, 1904). Based on her father's methodology and their study of serial sections, they determined the internal anatomy of these strange fish, which they then used to produce wax models that provided three-dimensional structure for the first time (Forey, 2005; Cunningham et al., 2014). This important work has been much cited even in recent times. Later study (e.g., by Carole Burrow and Zerina Johanson; Chapter 5) has resulted in various reinterpretations of *Palaeospondylus* as the larval stage of a fossil lungfish, a shark, and a hagfish, and now, based on the three semicircular canals and significant cranium, it is interpreted as an osteichthyan (Burrow, 2019). Also, with her father, Sollas published papers, mostly on invertebrates, but one was a description of the anatomy of the skull of the stem mammal *Dicynodon* using serial sections (Sollas and Sollas, 1912; discussed in Wyse Jackson and Spencer Jones, 2007). Like Dorothea Bate and Mary Wade (Chapter 4), Sollas became a Christian Scientist, which led her to question her animal experimentation and thus she ceased scientific work after 1916.

Austria. Karoline "Lotte" **Adametz** (1879–1966) was a self-taught paleontologist and prehistorian at the Natural History Museum in Vienna. After attending a commercial school, she became secretary and later director of the Geology–Paleontology Department at the museum from 1898 until her retirement in 1946. She conducted fieldwork with geologist Ernst Kittl (1854–1913) in the Austrian Central Alps and illustrated his publications using drawings and photographs. Later, she labeled fossils in the collection.

Adametz collaborated with prehistorians Julius von Pia and Franz Xavier Schaffer and provided many of the illustrations in Josef Bayer's (1927) *Man in the Ice Age* and Schaffer's (1924) *Lehrbuch der Geologie* (vol. 2). Beginning in 1919 she assisted in the excavations of Bayer and others. From 1914 to 1924 she led the Office of the Secretariat of the Austrian Geological Society and was also a member of the Anthropological Society in Vienna and the German Paleontological Society (Svojtka, 2011).

France. After Cuvier's time, Albert Gaudry (1827–1908) was the main vertebrate paleontologist in Paris. He influenced the following generations and was one of the first to use trees to reconstruct the phylogenetic history of fossil taxa taking their stratigraphic position into account (Cohen, 1994). However, he accepted theistic but not Darwinian evolution, rejecting natural selection. Gaudry does not appear to have had any French women paleontology students, but he was an important mentor for the Russian Maria Vasil'evna Pavlova (see later in this chapter). Women were forbidden to attend universities in imperial Russia until 1918, and those women interested in a university education had to struggle with bureaucracy; some, such as Pavlova, went abroad to European universities.

Most of the work by French women in vertebrate paleontology took place after World War II (see Chapter 5), but one of note is during this time but sadly she seems to have no obituary (cf. Falk, 2000; Jaussaud and Brygoo, 2004). Madeleine **Friant** (1897–197/8?) was the first vertebrate paleontologist and comparative anatomist who worked at the Muséum national d'Histoire naturelle, Paris (MNHN). Although she completed a doctorate in medical studies at the University of Strasbourg in 1927, she worked at the MNHN as a comparative anatomist and paleontologist on the anatomy and evolution of amphibians, birds, and various mammals, including cave bears, elephants, deer, insectivores, rodents, whales, and seals. British biologist D'Arcy Wentworth Thompson (e.g., 1945) supported her work on the relationship of rodent dentition to that of elephants (Friant, 1933). Her later research was focused on mammalian teeth, particularly the evolution of the teeth of beavers and other Tertiary mammals, and her numerous papers are listed in the BFV.

Germany. Johanna Gabrielle Ottilie "Tilly" **Edinger**'s (1897–1967) compelling story of pursuing research on fossil brains while enduring the Nazi regime and forced emigration from Germany to America is discussed in two parts: her early life in Germany (in this chapter) and her later years in the United States (e.g., Gould, 1980; Köhring, 1997; Chapter 4). Edinger was born in Frankfurt, Germany, the youngest of three daughters of a prominent

German Jewish family. Her father was Ludwig Edinger (1855–1918), an eminent physician and pioneer neurologist, who inspired her to consider science, and her mother, Anna Goldschmidt, was a social activist and descendant of a banking family. Edinger was tutored at home by English and French governesses and later attended the only Frankfurt girls' school. She studied at the universities of Heidelberg, Frankfurt, and Munich, and her postgraduate research, after World War II, on the endocast of a Mesozoic marine reptile, *Nothosaurus*, began her life's study of fossilized brains. She was the first female vertebrate paleontologist to receive a doctorate in Germany and pioneered the discipline of paleoneurology—the study of the evolution of brains (Buchholtz and Seyfarth, 1999, 2001; Köhring and Kreft, 2003).

Soon after graduation, Edinger took on unpaid assistant positions at both the Senckenberg Museum of Natural History and the Geological Institute of the University, Frankfurt, which gave her access to vast collections of vertebrate fossils. By 1926 she was named curator of fossil vertebrates. While in Frankfurt, Edinger had contact with two eminent vertebrate paleontologists, Friedrich von Huene (1875–1969) of Tübingen University and Louis Dollo (1857–1931) in Belgium. The latter urged that her work be considered a biological study rather than just for its association with geology (Buchholtz and Seyfarth, 2001). Dollo became Edinger's mentor and friend, and she was clearly influenced by him, writing later, "We must admire Dollo as a revolutionary who led V[ertebrate] P[aleontology] in new directions. . . . [He made] extinct vertebrates, so to say, re-enter life" (Edinger, 1964:42).

Edinger's first publication was her dissertation work on the Triassic *Nothosaurus* brain endocast (Edinger, 1921). Around 1923, in a letter to Dutch anatomist and neurologist C.U. Ariëns Kappers (1877–1946), Edinger wrote that she had noticed that a large literature exists about fossil "brains," distributed widely in all the journals of the earth, and she decided to give herself the assignment not only to collect but also to rework this material into a book, *Paleoneurology* (Köhring and Kreft, 2003; McNeill, 2018); her entire life was to be dedicated to this task. In 1929, Edinger published her first major monograph, *Fossil Brains*, based on a nearly decade-long postdoctoral research project. For this, Edinger had examined 280 papers that dealt with the brains and spinal cords of extinct vertebrates, but which no one had yet looked at holistically (Plate 12). Her achievement in synthesizing this work came through a framework of two seemingly disparate fields: geology and neurology. Emily Buchholtz and Ernst-August Seyfarth observe that Edinger had transformed paleoneurology into a discipline that was taxonomically, chronologically, and functionally informed (Buchholtz and Seyfarth 1999, 2001; see also McNeill, 2018).

During the Nazi rise in Germany in the 1930s, with the help of director Rudolf Richter, Edinger was able to retain her position as an unpaid "under-cover" curator at the Senckenberg Museum since it was a private institution (Schultze, 2007). But this did not wholly shield her from the threat of Nazi violence, which shaped her daily life and work. In March 1938, Ferdinand Broili (1874–1967), professor of paleontology and stratigraphy in Munich, asked Edinger to "voluntarily" give up her position on the editorial board of the journal *Neues Jahrbuch für Geologie und Paläontologie,* because some colleagues "would rather go without her cooperation"; Edinger obliged (Köhring, 1997:56). For five more years, she would strive to make herself as invisible as possible by entering through the side door of the museum, removing her name plate from her door, and staying out of sight. Despite the urgings of friends and colleagues, Edinger refused to leave, saying, "Frankfurt is my home, my mother's family has been here since 1560, I was born in this house. And I promise you they will never get me into a concentration camp. I always carry with me a fatal dose of veronal [the first commercially available barbiturate]" (quoted by Buchholtz and Seyfarth, 1999:355).

On November 9–10, 1938, Edinger's resolve changed. That night—the notorious Der Kristallnacht, or Night of Broken Glass (e.g., Kölbl-Ebert, 2017)—marked a frenzy of destruction in which Nazis systematically burned synagogues; destroyed Jewish businesses, homes, and institutions; and murdered and imprisoned Jewish people across the Reich. Edinger survived but was no longer permitted to enter the museum, and was left with little choice but to seek escape. As she awaited her quota number, the Emergency Association of German Scientists Abroad assisted her passage to London. She was allowed to take with her only 10 German marks (less than US$50 at that time) and two sets of cutlery, and so she set off to a new life. The US Immigration Act of 1924 allowed foreign ministers and professors to bypass immigration quotas if they could find work in American institutions, so Edinger's colleagues in Europe and the United States, especially Al Romer, rushed to help her secure a position. George Gaylord Simpson wrote to the American consulate praising Edinger as a research scientist of the first rank. . . . "She is so preeminent in this field that she may really be said to have created a new branch of science, that of paleo-neurology" (letter, December 31, 1938, quoted in Buchholtz and Seyfarth, 1999:355). Still, Edinger was not approved. In later life, Edinger (1961) looked back on the difficulty of being a German woman entering science in her assessment of one of the dominant professors of the time, Rudolf Virchow, who she noted was against the emancipation of women and against the theory of evolution; this was the milieu in which she and others had to succeed. The second phase of Edinger's life and work was spent in the United States.

Another German who faced the patriarchal regime, Erika Martha von Hoyningen-**Huene** (1905–1969), became a vertebrate paleontologist who specialized in Mesozoic fossil reptiles. Born in Tübingen, Germany, she, after Edinger, was the second female vertebrate paleontologist in pre–World War II Germany, working during the 1930s. After training with her father, Friedrich von Huene, and being mentored through her doctorate by paleontologist Edwin Henning (1935–2017) with the assistance of Otto H. Schindewolf (1896–1971), she received a PhD on April 1, 1933, the day Adolf Hitler came to power. As with Edinger, the rise of Nazism prevented her from obtaining a permanent position (Turner, 2009; Kölbl-Ebert, 2017). Schindewolf managed to give her contract work in Berlin during 1939–1945, but after the war, when the German economy had collapsed, she could not maintain the momentum (Turner et al., 2010a). Her work is little appreciated, overshadowed by her well-known father's, but among her few publications is one describing a tooth of the cynodont *Tricuspes tubingensis* that she collected from a quarry near Tübingen-Bebenhausen (Huene, 1933). She was the first to identify Triassic mammals and reptiles based on teeth in a microfauna, and Simpson (1935) noted her importance in this field. Other papers include description of a new rhynchocephalian from Somerset, *Pachystropheus rhaeticus* (Huene, 1935), later reidentified as an archosaur (Duffin, 1978; Turner et al., 2010a).

Hungary. Ilona **Nopcsa von Felsoe-Szilvas** (1883–1952), sister of Austro-Hungarian aristocrat and paleontologist Baron Franz Nopcsa (1877–1933), found an unusual skull and bones along a riverbank near the family's home in Hungary and collected them (Turner et al., 2010a). Her brother later described them while at the University of Vienna as the hadrosaur *Telmatosaurus transylvanicus*, the first dinosaur remains from Hungary. He went on to describe more than 25 genera of reptiles and five dinosaurs (Weishampel and Reif, 1984).

The Netherlands. Dutch paleomammalogist and Amsterdam-born Antje "Annie" **Schreuder** (1887–1952; Plate 13) began her biology studies in 1910 at the University of Amsterdam with teachers including geneticist Hugo de Vries (1848–1935) and Eugene Dubois (1858–1940). In 1916 she passed the state exam in botany, zoology, and geology and became research assistant to Dubois, remaining with him until 1929 (e.g., Hooijer, 1952; Kerkhoff, 1993;

Shipman, 2001). Schreuder reentered the University of Amsterdam in 1923 and gained her doctorate in 1928 with a study of Pleistocene beavers, and her last paper just before her death was a revision of the genus *Trogontherium*. Moving to an unpaid position at the Zoological Museum of the University of Amsterdam, through the 1930s and 1940s she became noted for her ability to sort out the chronology and fauna for the Pleistocene of the Netherlands. She became a pioneer in the field of micromammals, especially voles and desmans (water moles). Her fossils came mainly from bores, and she showed that this material can also be used for mammalian paleontology. Because of health problems, she gave up fieldwork in the late 1930s but encouraged colleagues to collect and send fossils to her (Det and Hoek-Ostende, 2002).

Poland. The first woman to study vertebrates in Poland was Russian-born Zinaida **Gorizdro-Kulczycka** (1884–1949). Born in Tashkent, Uzbekistan, she completed her education in zoology and paleontology under W. Szymkiewicz in the Higher Female Nature Course at Saint Petersburg (1903–1907). After graduation she worked in the department of geology at Saint Petersburg State University, where, under the direction of A.A. Inostrancewa, she did fieldwork around Tashkent and later did hydrological work in Turkestan and the Urals. In 1917 she married colleague Polish engineer Anton Kulczycki and went to work in central Asia for many years, doing pioneer paleontology for the Central and Asian State Museum, including work on Jurassic fish, which contributed to the Kara-Tau paleontological reserve. They went to live in Warsaw in 1925, and she went to collect in the Holy Cross Mountains, where her work on Devonian fish, under the mentorship of Moscow paleoichthyologist Dmitrii Vladimirovich Obruchev (1900–1970), on placoderms and lungfish (e.g., Gorizdro-Kulczycka, 1934), led her to become the head of the paleoichthyology lab at the Museum of the Earth in Warsaw. After a bad accident left her disabled and her untimely death in 1949, her remaining work was completed by her son (Malkowski, 1949).

South America

Mathilde **Dolgopol de Sáez** (1901–1957) has the distinction of being the first female vertebrate paleontologist in Latin America. She was born and worked in Argentina at the Museo de La Plata (Herbst and Anzótegui, 2016). She received a PhD in invertebrate paleontology in 1927 under the mentorship of Ángel Cabrera (1879–1960), who had just arrived from Spain as head of paleontology in the museum. She began publishing at that time, on both invertebrates and vertebrates including an early paper on a fossil bird (Dolgopol de Sáez, 1927). Later, her research focused on descriptions of fossil fish (e.g., Dolgopol de Sáez, 1940) and crocodiles (Dolgopol de Sáez, 1957).

United States

Annie Montague **Alexander** (1867–1950) was an America benefactress to and collector of paleontological and zoological specimens for the University of California, Berkeley (UCB). Her life and contributions have been recounted by Barbara R. Stein (2001). She was born in Hawaii, and her wealthy father provided her with opportunities to travel to Europe, the South Pacific, and Southeast Asia. She received a liberal education at Lasell College in Massachusetts and then entered UCB. Her lifelong interest in and support for paleontology began after hearing lectures in 1901 by John C. Merriam (1869–1945) about his discoveries of fossil mammals and reptiles in California and Oregon. Alexander offered to underwrite the costs of his summer collecting expeditions if she could participate in the fieldwork. When he agreed, Alexander went on to organize many field-collecting trips, including to Fossil Lake, Oregon, in 1901; to Shasta County, California, in 1902 and 1903; and to the West Humboldt Range, Nevada, in 1905 (Plate 14). This last trip, the "Saurian Expedition," revealed numerous Triassic ichthyosaur skeletons from Triassic deposits, including the then largest in the world and the most complete specimens in North America (Turner et al., 2010a). Alexander enjoyed the fieldwork and did much of the cooking and writing in 1905. She wrote, "We worked hard up to the last. My dear friend Miss Wemple stood by me through thick and thin. Together we sat in the dust and sun, marking and wrapping bones. No sooner were these loaded in the wagon for Davison to haul to Mill City than new piles took their places. Night after night we stood before a hot fire to stir rice, or beans, or corn, or soup, contriving the best dinners we could out of our dwindling supply of provisions. We sometimes wondered if the men thought the fire wood dropped out of the sky or whether a fairy godmother brought it to our door, for they never asked any questions" (Alexander, 1905:46). In 1906 Alexander began making monthly contributions to support research in the UCB Department of Paleontology. Over the years, she increased her donations, finally establishing an endowment for the Museum of Paleontology in 1934. Indeed, "without Alexander, paleontology would not have thrived at Berkeley" (Lipps, 2004:226). She made other donations to support faculty on sabbaticals, student research, and field expenses. In 1908 the UCB Museum of Vertebrate Zoology was established in part with funding from Alexander. She donated her own collections of recent mammals to the museum. During the years that followed, Alexander and her life partner, Louise Kellogg (1879–1967), collected both fossil and recent mammals for UCB museums.

Most important of the Americans in the late 19th century is Fanny Rysam Mulford **Hitchcock** (1851–1936) of New York (Plate 15). She is likely the earliest American woman to publish a scientific paper in vertebrate paleontology, in the *Proceedings of the American Association for the Advancement of Science* (Hitchcock, 1887, reported in the *American Naturalist* and *Geological Magazine* of 1888). At the time, she may have been an undergraduate at Columbia University, but she did not gain a degree there. She did return to university in 1890, and later she became the first woman to receive a PhD from the University of Pennsylvania in 1894, but in chemistry. Hitchcock had almost certainly wanted to study paleontology with American geologist and vertebrate paleontologist Joseph Leidy, but he had died in 1891 (Warren, 1998). It seems likely that she knew Leidy, a faculty member at nearby Swarthmore College, or perhaps she had met him and his wife, Anna Harden Leidy, who assisted her husband in his scientific work in the field and at the microscope (see Chapter 2), at an AAAS lecture. At the very least, Hitchcock was familiar with his earlier work on the enigmatic shark *"Edestus" vorax* Leidy, 1855, in which Leidy claimed the fossil was a series of teeth, later denied by Richard Owen in London and American paleoichthyologist John Strong Newberry (1822–1892) and others over the following decades (Ewing, 2017). Hitchcock was an early member of the AAAS, becoming a fellow later in life. Whether she attended the AAAS lecture in 1855 at which Edward Hitchcock (1793–1864; no relation) and Louis Agassiz debated the second "shark" specimen, which was found and sent to Owen for confirmation, is not known, but Hitchcock must have read about the new shark specimen just found in Australia—*"Edestus" daviesi*, later renamed *Helicoprion*—and her paper overturned ideas about these bizarre early sharks by correctly identifying the specimen as symphysial teeth, meaning that they were positioned in the midline of the jaw. She correctly compared them to the tooth "whorl" of another older bony fish, *Onychodus*, using classical comparative anatomy, but was later criticized (Ewing, 2017). This controversy over the identification and lifestyle of these fossils was debated over many years. More recently, scientists have solved the identity and agree with Hitchcock's interpretation, ultimately proving her correct (Ewing, 2017; see Chapter 5). Her contribution as the first female scholar in American vertebrate paleontology cannot be underestimated—even if she did not actually hunt a bone.

Ruth **Mason** (1913–1990) was a young amateur collector who discovered in the 1920s a dinosaur bone bed on her family ranch near Faith, South Dakota. Since then, tens of thousands of Cretaceous dinosaur fossils have been collected from the Ruth Mason Quarry, including the duck-billed dinosaur *Edmontosaurus annectens* and *Tyrannosaurus rex* teeth. Since 2003 the quarry has been the site for the Children's Museum, and many people have participated in excavations at this quarry and outdoor classroom (Turner et al., 2010a).

Mignon **Talbot** (1869–1950; Plate 16) was an American geologist who was born in Iowa City, Iowa. She earned a PhD in geology in 1904 from Yale University and went on to teach at Mount Holyoke College in Massachusetts from 1904 to 1935 (Aldrich, 1990). In 1908 Talbot became professor and chair of the Geology Department, and in 1929 she became chair of both the Geology and Geography Departments. She is likely the earliest university-educated American female to have contributed to vertebrate paleontology. By contrast, the earliest American male professionals, such as Edward Hitchcock, Joseph Leidy, and Richard Harlan (1796–1843), were working more than half a century before Talbot (Simpson, 1942; Warren, 1998). Talbot is best known as the first woman to find and describe an articulated skeleton of a dinosaur; she named the theropod dinosaur *Podokesaurus holyokensis* (Talbot, 1911), based on a partial skeleton that she and her sister located near the college (Weishampel and Young, 2001). Talbot had first identified it as an herbivore at a PS meeting in December 1910. Because she had no training in vertebrates, she contacted vertebrate paleontologist Richard Swann Lull (1867–1957), then a professor at Yale University, and he advised her to study and describe them. Lull suggested that the dinosaur was insectivorous or a wading form. When "dinosaur man" Friedrich von Huene was visiting Lull, he saw the specimen and based a new family of coelurosaurs on it. But its relationships are still uncertain because a fire destroyed the Mount Holyoke science building, which had housed the only known "type" specimen (Turner et al., 2010a). Talbot also had the distinction of being the first woman to become a member of the PS in 1904.

USSR

By the early 20th century, women were allowed to enter Russian universities. In 1906–1907 there were 1,949 women in attendance at all the principal universities. After several setbacks, with permission to attend rescinded in 1908, in 1911 there were 960 women in Russian universities. Following the 1917 revolution, many other women entered science; by 1929 some 5,000 women worked as researchers in many scientific institutions in the main centers of Moscow and Leningrad (former Saint Petersburg) (Nalivkin, 1970; Valkova, 2008). A few notable Russian women vertebrate paleontologists are known from this time.

Vera Isaacovna **Gromova** (1891–1973; Plate 17) was a Russian vertebrate paleontologist and paleomammalogist known for her research on the osteology and systematics of fossil and living ungulates and Quaternary mammalian faunas (Nalivkin, 1979). She grew up in the relative affluence of a lawyer's family in Orenburg, in southern imperial Russia, but rose to become a Soviet "First-class" scientist. She trained after the revolution of 1917, graduating from higher graduate courses in Moscow as a specialist in vertebrate zoology. By 1919 she had moved to Leningrad and joined the staff of the Zoological Museum of the Academy of Sciences, established 1724 in Saint Petersburg. In 1922 Gromova participated in expeditions to the Paleolithic of the Crimea and, unusual for women in this era, took along her young son, Igor Mikhailovich Gromov (1913–2003), who in turn mentored a generation of VPs (Abramson, 2003). For the next 20 years, while rising from the ranks of preparators to become head of osteology, she honed her knowledge working with the mammal collections, especially from eastern and southern Soviet states, until she was able in 1941 to make the first innovative systematic separation of ungulates. From 1942 to 1960 she was employed at the Paleontological Institute, where she was head of the mammal laboratory from 1946 to 1960 (Trofimov, 1961). Her publications include *The History of Horse (Genus Equus) in the Old World* (1949) and the Mammals part of *Fundamentals of Paleontology* (Orlov, 1964, 1968). In addition to her interest in horse evolution, she also proposed resolution of historical issues about the validity of the genus *Mammuthus* and the type species of *Elephas primigenius* (Gabunia and Trofimov, 1973; Sarjeant, 1978–1987), as well as working on the enormous indricotheres (Brinkman, 2013).

Anna Boleslavovna **Missuna** (1868–1922) was a Russian-born geologist who contributed to vertebrate paleontology (Ogilvie and Harvey, 2000; Lebedev, 2019) and, like Fanny Hitchcock, to unusual Late Paleozoic sharks. She was born in the Vitebsk Region (then in the Russian empire, now part of Belarus) to Polish parents. She was educated in Riga (now Latvia), where she learned to speak German, and in Moscow, where she had a higher education scholarship from 1893 to 1896. In 1906–1909 she became keeper of geological collections of the Moscow Society of Naturalists, working initially as an assistant, then as a teacher of petrography, paleontology, and historical geology at the Moscow Higher Courses for Women. In 1907–1908 she published papers in German and Polish in which she described a new shark species *"Edestus" (Protopirata) karpinskyi* from the Carboniferous Myachkovo Regional Stage near Kolomna, Moscow Region.

Maria Vasil'evna (Vasilievna) **Pavlova** (née Gortynskala) (1854–1938) was the first woman geologist in Russia and most noted female Russian paleontologist of the time, known for her research of fossil ungulates (hoofed mammals) and efforts to establish the Museum of Paleontology at Moscow State University (MSU), Moscow, Russia (Nalivkin, 1979; Deforzh, 2019). She was born in 1854 in Kozelets, Ukraine, to educated parents. She was homeschooled until the age of 11, when she entered the Kiev Institute of Noble Maidens, a finishing school that also included instruction in science. Following a brief marriage (her husband, Illich-Shishatskaya, her senior by many years, died in 1880) and encouraged by her father, she moved to Paris and studied natural history, graduating from the Sorbonne in 1884 (Nalivkin, 1979). Pavlova pursued research at the MNHN under the mentorship of Gaudry, who worked on Tertiary mammals, especially fossil ungulates.

In 1885 Pavlova returned to Russia, where she married geologist and paleontologist Alexei (Aleksii) Petrovich Pavlov (1854–1929), whom she had met in Paris. According to Olga Valkova (2008:150), Pavlova was allowed to work at MSU "only thanks to a personal authority of her husband [who already had a professorship there] and another professor." Pavlova initially studied collections without payment in the Geology Museum at MSU, including fossil invertebrate collections, publishing on Cretaceous ammonites collected by her husband. She kept in contact with Gaudry, as well as the American paleontologist Henry Fairfield Osborn (1857–1935) at the AMNH in New York, whose museum displays she much admired, and began to research Tertiary mammals in the collections, notably ungulates especially *Hipparion* (Pavlova, 1887–1906) and proboscideans (Pavlova, 1899) as well as evolutionary theory. She went on to publish with her husband a monograph (1910–1914) on the mammalian fossils of Russia based on fossils she collected, as well as those she observed in her visits to various museums from eastern Siberia to the northern Caucasus and Ukraine. She wrote more than one-half of the pre-1901 publications by Russian women geologists and is considered their most distinguished member (Creese and Creese, 1994; Bessudnova, 2006; Lyubina and Bessudnova, 2019). In 1897 Pavlova was one of two women invited to join the Organizing Committee of the International Geological Congress held in

Saint Petersburg for the first time. In 1910 she began teaching in the Geology Department at MSU, and in 1916 she received a PhD in zoology with honors. After the major shake-up of the revolution, she was promoted to profes-sor of paleontology in the Geology Department at MSU in 1919. That same year, within the new regime, she described a "painful situation" in which she had recommended a "talented" female student for a position as junior re-search officer at MSU. Esfir Falkova, the student, despite requisite skills, and publications, was ultimately turned down for the position. Describing the results in her memoir, Pavlova wrote, "Here is the result of the apprecia-tion of my scientific work from those closest to my scientific activities. [Here she names a professor in whom she had put 25 years of friendly trust] . . . Yes, today I'm feeling as if I returned from a very hard funeral" (memoirs of M. Pavlova, quoted in Valkova, 2008:156).

Pavlova's work throughout this time contributed to the paleontological foundation of Continental Neogene and Quaternary stratigraphy in west-ern Europe. She created a school of paleozoology with paleontologist-stratigraphers. Among her students were well-known scientists such as V.V. Menner, M.O. Bolkhovitinova, V.V. Teriaiev, M.I. Shulga-Nesterenko, and others (Deforzh, 2019). In 1925 she was elected a corresponding member of the Soviet Academy of Sciences and in 1930 an honorary member. In recog-nition of the research of Pavlova and her husband, the museum at MSU was named after them, the A.P. and M.V. Pavlova Museum of the Moscow Geo-logical Research Institute, and in 1926 both were awarded the Gaudry Medal by the French Geological Society. Material collected by Pavlova formed the basis for its fossil vertebrate collections in the paleontology section of the mu-seum (Nalivkin, 1979; Creese and Creese, 2015).

Clearly the number of women in Russian science began to grow during the early 20th century as the Soviet Union expanded. By 1936, the number of So-viet women to enter scientific work was 11,800 (Valkova, 2008; see Table 4.1 for VPs). Comparisons of women scientists in America reveal nearly a 10-fold decrease in number in the United States, with 450 in 1921 and 1,912 in 1938 (Rossiter, 1982). More about the women who entered the profession after World War II is revealed in Chapter 4.

Summary

By the late 19th and early 20th centuries, women were becoming university educated. Although most women fossil bone hunters were in Britain and the United States, other regions—France, Germany, South America, Russia, and even Australia—began to provide educational opportunities to women. The contributions of several women during these times are significant. Although not a VP, Marjorie Courtenay-Latimer's discovery of the coelacanth, a "living fossil," provided an important link in the transition of vertebrates from water to land. Another significant discovery was the presence of the vomeronasal organ, part of the olfactory system, in therapsid reptiles by British VP Helga Pearson. Field collecting was undertaken by several women, including British paleontologist Dorothea Bate, the first woman employed at the BMNH (although in an unpaid position), who excavated and identified mammal fossils from numerous sites in Europe and the Middle East. Annie Alexander was a benefactor and collector of paleontological and zoological specimens from the American West for the UCB museum. Women were gaining a foothold in natural history museum circles. Madeleine Friant has the distinction of being the first woman VP to work at the MNHN in Paris. Tilly Edinger spent her early life Germany at the Senckenberg Museum, where she began her landmark study of vertebrate fossil brains. Mathilde Dolgopol de Sáez was the first female VP in South America, working at the Museo de La Plata in Argentina on a range of fossil vertebrates. In the late 19th century in America there were two academics, Fanny Hitchcock, the first woman to publish a VP paper, and Mignon Talbot, the first woman to find and describe a dinosaur skeleton. Maria Pavlova, the most important woman Russian paleontologist at this time, was recognized for her research on fossil ungulates and efforts to establish the Museum of Paleontology at MSU, later named after her.

CHAPTER 4

Women in the Early Modern Years of Vertebrate Paleontology, Mid-20th Century (1940–1975)

When a person first starts out in vertebrate paleontology, the usual vision lurking somewhere in the mind is that of exotic field work with strange new worlds of fossils dripping from dramatic outcrops.

Louis L. Jacobs, former SVP president (Jacobs, 1997)

From the earliest days of our discipline there have been women who wanted to hunt for fossils. As recounted in Chapter 1, the first professional society devoted to the study of fossil vertebrates, the SVP, was founded in 1940, the year before America would join the war. As Margaret Rossiter observes, "The period (1940–1972) was a golden age for science . . . but a dark age for women in the professions" (Rossiter, 1982:xv). Women scientists were scarce and VPs even more so (Rossiter, 1995; Kölbl-Ebert and Turner, 2017). Among the 34 registrants at the December 1940 planning meeting at the Museum of Comparative Zoology (MCZ), Harvard University, Cambridge, Massachusetts, there were only three women present: Tilly Edinger, just joined as a research and staff member at the MCZ; Anne Roe Simpson, G.G. Simpson's wife and a trained psychologist; and Nelda Wright, Alfred Romer's research assistant (Simpson, 1941). Although only Edinger appears in the organization meeting photo (Plate 18), all three women signed the constitution.

World War II (1939–1945) affected paleontology around the world. As documented in the *SVP News Bulletin*, most of the North American male VPs went off to war, some not to return—that was also true in all Allied and opposing countries. When the Soviet Union was attacked in 1941, Aleksandra Paulinovna "Anna" **Hartmann-Weinberg** (1882–1942), Russian expert on large, herbivorous pareiasaurian reptiles, died during the terrible first winter of the Siege of Leningrad (Orlov, 1946; Watson, 1946; Chernov, 1989). Several women VPs, secretaries, and wives took over not only academic and museum jobs but also the role of reporting to the *SVP News Bulletin*. Edinger dealt with foreign news, followed by Hildegarde Howard and Priscilla Turnbull (after the war), who were the main correspondents, and Nelda Wright took on the huge task of editing. Reports on conditions of vertebrate paleontologists and museums in central Europe and Russia after the end of the war are found in the *SVP News Bulletin*. Among these was news of the destruction of the Munich vertebrate collections of Karl von Zittel and his students, as well as collections in Berlin, mainly in the Museum für Naturkunde, including "the better of two *Archaeopteryx* skeletons"; the fact that of three Bavarian universities, only one—Erlangen—survived the war; and the horrific bombing of the BMNH in London (e.g., Anon, 1945:6–8). In the United States a vertebrate paleontologist at the UCMP, V.L. Vanderhoof (1904–1964), remarked, "Attention has already been given to the preservation of types from destruction. Some thought has even been given to the preservation of paleontologists, but they, of course, can be replaced" (Lipps, 2004:230). Women scientists, including VPs, gained jobs in universities as a result but also diversified into other types of war work, e.g., Dorothy Bate (Chapter 3) took over moving collections out of London; Lore David got involved in radioactive minerals and the effects of radioactivity; Pamela Robinson went into munitions production in the United Kingdom (see later in this chapter); and Dorothy Hill went into strategic map reading in Australia (Campbell and Jell, 1998). Other contributions of women to VP during the war were as research assistants to paleontologist husbands or male colleagues.

When the war ended in 1945, there were immediate effects. For example, Charles Camp (1893–1975), third director of the UCMP, noted an astonishing increase in women's enrollment in his introductory paleontology course at UCB—of his 215 students, 90% were women (Camp, 1944). Similarly, in Europe and the Communist countries (Russia/Soviet Union, China post-1949), there was a tremendous surge of women in geology and zoology. The Soviet Union's system of education spread and women VPs appeared in the Soviet Socialist Republic (SSR) countries such as Estonia, Latvia, Lithuania, Poland, and Romania. The latter countries through this period particularly saw enormous numbers of women entering practical applied paleontology, with many VPs using vertebrates to determine the age and correlation of rocks (see later in this chapter).

Despite this period being one when women gained more advanced graduate degrees, this new turn of equality produced by war did not translate into jobs for women in the United States (Rossiter, 1995). In fact, there was a strong reversal of opportunities in the postwar 1950s. Places where women traditionally trained in VP might see work disappear. Rossiter (1995) presented evidence that universities, and colleges, closed their doors to younger women and rid themselves of older women until there was legislation to support affirmative action in the United States in 1972. Until then most US academic women were expected to leave their jobs if they married. Similar "rules" applied in other countries, persisting longer in some places, such as Australia. In the three decades before affirmative action, women VPs were still relatively rare. Edinger could say to a friend in 1948 that in the world there were only about 10 people making a living in VP (Köhring and Kreft, 2003:551)!

After World War II, paleontology programs regained their vitality and the collection and study of fossil vertebrates began again in earnest. Several SVP annual meetings and field conferences were held in the intervening years. The SVP annual meeting in 1948 (see Köhring and Kreft, 2003:fig. 22), held in conjunction with the GSA in New York City at the AMNH, better known as "the Natch," had eight women (some were VPs' wives) in attendance: Edinger, Jean Hough, Rachel Nichols, Dorothy Reed, Roe Simpson, Florence Wood, Wright (see later in this chapter), and a young student named Ruth Crozier. Given the importance of collections and field collecting to VP, organized field trips played an important role in the SVP and were well attended by women. Eighteen women joined the first in Nebraska in summer 1941 (including Roe Simpson; Marie Skinner, wife of Morris Skinner; and the wife of charter member R.G. Chaffee); the second, which was delayed because of the war until July 1947, was located in the Triassic deposits of northeastern Arizona, and three women participated (Roe Simpson again, Ruth Romer, and Barbara Hasting McKee, wife of VP Ed McKee).

SVP Charter Women and Early Supporters

We found so few women VPs in the 1930s and 1940s, especially in the United States, that we decided to investigate those who were among the founders and first supporters of the SVP, including the so-called charter members. As we have discovered, most were educated to college or university level or trained in VP before World War II and then often married fellow VPs. They were centered in California, especially Los Angeles, as well as Omaha, Nebraska; Cincinnati, Ohio; and New York City, New York. As noted earlier, among the founding participants at the MCZ was Edinger. Clearly, given her European training, she was the most experienced female VP in the United States at that time. She is also unique as a woman who presented research at the early SVP meetings. The first official SVP meeting, in December 1941, was held jointly with the PS and GSA in Boston, Massachusetts. Among the

150 members were 15 women recorded as charter members (Anon, 1941): Frances C. Baer, Carrie Adeline Barbour, Gretchen Burleson, Ida Wilson, Leigh Dodson, Edinger, Jane Everest, Edith Hellman Bull, Hildegarde Howard, Hazel Hunter, Jannette Lucas, Nichols, Roe Simpson, Wood, and Wright. Other women, early supporters of VP and SVP members, later in the 1940s, are also sometimes referred to as charter members, but the term is used rather loosely in the early issues of the *SVP News Bulletin*. We discuss these women here.

Carrie Adeline **Barbour** (1861–1942) is thought to be the first trained and employed VP woman in the United States. As a child in Springfield, Indiana, she collected fossils and became a nature lover. She went on to attend Miami University and Ohio College for Women and gained her BS, then studied art, model making, and wood carving at the Cincinnati Art Institute. In 1892 she joined the art faculty at the University of Nebraska–Lincoln, then resigned and went to help her brother Erwin Hinckley Barbour (1856–1947) in the Geology Department, where he was curator and fossil mammal collector at the University of Nebraska State Museum. Carrie was unusual in being a paid employee at that time. In 1912 she was named instructor in paleontology and during World War I was made assistant curator. She devoted nearly 50 years of service to the museum, beginning her career assisting in the paleontology lab in the Department of Geology and on the Morrill Geological Expedition field trips, collecting fossils in the Badlands of Nebraska and South Dakota. She made major contributions to museum preparation (Barbour, 1898, Ewing, 2017), preparing many of the fossils herself (Plate 19). She became a charter member at 79 but died the following year (Schultz, 1943), working in the museum to her last day (George Corner, pers. comm., 2019).

Gretchen Mary Lyon **Burleson** (later **Humason**) (1907–1991) was an American zoologist in the Department of Zoology, University of California, Los Angeles (UCLA). She joined the SVP in 1941 (as Burleson) and was a member at least until 1952. She wrote several papers on fossil pinnipeds (Lyon, 1937, 1941; Burleson, 1948).

Lore Rose **David** (1905–1985) was born in Opladen, near Cologne in the Rhine-Wupper district of northwestern Germany. After several years studying at various German universities, she received her PhD in embryology in 1932 from the Kaiser Wilhelm Institute in Berlin-Dahlem and began her scientific career as a zoologist. She wanted a new challenge, and so, on the advice of Fritz Haas, then at the Senckenburg Museum in Frankfurt, David began her studies in ichthyology. However, like Edinger, being from a German

Jewish family, she decided to leave as the Nazi regime took hold in 1933 (e.g., Kölbl-Ebert, 2017). That year, she obtained a position in zoology at the MNHN in the Jardin des Plantes in Paris, where she worked on fish from the French colonies. She moved to the Congo Museum at Tervuren, near Brussels in Belgium, in 1934 to further study Belgian Congo fish. David's parents then convinced her to immigrate to the United States, which she did in 1937. Once there, she joined her sister and brother-in-law in Pasadena, California, and began to work at the Los Angeles County Museum of Natural History (LACM). At the museum she became good friends with vertebrate paleontologist Chester Stock, who encouraged her to study fossil fish. Stock had her appointed as resident associate geologist at the California Institute of Technology, where she worked from 1938 to 1943 (Huddleston and Panofsky, 1988). David went on to work at the Paleontology Lab of Richmond Oil in California, spending time in the field collecting and using fish scales for biostratigraphy (David, 1942), work that inspired researchers in the late 20th century (coauthor of the present book S. Turner among them). David had a distinguished career with over 20 reports and pioneering papers on cartilaginous and bony fish from the Permian and the Cretaceous to the Tertiary. Her study of Miocene fish from southern California (David, 1943) is considered one of her most significant contributions to vertebrate paleontology. She joined with Californian colleagues Hildegarde Howard and others to review all western North American vertebrates (including Mexican) (Furlong et al., 1946). Her obituarists say that David was a charter member (so we place her here), but it seems she joined the SVP in 1943, and she may have dropped in and out as her employment waxed and waned. She reported regularly until eventually, in 1957, her failing eyesight forced her to give up her microscope work on fish scales, which had become her favorite subject. That same year, however, saw another major contribution by her to fossil fish in the *Treatise on Marine Ecology and Paleoecology* (David, 1957).

Leigh Marian **Dodson** (née Larson) (1907–1997) is listed as an SVP member from 1941 to 1944, then at Los Angeles City College. Larson graduated from UCLA with a BS in zoology in 1928 and taught for a while before marrying Harry Claiborne Dodson in 1930. She is listed as on the faculty at UCLA with an MS in 1933 and at UCB in 1936. She published a paper with Howard on Holocene bird remains from a Point Mugu shell mound in California (Howard and Dodson, 1933).

With her move to the United States, German-born Tilly Edinger became not only one of the most significant paleontologists of the 20th century but also a mainstay of the SVP. As noted in Chapter 3, Edinger created and led the field of paleoneurology. Before her, studies of the vertebrate brain were based on the soft tissues of brains of various vertebrates, from jawless fish to mammals including bats, equids, and sirenians. In 1940, Edinger was able to find work as a research associate at the MCZ with the help of the Emergency Committee in Aid of Displaced Scholars. She was only one of four women scientists granted such aid. Edinger had written presciently to one of her supporters, "One way (England) or the other (US), the fossil vertebrates will save me" (letter in German, December 18, 1938, Archives: Senckenbergische Naturforschende Gesellschaft, Frankfurt am Main, Germany). Referring to her book *Fossil Brains* (Edinger, 1929), as Emily Buchholtz and Ernst-August Seyfarth (1999) wrote, that book did in fact save her life.

Edinger eventually became a naturalized US citizen in 1945 with Alfred Romer among her witnesses. She was one of the three founding women of the SVP present at the organizational meeting of the society in 1940. She immediately took on the role of linking the society with Europe, reporting regularly on the state of affairs and giving foreign news, particularly of German colleagues (e.g., Edinger, 1951). Over the next decades she would return to Germany and help bring former colleagues back into the fold, eventually being honored by the German Paleontological Society. Edinger's unique contributions to paleontology were recognized by the award of honorary degrees from Wellesley College (1950), Giessen University (1957), and from the University of Frankfurt am Main in her hometown (1964) (Romer, 1967). She served in 1963 as the SVP's first female president. Edinger's letter in response to then–SVP president Sam Welles on her nomination as vice president of SVP in 1962 illustrates her wry sense of humor: "I trust that nobody who has seen me—as, for example, last year in Chicago—run around the floor from questioner to questioner can imagine that I could become a President. My middle ear deafness has now been joined by (old age) cochlear nerve decay so that I am unable to be a chair 'man'" (Edinger letter dated July 18, 1962). Edinger dedicated her later years to bibliographic work on paleoneurology, finished with the help of colleagues (Edinger, 1975). At the same time she was investigating the thickened bones of bony fish, which became part of her legacy in the term "Tilly bones" (Konnerth, 1966). Her deafness likely played a role in her death from head injuries in 1957, since she probably did not hear the approaching truck that hit her as she walked to Harvard's MCZ (Lang and Meath-Lang, 1995). As evolutionary biologist Stephen J. Gould wrote in the preface to Rolf Köhring and Gerald Kreft's (2003:7) book about Edinger, her life was "a lesson to all of us about one of the most remarkable natural scientists of the 20th century" (quoted in Leibing, 2004).

Jane **Gregory** (née Everest) (1915–2005) joined the SVP in 1941 as a student. During the war years, she worked for Union Oil company labs. She attended the 18th International Geological Congress in London 1948, where she met vertebrate paleontologist Joseph Gregory (1914–2007); they married in 1949. In the 1950s, she prepared a systematic catalogue of Peabody Museum types and published vertebrate fossils.

Edith **Hellman** Bull (1912–2015, aged 103!), of Far Rockaway, New York, became a charter member of the SVP beginning in 1941 along with her father, Milo Hellman (1872–1947), an orthodontist and dental morphologist involved in studies on the evolution of jaws and teeth of fossil anthropoids at the AMNH. Edith graduated from Skidmore College in New York in zoology and received a MS in vertebrate paleontology from Columbia University (Anon, 2015). Like her father, she worked at the AMNH but in the Vertebrate Paleontology Department, retaining her SVP membership until retirement. One of her jobs in 1945–1946 was recataloging the fossil fish collection, sorting out the Eocene Green River specimens. Her husband, John Bull, worked in the Ornithology Department; he and Edith collaborated on many books, including the popular identification guide *Birds of North America*, and co-led many AMNH expeditions around the world.

Hildegarde Wylde (née **Howard**, which is the name she mostly worked under) (1901–1998) is recognized as the world's first woman avian paleontologist. Her career spans much of the 20th century. She was inspired to study fossil birds by Loye Miller (1874–1970) (Howard, 1971), and she was a friend of fellow fossil bird lover Dorothea Bate (Chapter 3), but we place her here because she was a charter member of the SVP, joining in 1941. Howard was born in Washington, DC, and moved to Los Angeles with her parents in 1906. Her mother was a musician and composer and her father was a Hollywood scriptwriter. Howard attended the Southern Branch of the University of California (now known as UCLA) in 1920, planning a career in journalism. Her first instructor in biology, Pirie David, made the subject so interesting that she became a laboratory assistant for the class. She began her long affiliation with LACM as a student worker from UCLA. She met her husband, Henry Anson Wylde, at the museum, where he later became chief of exhibits in 1924. They were married from 1930 until his death in 1984. Howard worked for paleomammalogist Chester Stock, then professor of paleontology at UCB, sorting bones from the classic Rancho La Brea tar pits. In 1922

Howard completed her undergraduate degree at UCB, since UCLA was at that time a two-year school, and continued working for Stock. Once her degree was completed, she began research at LACM on the extinct California Turkey from Rancho La Brea, *Palapavo (=Melagris) californicus,* and she received credit toward an MS from UCB. This work resulted in her first major publication and was the beginning of a lifelong career in avian paleontology. She returned to UCB in 1925, earning an MS in 1926 and a PhD in 1928. Howard then returned to Los Angeles, where she began working full time at the museum in 1928. Her initial assignment as junior clerk was curation and research of the Rancho La Brea fossil bird collection (e.g., Howard, 1943) (Plate 20). It took 10 years before she was officially appointed curator of avian paleontology and she became chief curator of science in 1951, the first woman in the country to hold such a position. From 1957 to 1961, Howard served as editor of LACM publications. Even after retirement in 1961, she continued her research on fossil birds, publishing her last paper in 1992. In her seven-decade scientific career, as fellow avian paleontologist Ken E. Campbell (2000a:132) writes, "[Howard] described three families, 13 genera, 57 species and two subspecies of birds. Remarkably, of these, she described one family, seven genera, 34 species and one subspecies after she 'retired' from LACM." Also, as noted by Campbell (2000b:779), "Hildegarde undoubtedly faced many obstacles in her scientific career, beginning with being banned, as were all female students at that time, from class field trips in her early years at UCLA." Her appreciation for fossils is evident in vertebrate paleontologist Ted Downs's (1980:viii) reflection on her work: "Hildegarde has often declared that she preferred to see the birds without the feathers when identifying them."

She served as *SVP News Bulletin* regional editor from 1943 to 1952 (e.g., Howard, 1948), and she was awarded honorary membership in 1973 (Campbell, 2000a). Howard received many other awards for her research contributions (e.g., Campbell, 1980). She became the first woman to be awarded the American Ornithological Society Brewster Medal in 1953 for exceptional work on Western Hemisphere birds, as well as being the first woman president of the Southern California Academy of Sciences. In 1977 she was honored by having the Hildegarde Howard Cenozoic Hall, Southern California Academy of Sciences, named after her.

Hazel M. **Hunter** (1900–1996; née?), from Garden City, New York, was a member of the SVP for at least 25 years (1941–1964). She married AMNH archaeologist Fenley Hunter, who in 1953 discovered and researched the early man site at Tule Springs, Nevada, documenting Pleistocene fossils.

Idella **Kennedy** (née Carr) (1923–2003) joined in 1942 and remained a member at least until 1955. Kennedy was apparently a student at the University of Michigan in the early 1940s (listed as living in Detroit in 1946) and in the late 1940s was associated with the Carter County Geological Society and Museum in Ekalaka Montana (Kennedy family, pers. comm., 2019). Later she joined the PS (1952). By 1951–1952 Kennedy was at Western State College, Gunnison, Colorado.

Jannette May **Lucas** (ca. 1886–1965) was an American from New York who was an SVP member for at least 25 years (1941–1964), moving to Plymouth, Massachusetts. She wrote (or illustrated?) *Animals on the March* (1937) with W. Maxwell Reed (1871–1962), a pioneering US author of illustrated science books for children.

Catherine Ann **Moodie** (1913–1980), an American who worked at LACM, was the daughter of paleopathologist Roy Lee Moodie (1880–1934). She joined the SVP in 1941 but was only a member until 1942.

American Patricia May **Naughton** (1882–1943) seems to have been a teacher, as she is listed in 1942 when she joined the SVP as being at South High School, Omaha, Nebraska. She was the second woman charter member to die in 1944 (with a brief notice in *SVP News Bulletin* 11 but no obituary).

Rachel H. **Nichols** (née **Husband**) (1898–1985) joined the SVP in 1941. She was an assistant to visiting paleontologists at the AMNH and later served as guardian of the vertebrate collections and custodian of the Osborn Library. She received a BS from the University of Oregon. After graduating, she remained in Eugene, working as an assistant to paleontologist Earl L. Packard (1885–1983). She then moved to California, where she was employed at LACM from 1922 to 1924. It was during this time that she published on fossils, including toads from Rancho La Brea (Husband, 1924). On the advice of William Diller Matthew at the AMNH, she went next to the University of Kansas, where she obtained an MS in 1925. She then gained employment at the Natch as the museum cataloger in 1931. Here she befriended and put up the young Margaret Matthew (W.D. Matthew's daughter and the future wife of Edwin Harris "Ned" Colbert; Turner et al., 2010a), who joined the AMNH as a scientific illustrator (Elliot, 2000:46; see Chapter 6). Nichols arranged a meeting for Margaret with young Colbert and later married Margaret's brother-in-law John Nichols (Elliot, 2000:46).

Rachel Nichols became a mainstay of the *SVP News Bulletin* and BFV the early 1940s to the 1970s (see Colbert, 1985). She was made an honorary member of the SVP in 1982.

Katherine Evangeline Hilton **Palmer** (née Van Winkle) (1895–1982) was an American invertebrate paleontologist who joined the SVP in 1942 (Anon, 1942). She grew up in Oakville, Washington, with an interest in natural history, and she was the only woman in her school to go to college. Palmer received a PhD in zoology from Cornell University in 1925 before becoming the second director of the Paleontological Research Institute (Rossiter, 1995). Although her well-regarded research focused on Tertiary mollusks (Caster, 1983), she published two VP articles on fossil whales (Palmer, 1939, 1942). She was the first woman to be awarded the Paleontological Society Medal, that society's highest honor.

Anne **Roe Simpson** (1904–1991) was an American clinical psychologist who married her childhood friend, renowned vertebrate paleontologist George Gaylord Simpson (better known as "G.G."). She received her BS (1923) and MS (1925) degrees from the University of Denver and her PhD from Columbia University (1933). Early in their partnership in 1939, Roe Simpson went with G.G. on the first major paleontological expedition to Venezuela and collected fossil mammals (Simpson, 1980; Laporte, 2000). The couple collaborated on biometrics, evolution, and behavior. She is perhaps best known as coauthor with her husband of *Quantitative Zoology* in 1939 (2nd edition published 1960 with addition of coauthor Richard Lewontin) and *Behavior and Evolution* (1958). In 1946 she made use of 26 of her fellow SVP members (all were males) to do personality tests as part of her study of creativity at the 1945 annual SVP meeting. As Roe Simpson (1946:18–19) reported, "Among the results was the discovery that vertebrate paleontologists may be better adjusted than artists, but this is not conclusive because the differences are not statistically significant." Results of this survey, which at that time include only male respondents (Rossiter, 1995), contributed to her 1953 book *The Making of a Scientist*. In 1952 she moved to New York with Simpson then at the AMNH; the couple moved again in 1958 to the MCZ, and at Harvard they became a unique couple as the first married full professors (Rossiter, 1995; Laporte, 2000). The Simpsons hosted paleontologists at their summer home Los Pinavetes in the Jemez Mountains at La Jara, New Mexico. Roe Simpson handled visits with "grace and aplomb" and prepared meals using a slide ruler to calculate the time difference for cooking at high altitude (P. McKenna, written comm., August 2019). Although she was originally denied a tenure-track faculty appointment, Roe Simpson became the 9th woman in the history of the institution to become a tenured faculty member and the first woman tenured in Harvard's Department of Education. After retirement in 1967, the couple went to live in Tucson, Arizona, continuing work affiliated with the University of Arizona. In 1968 Roe Simpson was rewarded for her assessment of artists and scientists with the American Psychological Association Richardson Creativity Award.

Grace Evelyn **Russell** (née LeFeuvre) (1912–1998) was a Canadian fossil collector who married vertebrate paleontologist Loris Shano Russell (1904–1998) in 1938, a partnership that lasted 60 years (Plate 21). They joined the SVP together in 1941. As David Spalding (1993:163) noted, she "challenged orthodoxy by regularly accompanying Loris in the field. Although not a paleontologist, she has become almost as well known in the paleontological world as her husband." According to Kevin Seymour (2004:451), "they did everything together from hosting museum dignitaries . . . to working in the dirt and heat in the Badlands of Alberta collecting fossils." Many other stories about Grace Russell exist, including the following: "On . . . occasions I have listened to Grace lionize Loris, but often with incorrect facts about his history or achievements, only to hear Loris respond with polite and quiet corrections and Grace seemingly unaware that Loris had corrected her. They held each other in such high regard that there was no stress involved in such situations" (Churcher, 2000:269). Once married, Grace gave up her previous nursing career, and, as reported in issues of the *SVP News Bulletin* from the National Museum of Canada, Ottawa, or the Royal Ontario Museum, Toronto (ROM), their later base, Grace continued to be an active field assistant throughout their married life. Loris retired in 1971, but they continued fieldwork in western Canada until late 1992.

American Florence [E.]M. **Sunderland** (née Erford) (1885–1972) was living in Omaha when she joined and participated in the first SVP field conference in 1941 in Nebraska. In 1943 she moved to Portland, Oregon. According to her brief obituary (Schultz, 1972:31), "she contributed much, with her time and gifts, to VP in Nebraska and Oregon."

Ida **Wilson** (née DeMay) (1916–1990) joined the SVP in 1941 and remained a member for many years (1955 membership list). This American may have trained in zoology or vertebrate paleontology and been a research assistant to paleo-ornithologist Loye Miller at UCLA, because she published at least one VP paper with him, a major review of Californian fossil birds (Miller and DeMay, 1942).

Florence "Fran" **Wood** (née Dowden) (1898–1988) was an American zoologist specializing in crustaceans and the first woman to receive a PhD in zoology from Yale University, in 1925 (Anon, 1989). She married vertebrate paleontologist Horace Elmer Wood (1901–1975) and became both a research assistant to her husband and the illustrator for most of his publications (e.g., Wood, 1945). Together they made 15 expeditions to the US West, Mexico, and Canada, excavating fossils (*Princeton Alumni Weekly*, 1976). The fossil rodent *Franimys* was named for her by her brother-in-law Albert E. Wood. She was the New York reporter for the *SVP News Bulletin* in the 1950s.

Charter member and organizer Nelda E. **Wright** (1901–1992) was part of the SVP from the beginning, mostly in the background as research assistant and later "Girl Friday" to Alfred Romer at Harvard. There are conflicting stories of her origins—according to relations, she was born in Canada, but not much is known of her upbringing. She appears as photographer and illustrator for Richard S. Lull (Lull and Wright, 1942) at the Yale Peabody Museum of Natural History (YPM), New Haven, Connecticut, in the late 1930s (Turner et al., 2010a). Romer then employed Wright at the MCZ during and after World War II. Wright took on the role of *SVP News Bulletin* editor and, with Romer and Edinger, prepared the BFV on literature from the Middle Ages up to 1928 (Romer, 1954). She accompanied him in the field on occasion, reported on the MCZ in Romer's absence, and dealt with his publications after he died (Romer, 1974). Ned Colbert cited her for the drawing of tooth replacement in the hadrosaur *Kritosaurus* (Lull and Wright, 1942). She was recognized as an honorary member of the SVP in 1970.

There are two other early women SVP members whose contributions we could not verify; they remain in the realm of Falk's (2000) immortals. Frances C. **Baer** (191/192?–unknown) joined in 1941 and was still a member in 1969. She lived in Portsmouth, then Cleveland, Ohio, and later (post-1964) Washington, DC. She joined in 1941 and was still recorded as a member in 1969. We also know little about Helen H. **Lane** (1900–unknown), who joined in 1942 when she taught at Central High School, Omaha, Nebraska, and remained there at least until 1964. We do not know her connection to vertebrate paleontology, except that she might have been related to Henry Higgins Lane (1878–1965).

Early Women VPs into the 1950s

Other women joined the SVP later in the 1940s and into the 1950s, and some went on to notable professional careers in the following decades. Six women attended the 1954 SVP annual meeting in Los Angeles, including Sheila Brooks and Hildegarde Howard and VP wives Priscilla McKenna, Margaret Cook, and Jessie Camp. By the end of the decade, only 50 domestic women and only four "foreign" (i.e., overseas) women were listed (Anon, 1959, 1960), but still no honorary members. The influx of female overseas members was bolstered in 1952 (Anon, 1952) by Elizaveta Beljaeva and Vera Gromova from the Soviet Union, Dutch woman Antje Schreuder (Chapter 3; Table 3.1), and Dorothy Rayner in England (discussed later in this chapter).

An enthusiastic fossil collector was Helen Lucille "Lucy" Bailey **Baird** (1926–1963) (Sarjeant, 1978–1987; Sues et al., 2013), who married fellow student vertebrate paleontologist Donald "Don" Baird (1926–2011). After graduation from the University of Pittsburgh, she went to work at the Carnegie Museum as curator of paleontology Roy "Pop" Kay's secretary. She and Don received their MS degrees in 1947 from the University of Colorado; Lucy's was a thesis on museum presentation of VP specimens. Lucy then followed her husband across the country doing part-time educational, curatorial, or library work. They both joined the SVP in 1949 while in Cincinnati. Don became curator of the Princeton Natural History Museum, and Lucy, believing strongly that "a woman's place is in the field," continued as a keen collector and made many fossil discoveries in Ohio, West Virginia, Pennsylvania, and Nova Scotia, including Carboniferous Pennsylvanian reptiles and an amphibian skull in a lycopod tree at World Heritage Site Joggins Fossil Cliffs (Anon, 1963:35). Interestingly, the first jokes appear in the *SVP News Bulletin* in the early 1950s—all by women—and the first is by Lucy and Don's daughter Deidre Baird; others are by Dorothy Long and Grace Orton (see Anon, 1951).

Vivian (Vivienne) Leota **Castleberry** (née Fulton) (1907–1992), from a sheep-and-cattle-ranching family north of Ekalaka, Montana, joined the SVP in 1943. She graduated from the University of Montana State Normal College, Dillon, in May 1928 after a two-year teaching course and married soon after. Later she gained a BA in education in 1966 and was a teacher in the Ekalaka School District. According to her relatives (N. Askin, pers. comm., November 2018), there were fossils on her property in which she became interested. She became a member of the Carter County Geological Society (Anon, 1992).

Margaret Jean Ringier **Hough** (1903–1961) joined the SVP in 1947 as one of the few professional VP women in the 1940s and into 1950s. She worked mainly on the anatomy and evolution of carnivores. She was born in Saint

Louis, Missouri, and attended the University of Missouri. While there, she met and married Elliot Hough, then left the university and had a daughter. She gained her BS at the University of Chicago in 1929 and began postgrad work with Alfred Romer, juggling research with family responsibilities. She divorced in 1939 and became an assistant at the Field Museum of Natural History (FMNH) in Chicago—her first "real" job—and then an instructor at midwestern colleges. She then reentered graduate school at the University of Chicago and received both her MS (1942) and PhD (1946) from the Department of Zoology (Aldrich, 1990; Ogilvie and Harvey, 2000).

Hough was later employed by the AMNH and the US National Museum, where she examined collections and prepared her dissertation for publication—a widely read classic paper on the anatomy and evolution of the ear region of fossil and Recent carnivores (Hough, 1948). She was employed by the Paleontology and Stratigraphy Branch of the US Geological Survey from 1949 to 1960, and from 1960 until her death she was an assistant professor of biology at Long Island University (Patterson, 1961). Traveling to many sites across the United States to collect in Cenozoic deposits, she continued her work, especially on fossil cats, raccoons, bears, dogs, tenrecs, and so on, providing regular field reports for the *SVP News Bulletin* and announcements of her finds such as an Eocene tapir tooth (e.g., Dunkle, 1951). In 1960 she finally gained a faculty position at Long Island University, but within a few months she died suddenly (Patterson, 1961).

M. Dorothy **Reed** (1908–unknown) was an American who joined the SVP in 1946, when she was in charge of the VP collection in the Department of Geology and Paleontology at the ANSP. Reed was one of those who helped Edinger in her early days at Harvard (Köhring and Kreft, 2003:218). She remained in the SVP at least until retirement in 1964. Reed published one work, still well cited, on a new Eocene shark from New Jersey (Reed, 1946).

Another important woman in the history of the SVP from the beginning was Ruth **Romer** (née Hibbard) (1901–1992), wife of founding SVP vertebrate paleontologist Alfred Romer. She was a graduate of the University of Missouri and also studied at Bryn Mawr and the University of Chicago. She was working as a labor statistician in Chicago in 1923 when she met Romer, who had joined the university. They married in 1924 in Columbia, Missouri, and relocated to Cambridge, Massachusetts, when Alfred joined the MCZ and the faculty at Harvard University in 1934. She accompanied him on his travels at home and abroad, notably to Argentina in the late 1950s, and collected alongside him in the field, as well as organizing his academic and social life (Turner et al., 2010a).

Priscilla **Turnbull** (née Freudenheim) (1924–1985) was a paleontologist and zooarchaeologist who spent her career at the FMNH. Her education focused on geology and paleontology and she received a BS (1945) and an MS (1946), contemporary with Jean Hough. She then married vertebrate paleontologist William Turnbull (1922–2011), curator of mammals at the FMNH, and worked there as a scientific assistant and custodian (1946–1954) and a research associate (1974) until her early death. Priscilla Turnbull's field experiences were wide ranging and diverse. She collected fossils with shark specialist Rainier Zangerl (1912–2004) in the Mesozoic of Wyoming and with fossil fish curator Robert Denison (1911–1985) in the Devonian of Montana. Beginning in 1961 she specialized in the faunal analysis of Pleistocene mammals from Middle Eastern archaeological sites and Pakistan (Reed, 1986; Anon, 1985).

Several other women made important contributions by increasing public interest in paleontology. Marie L. **Hopkins-Healy** (1901–1983) was a vertebrate paleontologist who specialized in research on mammals, especially Plio-Pleistocene giant bison (*Bison latifrons*), mammoths, mastodons, ground sloths, horses, saber-toothed cats, musk ox, and deer (e.g., Hopkins, 1951). She was the first paleontologist allowed to conduct excavations on the Fort Hall Indian Reservation in Idaho in the early 1950s. She received a BS from the University of Chicago in 1923 and an MS from the University of Montana in 1933 and became and an SVP member in 1954. She was employed earlier than many women as a lecturer at Idaho State University (then Idaho State College) beginning in 1933 until her retirement in 1966 and conducted field-collecting parties with students over the years. During the fifth SVP field conference to Utah, some members had the "thrill" of a side trip to Pocatello, Idaho, to see Marie's giant buffalo (Frankforter, 1953). Charlotte P. Holton (see Chapter 6), one of her paleozoology students, was her field assistant in the late 1950s. The Idaho State Museum, Friends of Marie L. Hopkins, administers an endowment in her name for museum project support.

Emma Lewis **Lipps** (1919–1996) was a widely respected teacher with a PhD in botany at Shorter College in Rome, Georgia, for 44 years and very active in local VP projects (Holman, 1996). Most of her focus was on the Pleistocene vertebrates near Cartersville, Georgia, for which she annotated a bibliography (Lipps et al., 1988). She is best known for her involvement of many

Shorter College students in various Ladds Quarry, Georgia, projects over the years. She worked with paleontologist Clayton Ray (of the National Museum of Natural History, Smithsonian Institution, Washington, DC [NMNH]) in the early studies of the Ladds site and coauthored with him several papers on the site (Lipps and Ray, 1967; Ray and Lipps, 1968).

During this time several wives of vertebrate paleontologists were avid collectors, joining their husbands on fossil excavations. Two women in particular helped amass the Frick collection of fossil mammals at the AMNH: Shirley Marie **Skinner** (née White) (193?–2007) and Marian Mae **Galusha** (née Merchant) (1916–2004). Skinner worked as an assistant on equids in the AMNH. Recollections of both Marie and her husband, Morris Skinner, are included in a book by fellow Nebraskan Raleigh Emry—*Good Times in the Badlands* (2002). Once her children were grown, Galusha also worked at the AMNH, as a scientific secretary. She accompanied and documented collecting expeditions and mapping after she and her husband, Ted Galusha, retired in 1976. Priscilla **McKenna** (1930–) was born in Pomona, California, and graduated with a bachelor's degree in music (piano) from Pomona College in 1952. She was an enthusiastic participant in fossil-mammal-collecting trips alongside her husband, Malcolm McKenna (1930–2008). They also conducted fieldwork in many parts of the United States, as well as the Canadian Arctic, Argentine Patagonia, Mongolia (McKenna, 1964; Gallenkamp, 2001), and China. As she notes (P. McKenna, written comm., August 2019), "My involvement in Malcolm's paleontological life shaped my life and that of our children in a remarkable life-altering way—full of adventure and excitement and learning. We wouldn't have missed it for anything."

Africa

Mary Douglas **Leakey** (née Nichol) (1913–1996) was an English archaeologist and later paleoanthropologist whose work on human origins was overshadowed by and often erroneously credited to her husband, Louis Leakey. As a young child she became interested in art and humans after visiting French cave paintings with her artistic father. Her early schooling at a Catholic convent resulted in her expulsion for refusing to recite poetry, and she was later unsuccessfully tutored at home. While working on an archaeological expedition at the age of 17 with Caton Thompson, she caught the attention of Louis, who asked her to illustrate a book with him. Mary went to the field with him in Kenya and then Tanzania, where she first focused on the recovery of stone tools while he prospected for fossils. Mary went on to make significant vertebrate paleontological contributions, including her discovery in 1948 of the first skull of the Miocene primate *Proconsul*, a common ancestor of both apes and humans, and in 1959 of a skull of the fossil hominid *Paranthropus* (formerly *Zinjanthropus) boisei* at Olduvai Gorge in Tanzania (Plate 22). Among her most significant finds are the Laetoli hominid footprints in Tanzania, which provide solid evidence that hominids walked upright (Leakey, 1979, 1984; Wood, 1997). After Louis died in 1972, Mary continued excavating at Olduvai Gorge until 1984, often working with other members of the family (Morell, 1995; Diehn, 2019; and see Chapter 5). Her careful, methodical work earned her several awards, including the Linnean Gold Medal, and foreign membership in the US National Academy of Sciences.

Europe

France. In the postwar years, a new group of young women entered VP, mostly working for the MNHN or for the Centre national de la recherche scientifique in Paris.

Colette **Dechaseaux** (1906–1999) was a student of French vertebrate paleontologist Jean Piveteau (1899–1991) and later his assistant at Paris University. Her doctoral thesis was on the felids from the Oligocene phosphorites from Quercy, France. Despite disruption during World War II, Dechaseaux published over four decades, notably on mammal endocasts, building on Edinger's work on brains in the time before CT scans (e.g., Dechaseaux, 1962). She was particularly interested in bats and echolocation. Not long before her untimely death, Edinger wrote to her "dear sister" and colleague asking Dechaseaux to complete her work on camel brains in the event of going senile or her death (Edinger to Dechaseaux in November 1966, quoted in Köhring and Kreft, 2003:272). In the 1950s to 1960s Dechaseaux wrote for and did much editing work for the classic seven-volume *Traité de paléontologie* (Piveteau and Dechaseaux, 1952–1966), including for the period from the origin of vertebrates to primates. She retired in the early 1970s and then co-authored a textbook on vertebrate paleontology (Piveteau et al., 1978).

Jeanne **Signeux** (1902–1987) was a specialist on Cretaceous and Tertiary fish. For 25 years she was an assistant to vertebrate paleontologist Camille Arambourg (1885–1969) in the Institute of Paleontology, MNHN (Williams, 1953; Gaudant, 1990). For two decades, from 1949 until her retirement in the mid-1960s, she worked on Cretaceous to Lower Cenozoic sharks, including from Brittany and Morocco, naming several new genera (Ogilvie and Harvey, 2000).

Christiane Hélène **Mendrez-Carroll** (1937–1978) was a French vertebrate paleontologist specializing in the mid-1960s to the 1970s on morphology and systematics of mammal-like reptiles, especially South African therocephalians (Ogilvie and Harvey, 2000). She was educated in France and held a position at the Institute of Paleontology, MNHN. She visited South Africa to further her studies, attending the 2nd Gondwana Symposium in Cape Town and Johannesburg in 1970, meeting with Pamela Robinson (see Turner and Malakhova, 2010). In 1974 she married Canadian vertebrate paleontologist Robert "Bob" Carroll. Her promising career was cut short by a tragic death (Turner et al., 2010a).

Other women began their studies in this time. Germaine **Petter** (193/194?–) (retired from the Centre national de la recherche scientifique and the Institute of Paleontology, MNHN) studied various Tertiary to Pleistocene elephants and carnivores (viverrids, hyaenids, canids, mustelids). In the 1960s and 1970s she worked often in Africa (Ethiopia, Sahara, Madagascar) and in Spain, cooperating with Spanish colleagues. She named a fossil mustelid *Sabadellictis crusafonti* for her mentor vertebrate paleontologist Miquel Crusafont i Pairo (1910–1983). She has written with colleague Brigitte Senut a popular book about Lucy, the African hominid fossil (Petter and Senut, 1994), with her last paper in 2004 on Oldowan carnivores. Cécile **Poplin** (193?–) began as a research assistant at the MNHN, specializing in microtome technique (Poplin and De Ricqles, 1970). She gained her doctorate in 1974 on Carboniferous paleoniscoid bony fish from Indiana and Kansas, going on to study Paleozoic actinopterygians and xenacanth sharks from across France and around the world. Based later in the Département Histoire de la Terre, MNHN, she has worked with colleagues in the museum and overseas, including on Mongolian Cretaceous multituberculates by microtome serial section in collaboration with Zofia Kielan-Jaworowska and more recently with Dick Lund on fauna from Montana (e.g., Poplin et al., 1997; Lucas, 2001). Denise **Sigogneau-Russell** (193?–) is a leading expert on fossil frogs, therapsid reptiles, and Mesozoic mammals (e.g., Sigogneau-Russell, 2003) with groundbreaking studies on gorgonopsians and French dinosaurs. She is married to fellow VP Canadian Donald Russell and both are still working at the MNHN.

Germany. Gudrun **Corvinus** (1932–2006) was a Polish-born geologist, paleontologist, and archaezoologist. She completed her education at the University of Bonn and was affiliated with the University of Tübingen, Germany. Although primarily known as an archaeologist, she was trained in geology and was a member of the Afar expedition in Africa in the early 1970s that discovered the famous hominid skeleton dubbed "Lucy." While working in Namibia, she also discovered Miocene-aged mammalian fossils. She also did pioneer work on Paleolithic sites in India and Nepal. Corvinus died tragically, murdered in her home in India (Chauhan and Patnaik, 2008).

Minna (Wilhelmine) **Lang** (1891–1959) was a German physicist, teacher, and science journalist (Lang, 1960). She became the first woman to receive a PhD in physics from Frankfurt University in 1916. Given the difficulty that women with advanced degrees faced in finding employment in physics, she turned to teaching. In addition to her work as a high school teacher from 1919 to 1947, she worked as a journalist, writing about a variety of science topics, including paleontology (Müller, 1959). In 1945, after an air raid on Meiningen, she and a colleague rescued a natural science collection that later became part of the Meiniger Museum's Natural Science Department that she organized and maintained as curator. After World War II, she collected and curated mastodon remains from Meiningen. These specimens led to the establishment of the Mastodon Hall in the museum, which opened in 1953. Lang had a long and voluminous correspondence with the paleontologist Friedrich von Huene from 1947 to 1958, mainly on religious matters but also on historical topics and dinosaurs, and she wrote several papers with him, including one on Thuringian reptiles (Turner et al. 2010a).

The Netherlands. Margaretha "Greet" **Brongersma-Sanders** (1905–1996) was a Dutch vertebrate paleontologist, geobiologist, and oceanographer. She was born in Kampen, Netherlands, and raised in a scientific environment, encouraged by her scientifically trained grandfather and father, who became her biology teacher. She began her research career mentored by L.F. de Beaufort (1879–1968), receiving a PhD from the University of Amsterdam for a dissertation on fossil fish from Indonesia (Sanders, 1934) that is still considered an important reference since this part of the world is geologically and paleontologically so poorly known. As with Antje Schreuder, Brongersma-Sanders's first job was as assistant to Eugene Dubois, and here she met her husband, paleoherpetologist Leo Brongersma (Shipman, 2001). She may have been the first woman taphonomist, having been the first to categorize fish mass mortality, emphasizing algal blooms (red tides) and their importance in forming fish bonebeds (Brongersma-Sanders, 1957). During a honeymoon visit to South Africa, she had seen the astonishing mass fish kill on the coast of Namibia, which led to her inspirational work on the importance of fish mass assemblages to oil production, especially in open sea upwelling sites. This resulted in her appointment in the 1950s as a consultant with the Royal

Dutch Shell Oil Company. Much of her later work focused on geochemistry and oceanography, attempting to understand the black shales that preserve fossil vertebrates in fine detail (Sanders, 1948; Turner and Cadée, 2006).

Norway. Norwegian-born Natascha **Heintz** (1930–) was the daughter of Russian émigré and early vertebrate paleontologist Anatol Heintz (1898–1975). Like her father, she worked on early vertebrates, especially the Paleozoic agnathan Pteraspidiformes. Based at the Palaeontologisk Museet, Oslo, from the 1950s, she took part in arctic expeditions with the Norsk Polarinstitut, leading one in 1962 to retrieve dinosaur footprints found in Spitsbergen and Old Red Sandstone fish (Heintz, 1962). She participated in the 1968 Nobel Symposium, presenting on northern heterostracans (e.g., Miles, 1971).

United Kingdom. Doris Mary **Kermack** (née Carr) (1923–2003) was an English paleontologist and marine zoologist at Imperial College London. She received a PhD from UCL in 1953. She was a member of the thriving team of paleontologists at UCL that then included her husband, paleontologist Kenneth Kermack (1919–2000), and research assistant Frances **Mussett** (193?–) (Panciroli, 2019). Doris was the first to describe the stem mammal "symmetrodont" *Kuehneotherium praecursoris* and was lead author on her and her husband's major 1971 book *Early Mammals*. She was awarded the Linnean Gold Medal for outstanding service to the Linnean Society of London in 1988, where she had been a journal editor for many years.

Dorothy Helen **Rayner** (1912–2002; Plate 23) was an English geologist and vertebrate paleontologist who specialized in the anatomy and evolution of Mesozoic bony fish. She was also a recognized authority on the stratigraphy of the British Isles, writing a definitive textbook in 1971. Rayner attended the radical coed Bedales School, matriculating in 1931 at Girton College, University of Cambridge. Because the University of Cambridge did not admit women as full graduates before World War II, she was awarded her first-class BA (1935) from the college, becoming a University Harkness Scholar. From 1936 on, she was a Hertha Ayrton By-Fellow of Girton, working toward a PhD (1938) at Cambridge and at UCL, under the mentorship of D.M.S. Watson (see Chapter 3). With the outbreak of war in 1939, Rayner took up a teaching post as one of only three in the Geology Department of Leeds University. This was to remain her base all her working life; she became senior lecturer in 1960 but never attained a professorship. Her major analytical studies of Mesozoic "holostean" bony fish were produced during the 1940s and 1950s (e.g., Rayner, 1948), and she first received acclaim from the GSL for these with the Lyell Fund in 1950. Rayner went on to make paleoenvironmental studies of Paleozoic fish and tetrapods; recently, Giles and colleagues (2015) honored her with the creation of a genus of early bony fish, *Raynerius* (see Appendix 4). She gained an MA from Cambridge in 1948, and in further recognition of her contributions she was awarded a number of honors and medals, including the Lyell Medal from the GSL (Varker, 2004; Turner, in press).

Pamela Lamplugh **Robinson** (1919–1994) was an English vertebrate paleontologist who first specialized in Mesozoic fissure fill faunas of southwestern England and Wales and later taught geology and vertebrate paleontology at the University of London. Robinson was born in Manchester, and after early private and Manchester Girls' High schooling, she entered the University of Hamburg in 1938 for premedical studies. Her education was interrupted by World War II, when she worked at the Royal Ordnance Factory at Thorpe Arch, Yorkshire (1942–1945) and for the British Woollen Industries Research Association in Leeds. While there, she attended evening lectures in paleontology at Leeds University given by Rayner. Afterward she worked for nearly two years as librarian at the GSL before enrolling at UCL, where she was influenced by scientist J.B.S. Haldane (1892–1964), German vertebrate paleontologist Walter Kühne (1911–1991) who specialized in Mesozoic microfauna, and D.M.S. Watson. She graduated with first-class honors in 1951 and continued postgraduate research in the Zoology Department, gaining grants from the 1948 International Geological Congress Gloyne Fund in 1954–1955 for her work. Robinson remained a lecturer at UCL, receiving a PhD in 1957 for her study of the formation, stratigraphy, and faunal assemblages of the Triassic and Early Jurassic vertebrate-bearing fissure sediments from the Mendip Hills and Gloucestershire, published by the Linnean Society in 1957. In 1957 she made the first of many visits to India, where she was invited by Indian statistician Prasanta Chandra Mahalanobis (1893–1972) and their friend Haldane to establish the Geological Studies Unit at the Indian Statistical Institute in Calcutta; the institute, established in 1959, was later deemed a university and government ministerial visits were common to check progress of Robinson's educational program (Rao, 1973). There, she initiated research programs in Gondwana Mesozoic stratigraphy and vertebrate paleontology (e.g., Robinson, 1967). In Calcutta, she supervised the research of paleontologists Sankar Chatterjee, Sohan Jain, and Tapan Roy Chowdhury (1926–2013). The Geological Studies Unit team traveled widely, often on foot, in search of fossils and geological specimens (Colbert, 1980), and she set up a museum for dinosaur and other vertebrate material found that continues (Nigel Hughes, pers. comm., January 2020). Robinson was also important for her promotion of and attendance at the early Gondwana series of symposia (Robinson, 1971) and she was honored with the naming of several taxa (see Appendix 4). She supervised the research of vertebrate paleontologists L. Beverly Halstead (1933–1991), Barry Hughes, and Steven Rewcastle and in 1972 was an Alexander Agassiz Visiting Professor at Harvard. The following year she received the Wollaston Fund from GSL in recognition of her work promoting and establishing vertebrate paleontology in India (Milner and Hughes, 1995; Turner et al., 2010a; Turner, in press).

Early Women VPs into the 1960s and Mid-1970s

As Rossiter (1995) notes, women's "consciousness" was on the rise in the 1960s and many former practices in science began to change with the fall of major barriers to gaining jobs. Eventually, legislation for equal pay (1963 in the United States) and affirmative action was realized in several countries (Kelly, 1993; Weatherford, 1994; Connelly, 2015). Women scientists, especially in biological fields, and VPs increased in numbers during the 1960s. More young women trained in VP and joined the SVP during this period. Fourteen women attended the 1971 SVP meeting in Washington, DC, including Zofia Kielan-Jaworowska, Elaine Anderson, Barbara Waters, Margery Coombs, Anna "Kay" Behrensmeyer, Sylvia Fagan Graham, Susan Berman, VP wives Ruth Romer, Mrs. William Gonyea, Mrs. Walter Greaves, Marian Galusha, Priscilla McKenna, and Mrs. John Wilson, and one unknown, some of whom we discuss further shortly.

These decades saw renewed international cooperation, with major joint expeditions, especially in Asia, and new work in Africa, India, and the polar regions, which brought forth new dinosaurs and mammals in particular (e.g., Rozhdestvensky, 1977; Leakey, 1984; Morell, 1995; Boishenko, 2001). Most important were the new ideas regarding continental drift, plate tectonics, phylogenetic systematics, and vicariance that were to overturn so many earlier ideas about evolution and the paleogeographic distribution of vertebrates. Several of the women during this era were contributing to the new paradigm (e.g., Robinson earlier; Turner, 1970; and see the discussion of Mary Dawson later in this chapter).

Africa

Shirley Cameron **Coryndon** (1926–1976) was an English paleontologist who worked on mammals, especially hippopotami, mostly in Africa. Her first work on bones was as a radiographer in London after the war. There she met and married Roger Coryndon, son of a former governor of Kenya. She moved to Kenya and began working at the Coryndon Museum (now the Kenya National Museum), where she met the Leakeys. Here Coryndon became interested in mammalian paleontology, and she became the Leakeys' research assistant (Leakey, 1984; Morell, 1995). She studied paleontology with Donald MacInnes at the Museum in Nairobi. She worked at most of the Miocene and Pleistocene sites in Kenya and Tanzania, including Olduvai Gorge. On the death of her husband in 1963, Coryndon returned to London and worked at the BMNH, continuing her work on African Tertiary faunas and contributing to the *Fossil Vertebrates of Africa* (e.g., Savage and Coryndon, 1976), a series she began in 1970 with Louis Leakey and her second husband. In 1969 she married fellow vertebrate paleontologist Robert "Bob" J.G. Savage (1927–1998), whom she had met in Kenya in 1955, and moved to Bristol, where she was involved in organizing the Bristol VP collections until her untimely death (Dineley, 1977).

Marjorie **Greenwood** (née George) (1924–2006) was born and educated in Johannesburg. In 1944, the first in her family, she entered the University of Witwatersrand to study zoology, after which she gained a position as a "demonstrator" in the Zoology Department, working alongside Raymond Dart (1893–1988) and paleontologist James Kitching (1922–2003). Greenwood joined Kitching on field trips to the Karoo and elsewhere in southern Africa, collecting Permian to Triassic bones for the Bernard Price Institute. She assisted and published with Robert Broom on the therapsids (e.g., Broom and George, 1950). After marriage to ichthyologist Humphry Greenwood (1927–1995) in 1950, she concentrated on family but did investigate Dart's material from the Transvaal and specialized in small mammals such as hyraxes. When the family moved to England in 1957, she began working at the BMNH, where she joined forces with Percy Butler (1912–2015) to work on a range of Plio-Pleistocene shrews, hedgehogs, and bats from the Leakeys' Olduvai Gorge (e.g., Butler and Greenwood, 1965) and elsewhere. In all, she named many taxa (Jenkins and Hutterer, 2017).

Asia

In terms of women, the history of VP in China does not begin until after World War II (Lucas, 2001). China emerged from the war in economic distress after Japanese invasion, the civil war that followed, and the major shift to the People's Republic in 1949. Political change eventually led to upheaval with the Great Leap Forward in 1958, and then China was plunged into a decade of the Cultural Revolution in the mid-1960s. VP scientists in Communist China fared in different ways (e.g., Lucas, 2001; Kölbl-Ebert and Turner, 2017). As did others in the scientific community, members suffered purges and sometimes were arrested as "spies," "capitalist roaders," or "revisionists" (Chang, 1969). Young students were ordered to join the Red Guard (Hóng Wèibīng), a mass student-led paramilitary. Universities were closed from the summer of 1966 through 1970, when they reopened for undergraduate training but with greatly reduced admissions and a heavy emphasis on political training and manual labor. This decade affected the studies of many who later became VPs. Political rectitude was more important than academic talent. All scientific journals ceased publication in 1966, and subscriptions to foreign journals lapsed or were canceled. For most of a decade, China trained no new scientists or engineers, and VP was essentially cut off from foreign scientific developments. Women VPs in this era had to be highly resilient.

CHINA

Chang-Kang **Hu** (1928–2011) was 20 when the new People's Republic of China was founded. She studied Quaternary mammal fossils excavated at Choukoutien, near Beijing, the famous site of *Homo erectus* (*Sinanthropus pekinensis*), and often worked in conjunction with Russian colleagues such as Inessa Vislobokova (Chapter 5) on fossil deer. Hu was born in Shaoxing, Zhejiang, in southern China. She attended the Sino-French University in Beijing from 1947 to 1950 and majored in biology. She then transferred to Nankai University in Tianjin, where she received her BS degree in biology in 1951. After graduation, she was assigned by the Central Government to the Cenozoic Research Laboratory, under the Committee of Chinese Geological Planning (later Ministry of Geology), as a research assistant to C.C. Young, the pioneer of Chinese VP (e.g., Lucas, 2001) and an early correspondent with the SVP. In 1953, after he had founded the new Laboratory of Vertebrate Paleontology (the precursor of the IVPP) in the Academia Sinica, Hu followed him to the new institute. Soon she was publishing on mammal fossils from Choukoutien and other localities. She worked with German H.D. Kahlke on the distribution of giant deer in China. In the 1950s and 1960s, Hu joined many important field expeditions organized by the IVPP, most notably one of the Sino-Soviet Paleontological Expeditions, for which she was the deputy leader. Often there were limited resources and poor working conditions, but she produced important work, including "Mammalian Fossils Associated with the Hominid Skull Cap of Lantian, Shaanxi" (with Minchen Chou and Y.C. Lee in 1965) and a monograph with Tao Qi on Gongwangling Pleistocene mammalian fauna of Lantian, Shaanxi, in 1978. As with other scientists in her generation, Hu's life was not easy. She had to finish high school under the threat of air raids during the day in the war against the Japanese (1937–1945). She transferred to another university because her university was closed after the Chinese civil war. In the late 1970s, when China resumed more normal conditions, Hu eventually continued her work and was promoted to an associate research professor at the IVPP in 1978. In the 1980s, she led a small team from the IVPP and visited the United Kingdom and the United States, including the Smithsonian Institution, the University of Pittsburgh, and the University of Kansas, and joined field excavations in Denver, Colorado. As the senior scientist, she participated in the successful introduction of the endangered Père David's Deer (*Elaphurus davidianus*) from Woburn Abbey United Kingdom, to China in 1985–1987.

In the late 1980s, the Chinese Academy of Sciences underwent large-scale restructuring and a sizable number of senior scientists faced early retirement, including Hu in 1989. She then assisted her husband, Tung-Sheng Liu, until his death in 2008. Hu died in Beijing.

Yu-Ping **Chang** (**Zhang**) (1934–1985) was also a VP at the IVPP. She was born in Tienjing and graduated from the Beijing College of Geology with a degree in geological engineering in 1956. She then joined the IVPP, where she remained until her death. She conducted fieldwork in nearly every province of China. Among her major contributions to VP was her discovery and collection of mammals (e.g., tillodonts, suids) from the Paleocene in the Nanxiong area of South China. In 1957 she worked at Late Eocene mammal localities in Luishiu, Hunan Province, and in 1959 she co-led with Irina Aleksandrovna Dubrovo (1922–2011) an expedition to Ula Usu (Eocene) in Shara Murun, Inner Mongolia, as part of the Chinese-Russian expeditions to this area (Kurochkin and Barsbold, 2000). From 1970 onward she focused on early Tertiary deposits in South China and those at the Cretaceous–Paleocene boundary (e.g., Zhang, 1989). Before her untimely death in 1985, she was working on a bilingual dictionary of paleontological terms and establishing a computer-based translation facility for Chinese vertebrate paleontologists (see Rich et al., 1994). She spent 1982–1983 in Australia working on this project (Rich and Rich, 1985).

Another who grew up in the post–World War II era was Meeman **Chang** (**Zhang**) (1936–), a Chinese vertebrate paleontologist at the IVPP, Chinese Academy of Sciences, Beijing. She has spent her research career studying fish faunas and environments in China's eastern coastal provinces in diverse areas, including taxonomy, phylogeny, zoogeography, paleoecology, and bio-stratigraphy. Her work has helped clarify the fossil record of fish, notably the links between Paleozoic fish and the evolution to land-based vertebrates (e.g., Elliott et al., 2010). She has played a key role for years in bringing little-known Chinese fish fossils to the attention of the scientific world. Along with her many successes, she also faced significant difficulties at the start, because of the shifting political landscape of her homeland. Although Chang's father, a pathologist from Nanjing, wanted her to become a physician, love of her country led her to choose geology instead. In 1958, during the Great Leap Forward, she was among those who heeded the call of Vice President Liu Shaoqi to study the earth so that China might exploit its natural resources. She earned her undergraduate degree in 1960 at the Lomonosov Moscow State University, where she became fluent in Russian. In 1965, she was chosen to do graduate research at the Swedish Museum of Natural History, where she began to study with Erik Stensiö (1891–1984), who had made Stockholm a leading fossil fish research center. Her time there proved short, as the Cultural Revolution swept China in 1966. Chang, ever the patriot, halted her studies and returned home. In Beijing, she was confronted by the Red Guard, who were "purifying" China by isolating and punishing the academic classes. Like other academics, she was sent to the country and had to haul rocks and build a dam but eventually was able to return to science, conducting practical research on mosquitos. Only when the Cultural Revolution ended was she able to resume her work, finally defending her doctorate, entitled "The Braincase of *Youngolepis*, a Lower Devonian Crossopterygian from Yunnan, South-Western China," in Stockholm in 1981.

Chang's contribution was recognized in 1983 when she became the first woman to head the IVPP (Miao, 2010). Patricia Vickers-Rich and Susan Turner both went to work in China for the first time in the 1980s and Chang was Turner's host at IVPP and treated her as a little sister; this family relationship with colleagues has continued through time. Her role as director of an Academy of Sciences institute was significant not only because of her gender but also because it marked the IVPP's move away from political appointments to those based on merit. Chang served two terms as director, ending her tenure in 1990. She helped shepherd the institute from the days when whole families were living on an upper floor of the research building, moving it to a new facility that included modern laboratories, and promoted young researchers to work both in China and overseas. She became an academician of the Chinese Academy of Sciences in 1991. Chang was elected to honorary membership in the SVP in 1997 and awarded the Romer-Simpson Medal in 2016. In 2018 she was given the L'Oréal-UNESCO for Women in Science Award for her pioneering work on fossil records leading to insights on how aquatic vertebrates adapted to life and land (e.g., Jiang, 2018).

Australia

Female paleontologists are few in Australia in the post–World War II period. The important pioneer Dorothy **Hill** (1907–1997) earned a PhD at Cambridge University in the 1930s and returned to a job in the Geology Department of the University of Queensland (e.g., Campbell and Jell, 1998). As one of the few academic geologists at the time, she undertook many different kinds of geoscience work besides her own specialty of coelenterates. During pioneering mineral and oil surveys of Australia, fossil vertebrates were found, and she always made efforts to identify these. She founded important publications, such as a series of guides to fossils of Queensland through time (e.g., Hill et al., 1970) that included vertebrates and the journal *Alcheringa*, which included the first papers on fossil Australian vertebrates (e.g., Turner, 2007). Joyce **Gilbert-Tomlinson** (1916–1981) was another who contributed to discovering early vertebrates in Australia, notably some of the first Devonian fish and one of the oldest in *Arandaspis* from the Ordovician (Rich et al., 1985; see Appendix 4). She was a BSc graduate of the University of Sydney and later worked for the national geological survey (Dickins, 1982; Turner, 2007).

By the 1960s and 1970s, vertebrate paleontologists trained overseas were coming to work in Australia (see Chapter 5).

South America

From the end of the 1930s, Noemi Violeta **Cattoi** (1911–1965) held a position in the then Paleozoology Section (Vertebrates) at the Museo de La Plata, Argentina (Herbst and Anzótegui, 2016). She took over at the beginning of 1960 as head of the Paleozoology Division. We know little of her decision to study VP but her research focused on Cenozoic mammals (notoungulates, tapirs) (Cattoi, 1943, 1951) and birds (e.g., Cattoi, 1957; Cucchi, 2016). She was a professor of vertebrate paleontology at the Faculty of Natural Sciences and Museum of the National University of La Plata, a position she resigned to join Consejo Nacional de Investigaciones Científicas y Técnicas (CONICET; the National Scientific and Technical Research Council) in Argentina, founded in 1958, becoming the first female paleontologist researcher in that organization (Tonni, 2005).

United States and Canada

Elaine **Anderson** (1936–2002) was a vertebrate paleontologist and pioneer zooarchaeologist who specialized in Pleistocene mammals. She was born in Salida, Colorado, and grew up in Denver. She graduated with a BS from the University of Colorado in 1960. She joined the SVP in 1962. For her PhD she went to Finland as a Fulbright Scholar and studied under the mentorship of the Finnish vertebrate paleontologist Björn Kurtén (1924–1988), whom she described as "inspiring, helpful and demanding" as a teacher (Anderson, 1992:124). Anderson worked primarily at the University of Colorado Museum of Natural History in Boulder, undertaking fieldwork and researching Pleistocene mustelids from Little Box Elder Cave, Wyoming, and briefly at the Idaho Museum of Natural History and the Maryland Academy of Sciences. She gave up her scientific life in 1972 to care for her ailing mother, returning to full-time work in 1984 teaching osteology at the Denver Museum of Nature and Science. Anderson is best known for her major book *Pleistocene Mammals of North America*, coauthored with Kurtén in 1980 (Donner, 2014) and illustrated by Margaret Lambert (see Chapter 6). In her final years, she was collaborating with Russ Graham on specimens from Porcupine Cave in the Colorado Rockies (Carpenter, 2002). She was elected an SVP honorary member in 2002.

Mary R. **Dawson** (1930–) is a vertebrate paleontologist with research focused on the evolution of mammals, especially Cenozoic rodents and rabbits. She was born and grew up on the Upper Peninsula of Michigan and received her undergraduate BS degree from Michigan State University (formerly Michigan State College) in 1952. After graduating, Dawson won a Fulbright scholarship to study at the Institute of Animal Genetics, University of Edinburgh, Scotland. Her formal training in vertebrate paleontology took place under the guidance of Robert Warren Wilson (1909–2006) at the University of Kansas. She studied the evolutionary history of lagomorphs (rabbits, hares) in North America and received a PhD. Following a brief faculty appointment to teach zoology at Smith College, Massachusetts, a research year at the YPM, and two years of service as program director at the National Science Foundation, Washington, DC, Dawson joined the curatorial staff at the Carnegie Museum in Pittsburgh in 1962. Dawson has spent most of her career at the museum as curator of vertebrate paleontology until she retired in 2003. When she first arrived, she was told by the museum director that, as a woman, she'd never be made curator. Ten years later she was appointed by the same director to that post at the museum (Carpenter, 2000). Dawson's pride in her work at the Carnegie Museum is evidenced by her belief that a museum's real goal is to encourage children's interest in science.

In 1981, Dawson and paleontologist Robert "Mac" West, who began their research on Ellesmere Island in the mid-1970s, were awarded the National Geographic Society's Arnold Guyot Memorial Award for their outstanding contributions to High Arctic geology and paleontology. From 1973 to 2002 the National Geographic Society supported 10 of Dawson's Arctic field expeditions. As Jaelyn Eberle and Malcolm McKenna write (2007:7), "Over the span of 11 field seasons, Dawson and her colleagues were responsible for one of the greatest contributions to vertebrate paleontology—the first and best record of Paleogene terrestrial vertebrates from within the Arctic circle." She hypothesized that rocks on several islands in the Canadian Arctic held clues to the land bridge that once connected northern Europe to North America, allowing dispersal of land mammals across the North Atlantic. Earlier G.G. Simpson had hypothesized that faunal exchange between Europe and North America occurred but immigrants came the long way around through Asia across the Bering land bridge (Beringia). Geologic and geophysical evidence, including new data from the then-unfolding theory of plate tectonics, supported a North Atlantic bridge, with fossil vertebrate evidence provided by the discoveries of Dawson and colleagues.

Dawson relished fieldwork, and she continued work in the Arctic well into her 80s (Beard, 2007). During a rare encounter during field season 1977, she faced a pack of wolves while exploring for fossil vertebrate bones. Although unhurt, Dawson's brush with the lead wolf grazed her cheek, leaving it wet with saliva (Munthe and Hutchison, 1976). Dawson's other field expeditions are no less noteworthy. She was the first person to find Eocene fossil mammals in Montana's Glacier Park. Her discoveries helped document the park's geology—with older Precambrian rocks overthrust onto younger rocks—and record the rich faunal diversity, ranging from small rodents and shrews to giant brontotheres. She carried out expeditions to China in the 1990s, and in 1995, along with colleague Christopher Beard, Dawson found fossils of a "dawn monkey," prompting colleagues to declare their discovery as humankind's "Garden of Eden" (Carpenter, 2000).

Dawson has been a member of the SVP since 1954 and an honorary member since 1999. She was only the second woman to serve as SVP president, from 1973 to 1974. In 2002 the SVP awarded her—the first American woman—the Romer-Simpson Medal, its highest honor. Dawson was intro-duced as the recipient of this medal by then-SVP president and curator of paleontology at the Carnegie Museum Hans-Dieter Sues, who remarked that when his daughter was queried about her future career, she stated that she wanted to be a paleontologist "like Mary." A $50,000 predoctoral fellowship was endowed in Dawson's name by the SVP in 2011, providing individual $3,000 grants annually to students.

Sylvia **Fagan** (née Robinson, later Graham) (193?–unknown) was an American paleomammalogist working with Malcolm McKenna and Horace and Albert Wood at the AMNH in the 1960s. She gained an MA in zoology at the University of Kansas in 1959 on the Miocene fossorial rodent *Myl-agaulus laevis* from Colorado (Fagan, 1960) and then went to work in the Biology Departments of Newark College and Rutgers University, New Jersey. She joined the SVP in 1958 and went into the field to Wyoming and Montana on the 8th SVP Field Conference. In 1960 she was part of McKenna's team prospecting the Cretaceous Lance Formation for small vertebrates, working on cranial morphology of *Peratherium* and allied marsupials.

Irene **Vanderloh** (1917–2009), a Canadian amateur VP, was born in 1917 in Steveville, Alberta, near what became in 1955 Dinosaur Provincial Park (Gross, 1998; Spalding, 1999; Tanke, 2010, 2019). She was the second oldest of 10 children, and she became interested in vertebrate paleontology at a young age in the early 1920s. She likely met the bone-hunting Sternberg sons, either Charles, who surveyed the area in the mid-1920s (Spalding, 1999), or George when he was camped close to Steveville collecting dinosaur fossils for the University of Alberta in Edmonton (Tanke, 2010). As she grew older, Vander-loh became increasingly interested in paleontology not only as a hobby but also as a potential career, but she was told paleontology "wasn't for girls" (Vanderloh, 1986:144). So she did not pursue it as a career and later became a hairdresser and did odd jobs in the area. After a first marriage and a di-vorce, she returned to Alberta and married a rancher, Vic Vanderloh, and raised two children. Her desire to contribute to paleontology grew from that time. The Vanderloh homestead was near the Dinosaur Provincial Park bad-lands, so Irene frequently explored the Upper Cretaceous buttes and cou-lees and developed a keen eye for finding small theropod bones (Spalding, 1993). Thus, she rose to prominence in Canadian paleontology during the heyday of her fossil collecting in the badlands during the 1960s and 1970s, often collecting fossils with Hope Johnson (1916–2010), another female amateur collector who began work in 1949 and later worked for the Dinosaur Provincial Park, popularizing dinosaurs (e.g., Johnson and Storer, 1974; Spalding, 1993; see Chapter 6). Likely Vanderloh shared her knowledge of the badlands with Johnson, providing the fossil identifications.

One of Vanderloh's most significant discoveries was a small theropod skeleton, described by vertebrate paleontologist Dale Russell (1937–2019) as the holotype of *Stenonychosaurus inequalis* (Russell, 1969), later found to be the junior synonym of *Troodon formosus*, which remains the best, most complete associated skeleton of *Troodon* found in Alberta. Other significant discoveries by Vanderloh include the articulated pelvis and hind limbs of a subadult tyrannosaurid (a probable carnosaur), and another skeleton that became the holotype of the maniraptoran *Saurornitholestes langstoni* (Sues, 1978), and she was likely involved in locating several other dinosaur skulls and skeletons (including *Centrosaurus*) (Tanke, 2010). The Royal Tyrrell Museum of Palaeontology and the Royal Alberta Museum have received at least 18 mostly rare small theropods in donations from Vanderloh's collecting. Vanderloh died in Brooks, Alberta, and was interred near the Dinosaur Provincial Park, her favorite collecting ground (see Tanke, 2010).

Europe

UNITED KINGDOM

The 1960s were a time of new vigor and increased jobs and opportunities for women with university educations, and several institutions were teaching VP. Several of the current VPs in Britain had begun their research, including Sheila Mahala Andrews and Anne Howie (later Warren, see Chapter 5) (Cambridge), and later Elizabeth Loeffler (Bristol), Janice Collins, Susan Evans (London), Susan Turner and Cherrie Bramwell (Reading), the late Jennifer Agnew-Clack, Eileen Beaumont-Tilley, and Angela Milner (Newcastle upon Tyne) began as students (see Chapters 5 and 6 and online database).

One who typifies this era is the English vertebrate paleontologist Cherrie Diane **Bramwell** (1945–), who specialized in (ate, slept, and dreamed) fossil pterosaurs. She was born in eastern London (Essex) and obtained her BSc degree in zoology via evening classes from Birkbeck College, University of London. She went on to PhD training beginning in 1968 under the mentorship of L. Beverly Halstead at the University of Reading (gaining it later in the 1980s). Because she considered the bat wing to be an analogue for a pterosaur wing, as a graduate student she bought herself a fruit bat and later gained a medical grant to maintain a colony of fruit bats. During her studies she worked with University of Reading aeronautical engineer George R. Whitfield, who collaborated with her on the biomechanics of pterosaur flight, and together they achieved much in promoting paleoengineering (Whitfield and Bramwell, 1971). In 1975 they won the first Trident Award for Communication in Science. With designer Steven Winkworth, she created a life-size pterosaur model that flew over the cliffs at Dorset in England, featured in BBC television program *Pterodactylus Lives* in 1974 (see "Flight from 70 Million Years B.C." in Bibliography Sources and Further Reading). Bramwell was notable for her media presence through the 1980s and 1990s, popularizing science on children's television programs (Turner et al., 2010a, but the rumors of her death they cite proved false).

USSR

In the Soviet Union many women trained in geology and then applied paleontology to biostratigraphy. There were vertebrate specialists notably using mammals and then fish to map the vast northern areas to aid in the search for resources. From these field studies came paleobiological and paleoclimatologic research (e.g., Schultze et al., 2009). Our Women in VP database and sources such as the BFV indicate that there were more than 50 women VPs in the postwar Soviet Union (Table 4.1). We have looked at Vera Gromova's career in Moscow (Chapter 3; see also Haines and Stevens, 2001). Other colleagues at Paleontological Institute (PIN) of the Russian Academy of Sciences, Moscow, were paleomammalogists, such as Elizaveta Beliaeva (see discussion that follows) and N.M. Ianovskaia (194?–198?). The latter was involved in Soviet-Mongolian paleontological expeditions in the mid-1950s (Barsbold, 2016). Homegrown Mongolian women VPs did not come until later in the 20th and 21st centuries (see Chapters 5 and 6).

Elizaveta Ivanovna **Beliaeva** (aka Beljaeva/Belyaeva) (1894–1983) was a Soviet paleontologist who studied fossil mammals and other vertebrate faunas of the Tertiary and Quaternary of the USSR and Mongolia. Her research focused on the systematics, morphology, phylogeny, and paleozoology of proboscideans, perissodactyls, and artiodactyls. She received her graduate education in Saint Petersburg. She began her scientific career as an assistant curator in the Paleontology Department of the Museum of the Mining Institute, Saint Petersburg. In 1921 she continued her scientific activity at the Osteological Department of the Geological Museum of the USSR Academy of Sciences and then in 1930 moved to the newly formed Paleontological Institute of

TABLE 4.1

A selection of women VPs working during the Soviet regime from 1917 to 1992 (Russian unless stated otherwise)

Name	Dates	Location	Research	Specialty
N.A. Alekperova	193/4?–unknown	Azerbaijan	Mammals	Deer
L.P. Aleksandrova	193/4?–unknown	Moldovia	Mammals	Small mammals
E.V. Alekseeva	193/4?–unknown	Russia	Mammals	Mammoths
Ludmila Ivanovna Alekseeva (Alexeeva)	1927–	Khar'ov; Tajikistan	Mammals (primates)	Quaternary monkeys, camels, mastodons
Jelena Sergaja Andelkovic	193/4?–unknown	Yugoslavia; Serbia	Fish	Salmonid fish
Sevil' Mutalibovna Aslanova	193/4?–unknown	Russia	Mammals	Miocene vertebrates, deer
Piruza Ablayeva Aubekerova	193/4?–unknown	Alma Alta, Kazakh	Mammals	Fieldwork, stratigraphy
V.T. Balakhmatova	1905–1997	Russia	Mammals, ungulates	Anthracotheres *Hipparion*
Virginia **Barbu**	1894–1983	Romania	Mammals	Proboscidians, rhinos
Elizaveta Ivanovna **Beljaeva** (**Beljaeva**)	193/4?–	Russia	Mammals	Quaternary
Anna Aleksandrovna Boboyedova	1937–2017	Russia	Mammals	Cenozoic
Margarita Borisovna Borisoglebskaya	1931–	Central Asia	Mammals	Artiodactyls, carnivores
Yekaterina Leonidovna Dmitrieva	1922–2011	Central and East Asia	Mammals	Rhinos, mammoths
Irina Aleksandrovna Dubrovo	193/4?–		Mammals	
Zhanna Dzhalil'yevna Dzhafarova	192?–	Azerbaijan	Fish, mammals	
V.V. Feniksova	1920–2002	Russia	Mammals	Permian, Tertiary, Quaternary
Anna Yakovlevna Godina		Mongolia	Mammals	Miocene giraffes
Zinaida **Gorizdro-Kulczycka**	1884–1949	Turkestan; Holy Cross Mts., Poland	Fish	Jurassic fish, Devonian placoderm, lungfish
N.I. Grechina	194/5?–	Kamchatka; central Asia	Fish	Tertiary teleost
Vera Isaacovna **Gromova**	1891–1973	Russia	Mammals	Carnivores, horses, tapirs, rhinos
Anna **Hartmann-Weinberg**	1882–1942	Leningrad	Reptiles	Pareiasaurs
N.M. Ianovskaia	194?–198?	Mongolia	Amphibians	Labyrinthodonts
Valentina Nikiforova **Karatajūtė-Talimaa** (Lithuanian)	1930–	Lithuania; Arctic; Russia; Siberia	Early vertebrates, fish	Paleozoic thelodonts, sharks, placoderms
Alvina Aleksandrovna Kazantseva-Seleznova	1929–	Russia	Fish	Permian–Triassic bony fish
I.M. Kiebanova	192/3?–	Zaisan basin	Mammals	Notoungulata
Zofia **Kielan-Jaworowska** (Polish)	1925–2015	Mongolia	Reptiles	Cretaceous dinosaurs
Yelena Dometyevna Konzhukova	1902–1961	Russia	Tetrapods (reptiles, mammals)	Labyrinthodonts, cynodonts

(continued)

TABLE 4.1 (continued)

Name	Dates	Location	Research	Specialty
Irina E. Kuzmina	1918–2000	Urals	Mammals	Mammoths, horses
Natalya Sergeyevna Lebedkina	1932–2003	Russia	Amphibians, reptiles	Amphibians
Lyubov' Anatol'yevna **Lyarskaya**		Gauja River and Lode Quarry, Latvia	Early vertebrates, fish	Devonian agnathans, placoderms
A.M. Malega	194/5?–	Kyzzyinur	Mammals	Eocene
S.P. Malinovskaya	194/5?–	Russia	Fish	Devonian placoderms
Elga **Mark-Kurik** (Estonian)	1928–2016	Estonia; Arctic	Early vertebrates; fish	Devonian placoderms
Teresa **Maryańska** (Polish)	ca. 1937–	Poland; Mongolia	Reptiles	Cretaceous dinosaurs
Anna Vasil'yevna Minich	194?–	Urals	Fish	Permian–Triassic paleoniscoid fish
Anna Boleslavovna **Missuna** (Baltic)	1868–1922	Russia	Fish	Carboniferous shark
Larissa Illarionovna Novitskaya	1934?–	Novaya Zemlya	Early vertebrates, fish	Devonian amphiaspids
Olga Pavlovna Obrucheva	1920–2008	Russia	Fish	Devonian placoderms
Halszka **Osmólska** (Polish)	1930–2008	Mongolia	Reptiles	Cretaceous dinosaurs
Maria Vasil'evna **Pavlova** (Pavlowa)	1854–1938	Russia; Ukraine	Mammals	Tertiary ungulates
Nina Semenova **Shevyreva**	1931–1996	Ciscaucasus; Mongolia	Mammals	Paleogene rodents, lagomorphs
Marina Vladimirovna Sotnikova	195/6?–	Russia	Mammals	Carnivores (cats)
A.A. Svichenskaia	192/3?–	Russia	Fish	Mullets
Yevgeniya Konstantinovna Sychevskaya	1936–	Central Asia; Mongolia	Fish	Teleost fish
S.A. Tsyganova	194?–	Russia	Mammals	Mammoths
Tan'ia Tumanova	194?–	Russia	Reptiles (dinosaurs)	Dinosaurs ankylosaurs
Galina Ivanova Tverdochlevoba	194?–	Cis-Urals	Amphibians, reptiles	Permian–Triassic vertebrates
Inessa Anatolyevna **Vislobokova**	1947–	Siberia	Mammals	Pleistocene horses, artiodactyls
Emilia Ivanovna **Vorobyeva**	1934–2016	Russia; Latvia	Fish	Devonian sarcopterygians
Lyubov' Vasilyevna V'yushkova	194?–	Novosibirsk; central Asia	Fish	Devonian acanthodians, placoderms
Natalia Mikhaylovna Yanovskaya	1912–1979	USSR; Mongolia	Mammals	Brontotheres
Galina V. Zakharenko	1961–	Russia	Fish	Devonian placoderms
Galina Aleksandrovna Zerova	195?–	Ukraine	Reptiles	Miocene snakes, lizards

Sources: Bibliographies of Vertebrate Paleontology (BFV) and BFV online.
Note: Where names are in bold, see more information in the text.

the USSR Academy of Sciences, founded by her mentor A.A. Borissiak (1872–1944) (Trofimov and Reshetov, 1983; Bodylevskaya, 2007). He encouraged her from afar during the war years when he moved the collections out to Kazakhstan in the east. After World War II, she worked mostly on Tertiary and Quaternary faunas from the eastern Soviet states, contributing several chapters to the classic *Fundamentals of Palaeontology*, and she continued her research beyond her retirement in 1964 until about a decade before her death.

Emilia Ivanovna **Vorobyeva** (1934–2016) was a developmental biologist and paleontologist at the Russian Academy of Sciences. She was born in the town of Sloboda, Vyatka (now Kirovsky) region. In 1954 she graduated from the Department of Vertebrate Zoology at MSU, then entered the postgraduate school of the Paleontological Institute of the USSR, working with well-known paleoichthyologist D.V. Obruchev, who had mentored a succession of women students to work on fossil fish, including his wife Olga Pavlovna Obrucheva (1920–2008) (Schultze et al., 2009; Obroucheva, 2014). He offered Vorobyeva the study of previously poorly known groups of sarcopterygian fish related to the origin of tetrapods. She defended her thesis, "Rhizodont Crossopterygian Fishes of the Main Devonian Field," published as a monograph in 1962. In the late 1960s and 1970s Vorobyeva began long-term study of a group of Devonian sarcopterygians, which she named Panderichthyidae. She joined the Severtsov Institute of Evolutionary Morphology and Ecology of Animals (now the Severtsov Institute of Problems of Ecology and Evolution, Russian Academy of Sciences) and began experimental work on modern vertebrate embryos as a means of understanding tetrapod limb development, conducted in collaboration with English embryologist R. Hinchcliffe. Detailed studies of the ontogeny of extant vertebrates and thorough morphological study of fossil fish, including *Panderichthys*, led Vorobyeva to the origins of tetrapod limbs from the fins of sarcopterygians (Borisyaka and Severtsova, 2017). Vorobyeva is the author of about 400 publications, including seven monographs on herpetology, ichthyology, paleoichthyology, evolutionary morphology, phylogenetics, and systematics of lower vertebrates, notably with Obruchev the Osteichthyes part of *Fundamentals of Paleontology* (Vorobyeva and Obruchev, 1964; Rozhanov, 2018).

Romania. The study of fossil mammals also had a long tradition in Romania with 19th-century male scholars, but the first woman VP was Virginia **Barbu** (1905–1997), who was for 30 years professor of paleontology at the Institute of Petrol, Gas and Geology in Bucharest. She researched the distribution of the horse *Hipparion* in Romania and Eastern Europe (e.g., Radulescu,

1999; Grigorescu, 2004). In 1959 Barbu published rich *Hipparion* remains from Cimislia, now Moldavia. She went on to investigate anthracotheres in Tertiary coals.

Probably the most important new schools for women in vertebrate paleontology outside Russia were in Poland (for reptiles, especially dinosaurs) and in the Baltic States (various fish studies).

Estonia. Elga Yuliusovnao **Mark-Kurik** (née Mark) (1928–2016; Plate 24) was an Estonian geologist and vertebrate paleontologist who specialized in Devonian fish and stratigraphy. She was born in Tartu, Estonia, into an academic and artistic family. She graduated with a degree in geology from Tartu University in 1952 and became interested in placoderms in her honors year. She joined the Institute of Geology of the Estonian Academy of Sciences in Tallinn and remained a successful member for 64 years (Kaljo, 2018). She received a PhD in geology and mineralogy there in 1955 with a dissertation on Devonian agnathan psammosteids of Estonia and was one of Obruchev's students, often mentored long-distance from Moscow. She was employed at the Institute of Geology until 1988 when the Estonian Academy of Sciences developed financial difficulties after the breakup of the USSR, and then she survived as an independent researcher on grants at the Institute of Geology, which was transferred in 1998 to Tallinn Technical University. She did not join the SVP until 1964 but then worked in the field in the United States and elsewhere on several occasions, joining forces with colleagues in many countries to work on her fish right up to her death (e.g., Mark-Kurik et al., 2018).

Her research interests included the morphology, taxonomy, evolution, paleoecology, and biostratigraphy of Silurian and Devonian fish, including agnathans and arthrodires. She participated in excavations in Ukraine, Russia, Latvia, and Siberia, as well as in her native Estonia (Schultze et al., 2009). Among her most significant discoveries was a bone fragment found in a collection that she sent to paleontologist Per Ahlberg, who determined that it was from a lobe-finned fish, *Livoniana*, a transitional fossil between fish and tetrapods (Ahlberg et al., 2000). In 1972 Mark-Kurik participated in an excavation near Lode Quarry, Latvia (e.g., Kuršs et al., 1998), where a large layer of fish skeletons was found, including *Asterolepis*, *Laccognathus*, and *Panderichthys* (see also the discussions of Lyubov' Lyarskaya and Vorobyeva). Fragile specimens of the last have been studied more recently using a CT scanner by then–PhD student Catherine Boisvert (see Chapter 5) at Uppsala University (Boisvert et al., 2008). Although Mark-Kurik strictly did not have a university position, she was very supportive of young paleoichthyologists and mentored several, including Tiiu Märss (Chapter 5) and many of the living Paleozoic fossil fish workers (see Festschrift Kaljo, 2018). One important contribution was organization of the 1st Middle Paleozoic Fish Symposium in Tallinn in 1976, which led to the eventual worldwide project Palaeozoic Microvertebrates (a UNESCO and the International Union of Geological Sciences International Geoscience Programme-IGCP 328) and later projects in which she took part in the 1990s to 2000s.

Latvia. Lyubov' Anatol'yevna **Lyarskaya** (1932–2003) was born in Siberia and is another vertebrate paleontologist from the 1950s Moscow Obruchev lab. She worked at the Institute of Marine Geology, Riga, during the Soviet era and was most active in the 1970s to early 1990s working on Paleozoic fish, especially Late Devonian heterostracans and placoderms from the classic Gauja River site and Lode Quarry, the latter often in cooperation with Mark-Kurik (Lyarskaya and Mark-Kurik, 1972).

Lithuania. Valentina Nikiforova **Karatajūtė-Talimaa** (née Karatajūtė) (1930–) is a vertebrate paleontologist who has devoted her research career to the study of Paleozoic agnathans and early fish. She was born in Lazdijai, southern Lithuania. As a geology student, her study of armored fish (placoderms) at various outcrops in Lithuania was encouraged by one of her professors, Jouza Dalinevicius. She graduated from Vilnius University with a degree in geology in 1954 and began postgraduate studies at the Moscow Paleontology Institute in Moscow. Her PhD on Devonian placoderms (asterolepids) was completed in 1958, mentored by Obruchev. She returned to Vilnius to the then Institute of Geology and Geography of the Lithuanian Academy of Sciences, SSR, where she remained throughout her career. During the 1960s she began research on fish microremains (especially thelodont, acanthodian, and shark scales) and their stratigraphic distribution. During her career she made collecting trips to various outcrops, including those in Lithuania, Mongolia, Latvia, Spitsbergen, and Siberia (Grigelis and Turner, 2006; Schultze et al., 2009). In recognition of her valuable contributions to vertebrate paleontology, she was elected an honorary member of the SVP in 2002.

Poland. Zofia **Kielan-Jaworowska** (1925–2015), affectionately known as Zosia, was born in Sokolow, central Poland. As a schoolgirl of 14 during the Nazi occupation in World War II, with higher education banned and punishable by death under Adolf Hitler, she contributed to her country's war effort by joining the Grey Ranks, an underground paramilitary group organized from scouting troops. She began her studies by attending clandestine lectures held in private apartments. A year later she received training and served as a medic in the Polish resistance until 1944 and the eventual retreat of Nazi forces (Kielan-Jaworowska, 2005; Cifelli et al., 2015). Her family moved to Warsaw in the aftermath of World War II. As a student in 1946, she learned about paleontological expeditions to the Gobi Desert by Roman Kozlowski (1889–1977), founder of Polish paleontology and ultimately the Institute of Paleobiology. As she writes, "I was fascinated by . . . American expeditions to the Gobi Desert and never in my wildest dreams did I expect that I, too, would someday go there" (Kielan-Jaworowska, 1969:3). Sixteen years later, in 1962, Kielan-Jaworowska organized the first Polish-Mongolian paleontological expedition to the Gobi Desert, becoming the first woman to lead a dinosaur excavation expedition. She returned seven times (1963–1971), collecting Mesozoic dinosaur and mammal fossils, which has been considered her "greatest achievement" (Cifelli, 2015). Her field teams included other important Polish women (see later in this chapter and Lavas, 1993; Spalding, 1993; Turner et al., 2010a).

Kielan-Jaworowska was a founding member and later professor of the Institute of Paleobiology, Polish Academy of Sciences, at the University of Warsaw, earning an MS in zoology in 1949 and PhD in paleontology in 1953. During a mountaineering trip in 1950, she met Zbigniew Jaworowski (1889–2011), whom she married eight years later. He was to be an inseparable part of Kielan-Jaworowska's life, and his expertise in radiation medicine led to the discovery that part of her collection of dinosaur bones from Mongolia was highly radioactive (Cifelli et al., 2015).

Although she is best known for her research on fossil vertebrates, Kielan-Jaworowska's early work was focused on Paleozoic invertebrates from Poland and adjacent countries. With the retirement of Kozlowski in 1960, she was appointed to succeed him as the director of the Institute of Paleobiology, a position that she held until 1982. She then stepped down from the directorship to accept a visiting professorship at the MNHN, where she expanded her research on multituberculates (see the discussion of Poplin earlier in this chapter). Soon after returning to Warsaw, Kielan-Jaworowska was appointed professor of paleontology at the University of Oslo, a position she held from 1986 to 1995. She again returned to Warsaw and was appointed professor emerita at the Institute of Paleobiology. In recognition of her many significant contributions to vertebrate paleontology, she was elected as an honorary member of the SVP in 1975. In 1995, Kielan-Jaworowska was awarded the Romer-Simpson Medal, the SVP's highest honor, recognizing sustained and outstanding excellence in the discipline of vertebrate paleontology. She was the first women to receive this award.

Kielan-Jaworowska's (1969) book *Hunting for Dinosaurs* recounts her experiences leading the Polish-Mongolian expeditions. Her narrative includes vivid descriptions of fieldwork, including blinding sandstorms, swarming mosquitos, and critical water shortages. Several Mongolian women and two other female scientists were on these expeditions. Teresa Maryańska (see later in this chapter), then an assistant of the Museum of the Earth in Warsaw, and Maria Magdalena Borsuk-Białynicka (Chapter 5), an assistant at the Paleozoological Institute at the University of Warsaw, were members of the first 1964 expedition. Halszka Osmólska (see later in this chapter), a member of the 1965 expedition, regarded Kielan-Jaworowska as "a mammal collector of enormous patience and concentration" (Borsuk-Białynicka, pers. comm., September 2009).

As noted by paleontologist and longtime collaborator Richard Cifelli in his obituary of her, "Much of what we know about the origin and early evolution of mammals stems, directly or indirectly from the work of Zofia Kielan-Jaworowska. Her style was, at times, unapologetically exacting—an apprenticeship with her was akin to martial-arts training with a Buddhist monk—but she pushed the rest of us to reach for better science" (Cifelli, 2015:158). As Priscilla McKenna (written comm., June 2015) notes, "My friendship with Zofia has been one of the great privileges of my life. With her strong and enduring courage, unquenchable spirit, brilliant and talented mind, boundless energy and perseverance, personal and scientific integrity, warm personality, and her passion for scientific work, she set an example for all of us." Kielan-Jaworowska died on March 13, 2015, just weeks before her 90th birthday.

Teresa **Maryańska** (ca. 1937–) is a Polish vertebrate paleontologist who has specialized in Mongolian dinosaurs, particularly pachycephalosaurians and ankylosaurians. She was a member of Kielan-Jaworowska's 1964, 1965, 1970, and 1971 Polish-Mongolian expeditions to the Gobi Desert. Based for many years at the Museum of the Earth in Warsaw, she published mainly in tandem with Osmólska on numerous dinosaurs (particularly pachycephalosaurians and ankylosaurians) collected during those expeditions (e.g., Maryańska and Osmólska, 1974) and is regarded not only as one of the leading women in the field of dinosaur research but also as one of the first women to contribute to dinosaur taxonomy and systematics (Dodson, 1998; Turner et al., 2010a).

Halszka **Osmólska** (1930–2008), a Polish vertebrate paleontologist, was a student during the Nazi regime and studied biology at Poznan University in 1949, moving to the University of Warsaw in 1952. Despite an interest in dinosaurs, she completed her MS (1955) and PhD (1962) theses on Devonian and Carboniferous trilobites, similar to Kielan-Jaworowska's early years. As an undergraduate in 1953, she joined Kozlowski at the Laboratory of Paleozoology (now the Institute of Paleobiology, Polish Academy of Sciences) and was promoted to professor of paleontology in 1983 and later to director (1983–1989). Beginning in 1969, Osmólska changed her research focus and began to study Late Cretaceous dinosaurs and other reptiles from the Gobi Desert. She went on to become the most well known of Polish researchers on dinosaurs and early crocodilians. She took part in the Polish-Mongolian dinosaur expeditions to the Gobi Desert in 1963–1971 under the leadership of Kielan-Jaworowska (Kielan-Jaworowska and Smith, 1965). Her VP contributions include descriptions of theropod dinosaurs such as *Gallimimus, Oviraptor, Borogovia,* and *Tochisaurus* (e.g., Osmólska et al., 1972; Osmólska, 1976, 1987), all studies important in unraveling bird origins. Later, she collaborated with her Gobi Desert colleague Maryańska on anatomical and phylogenetic studies of ornithischian dinosaurs (Kielan-Jaworowska, 2008; Turner et al., 2010a).

Australia

Judith Eveleigh **King** (1926–2010) was a British zoologist who specialized in the anatomy and taxonomy of seals. After receiving an honors degree from the University of London, she remained at the BMNH for 20 years. From 1964 to 1984, she worked in the Zoology Department of the University of New South Wales, where she published papers on a Pleistocene seal and sea lion (King, 1973, 1983).

Summary

By the mid-20th century, there was a dramatic increase in women studying geology and zoology in the United States and around the world, especially in Europe and the Communist countries (i.e., the Soviet Union and China). In the United States, the first exclusively professional vertebrate paleontological organization, the SVP, had been organized. Although only a handful of women presented research at the SVP annual meetings until the mid-1970s, women were active participants, many as *SVP News Bulletin* and later BFV editors (e.g., Rachel Nichols and Nelda Wright, see BFV). Among charter members, Carrie Barbour at the Nebraska State Museum was the first trained and employed VP woman. Other notable charter members include fish paleontologist Lore David, Tilly Edinger (who was now residing in the United States), and Hildegarde Howard, the first woman avian paleontologist. In Africa, British paleoanthropologist Mary Leakey made significant discoveries of fossil apes and hominids. Among women VPs in Europe are fish paleontologists Jeanne Signeux in France and Margaretha Brongersma-Sanders in the Netherlands; the latter may have been the first taphonomist, based on her work categorizing a fish mass mortality off the coast of Namibia. In the United Kingdom, notable women VPs include Pamela Robinson, who specialized in vertebrates from Mesozoic fissure fills in England and Gondwanan faunas in India. By the 1960s, women VPs had significantly increased in number and spread around the world. Notable Chinese scientists include fish paleontologist Meeman Chang, also the first woman to head the IVPP. In the United States, mammalian paleontologist Mary Dawson was just beginning her VP career at the Carnegie Museum, studying Cenozoic rabbits and rodents. In the USSR and allied countries, several noted women VPs applied fossils to biostratigraphy, including Elizaveta Beliaeva and Emilia Vorobyeva. Two other noted fish paleontologists during this time are Elga Mark-Kurik in Estonia and Valentina Karatajūtė-Talimaa in Lithuania. Prominent Polish mammalian paleontologist Zofia Kielan-Jaworowska organized and led a series of Polish-Mongolian expeditions to the Gobi Desert from 1961 to 1971, collecting an impressive assemblage of Mesozoic dinosaur and mammal fossils.

CHAPTER 5

Women in Vertebrate Paleontology, Late 20th to Early 21st Century (1976 to the Present)

Exploration is an obsession. The more I discover, the more I want to know. Unfortunately, I will never discover everything I want. One can never, never find all the answers.

Maeve Leakey, in Hemming, 1998

For more than 100 years, vertebrate paleontology in the United States and elsewhere was focused primarily on the collection and description of fossil vertebrates and the rocks that contained them. Beginning in the 1960s and 1970s, and continuing to the present, isotopic dating methods (i.e., K-Ar dating) and paleomagnetic stratigraphy were used together with fossils (biostratigraphy) to provide the ages of rock units (geochronology) and as the basis for correlation of faunas (e.g., Julia Sankey, Susannah Maidment). There is a long history of using vertebrate fossils to relatively date rocks—ranging from using fish scales and teeth (e.g., Olga Afanas'ieva, Valentina Karatajūtė-Talimaa, Susan Turner, Claire Derycke) to using small mammals (e.g., Erika von Huene, Marie Hopkins, Karen Black). The study of fossils in the making, taphonomy, emerged as a subdiscipline of paleontology in the mid-20th century and gained more common practice in the mid-1980s, carrying on to the present with both field and laboratory studies (e.g., Anna "Kay" Behrensmeyer, Sarah Gabbott). The related study of trace fossils began with study of footprints and trackways (e.g., Orra Hitchcock: Chapter 6, Mary Wade) and continues today with the use of new technology, such as that providing analyses of molecular traces of vertebrates (e.g., Karen Chin, Mary Schweitzer).

As the late 20th century unfolded, paleontology maintained a geological thrust but significantly expanded its biological emphasis. Many women VPs now followed past practices and pursued expertise in a chosen group of vertebrates incorporating one or more subdisciplines, such as functional anatomy, systematics, development, taphonomy, paleoecology, and biostratigraphy (see also Chapter 1). Vertebrate paleontology research specialties have expanded and become more interdisciplinary, with researchers employing modern techniques for the study of fossil vertebrates both in the field and in the laboratory (e.g., Servais et al., 2012). Studies of biodiversity dynamics (i.e., species origination, extinction) and morphological diversification, as documented by the fossil record, have begun to be explored, including through reconstructions of fossil communities and ecosystems (e.g., Felisa Smith, Catherine Badgley, Blaire Van Valkenburgh). The emergence of phylogenetic systematics and the cladistics revolution began to affect vertebrate paleontology in the late 1970s (e.g., Forey et al., 1991) with improvements in statistical methods, computer algorithms, and the availability of DNA sequences having facilitated evolutionary studies integrating fossil and extant taxa (e.g., Maureen O'Leary, Annalisa Berta). The discovery and analysis of fossils together with comparative developmental data from extant forms have informed the origin of evolutionary novelties such as eye musculature, squamation and teeth (e.g., Carole Burrow), paired fins (e.g., Zerina Johanson), the tetrapod limb (e.g., Jennifer Clack), feathers (e.g., Julia Clarke), turtle shells (e.g., Ann Burke), and eggs (e.g., Frankie Jackson, Kristi Curry Rogers). Broader "paleoinformatics" studies have been initiated utilizing various databases (e.g., the online Paleobiology Database, MorphoBank, Neogene of the Old World; see Uhen et al., 2013). Stable isotope geochemistry is employed to help understand climate and diet changes (e.g., Penny Higgins, Kathryn Hoppe, and Kena Fox-Dobbs). For the last three decades, digital visualization, including CT scanning and related imaging techniques, has opened up exciting new directions in the investigation of the hard and soft anatomy of fossil vertebrates (e.g., Anjali Goswami, Rachel Racicot, and Suzanne Strait-Holman) and allowed a deeper look at the structure of microfossils (e.g., Tiiu Märss). Other techniques such as finite element analysis have helped understand the feeding and locomotor mechanics of animals (e.g., Rayfield). Histological techniques have had a long tradition of use in paleontology and, in today's use, have benefited from technological advances that have permitted the recovery and analysis of DNA, bone, and soft-tissue structure and their preservation (e.g., Burrow, Anusuya Chinsamy-Turan, Kristi Curry Rogers, Alexandra Houssaye, and Katherine Trinajstic).

International activities have been encouraged—such as through joint expeditions, the United Nations Educational, Scientific, and Cultural Organization (UNESCO), and scientific unions—and new collaborations forged, many by women VPs, that continue to uncover vast numbers of new or little-known extinct species. These include Trinajstic's Devonian fish mothers in Australia; Mary Dawson, Julia Clarke, and Jaelyn Eberle's High Arctic field expeditions;

the Leakeys' African early hominid expeditions; Suzanne Hand's Riversleigh Australia bats and marsupials; and Jingmai O'Connor's field research on early birds in China and Myanmar. More recently, new directions have evolved using paleontological data (including paleogenomics) to understand and manage climatic and other environmental changes caused by humans—that is, conservation paleontology (e.g., Catherine Forster, Elizabeth Hadly, Beth Shapiro; Bottjer, 2006; Conservation Paleobiology Workshop, 2012). Thus, employment opportunities for paleontologists have diversified in the 21st century, including the more traditional academic positions in colleges and universities and those as museum curators and collection managers, as well as jobs with consulting firms and government agencies—for example, the National Park Service, the Bureau of Land Management, and UNESCO Global Geoparks.

The last 50 years have also seen a dramatic increase in the number of women vertebrate paleontologists. Knowing it is not possible to include all women, we focus on two groups (Tables 5.1–5.3): (1) those mostly active in the late 20th century, including those whose research activity is more recent (late 1970s) and, for most, continuing today (Table 5.1), and (2) young scientists who have just begun their careers (21st century); their history is still unfolding, and they represent the future of the field (Table 5.2). For both groups, we discuss the careers of some women in more detail to highlight underrepresented ethnicities and regions or special contributions to VP. Others are more briefly discussed in terms of research specialty and educational background following our geographic approach (e.g., Chinese VPs in Table 5.3) and research subdiscipline.

Late 20th Century (1970s–1999)

Africa

Anusuya **Chinsamy-Turan** (1962–) is a South African vertebrate paleontologist and professor at the University of Cape Town. Chinsamy-Turan's research specialization is the microstructure of fossil and living vertebrates, especially nonavian dinosaurs and dicynodonts, birds, and dinosaurs (e.g., Chinsamy-Turan, 1995, 2002). She received both a BS (honors) in 1983 and a PhD in 1991 from the University of Witwatersrand. She has authored several books, including *The Microstructure of Dinosaur Bone: Deciphering Biology through Fine-Scale Techniques* (2005) and *Forerunners of Mammals: Radiation* (2011), a children's book, *Famous Dinosaurs of Africa* (2008), and a popular book, *Fossils for Africa* (2014). In recognition of her research, she has been elected fellow of the University of Cape Town and fellow of the Royal Society of South

TABLE 5.1
Women VPs active in the late 20th century

Name	Dates	Country	Research	Major contribution
Olga **Afanas'ieva**	194/5?–	Russia	Early vertebrates	Systematics, paleohistology
Maria **Alberdi**	1949–	Spain	Mammals	Fieldwork
Andrea **Arcucci**		Argentina	Reptiles (dinosaurs)	Systematics
Gloria **Arratia**		Chile, United States	Fish	Fieldwork, paleoecology
Catherine **Badgley**		United States	Mammals (faunal studies)	Anatomy, systematics
Ana **Báez**	1942?–	Argentina	Amphibians	Anatomy, systematics
Nathalie **Bardet**		France	Mesozoic marine reptiles	Taphonomy
Anna **Behrensmeyer**	1946–	United States		Fieldwork
Rachel **Benton**		United States		
Annalisa **Berta**	1952–	United States	Mammals (marine)	Anatomy, systematics, evolution
Maria **Borsuk-Białynicka**	1940–	Poland	Amphibians and reptiles	Functional anatomy
Wendy Bosler	1952–	Africa, Australia	Mammals (primates)	Anatomy
Elizabeth Brainerd		United States	Fish and reptiles	Anatomy and biomechanics
Laurie **Bryant**		United States	Reptiles (dinosaurs), mammals	Fieldwork
Emily **Buchholtz**		United States	Mammals	Anatomy, evo-devo
Ann **Burke**		United States	Reptiles (turtles)	Evo-devo
Carole **Burrow**		Australia	Fish	Systematics, paleohistology
Yael **Chalifa**	1940–2006	Israel	Fish	Systematics
Karen **Chin**		United States	Reptiles (dinosaurs)	Trace fossils
Brenda **Chinnery**		United States	Reptiles (dinosaurs)	Fieldwork
Anusuya **Chinsamy-Turan**	1962–	Africa	Reptiles (nonavian dinosaurs)	Paleohistology
Jennifer **Clack**	1947–2020	England	Early tetrapods	Fish-tetrapod transition
Julia **Clarke**		United States	Reptiles (birds)	Fieldwork, anatomy, evolution
Jane **Colwell-Danis**	1941–	Canada	Mammals	Fieldwork, curation
Margery **Coombs**	1945–	United States	Mammals (chalicotheres)	Fieldwork, taphonomy, paleoecology, anatomy
Penelope **Cruzado-Caballero**		Argentina	Reptiles (dinosaurs)	Evolution, paleobiology
Kristi **Curry Rogers**		United States	Reptiles (dinosaurs)	Paleohistology, anatomy, systematics
Lyndall Joan Dawson	1940–	Australia	Mammals (marsupials)	Fieldwork, taphonomy
Kym Dennis-Bryan	1942?–	England	Fish (placoderms)	Systematics, anatomy

(continued)

TABLE 5.1 (continued)

Name	Dates	Country	Research	Major contribution
Claire **Derycke**		France	Early vertebrates, fish	Paleohistology, biostratigraphy
Jaelyn **Eberle**		United States	Mammals	Fieldwork, evolution, paleoclimate
Vera Eisenmann		France	Mammals (equids)	Systematics
Margarita **Erbajaeva**	1938–	USSR	Mammals	Fieldwork, systematics, paleoecology
Susan **Evans**		England	Amphibians and reptiles	Anatomy, systematics, paleoecology
Marta **Fernandez**		Argentina	Fish, reptiles (crocodyliformes, ichthyosaurs, dinosaurs)	Anatomy
M. Eileen Finch	194/5?–	Australia	Mammals (marsupial lion)	Anatomy
Ann-Marie **Forstén**	1939–2002	Finland	Mammals (horses)	Anatomy
Catherine **Forster**		United States	Reptiles (dinosaurs)	Systematics
Sarah **Gabbott**		England	Early vertebrates and fish	Taphonomy
Zulma **Gasparini**	1944–	Argentina	Reptiles	Fieldwork
Joyce Gilbert-Tomlinson	1916–1981	Australia	Early vertebrates, fish	Fieldwork
Naomi **Goldsmith**		Israel	Mammals (rodents)	Systematics
Katia Gonzalez-Rodriguez		Mexico	Mesozoic fish	Anatomy
Anjali **Goswami**		United States, England	Mammals	Anatomy, evolution, development
Elizabeth **Hadly**		United States		Paleoecology, paleoclimatology
Suzanne **Hand**		Australia	Mammals (marsupials), bats	Fieldwork, evolution, paleoclimate
Sue Hirschfeld	194?–	United States	Mammals (ground sloths)	Anatomy, systematics
Ella **Hoch**	1940–	Denmark	Tertiary–Holocene vertebrates (birds)	Fieldwork, historian
Jeanette **Hope**	1942–	Australia	Mammals	Faunal remains at archaeological sites
Nina Jablonski		United States	Mammals (primates)	Anatomy
Frankie **Jackson**		United States	Reptiles (dinosaurs)	Eggs, paleobiology
Christine **Janis**		England, United States	Mammals (ungulates)	Anatomy, evolution, paleoecology
Anna Jerzmanska		Poland	Fish (actinopterygians, sharks)	Systematics
Joyce Joffe	194?–	Israel	Reptiles (crocodiles)	Research
Zerina **Johanson**		England	Early vertebrates	Development
Gillian **King**	194?–	England, Africa	dicynodonts	
Irina **Koretsky**	194?–	Russia, United States	Mammals (marine)	Anatomy, systematics

France de Lapparent de Broin	1938–	France	Reptiles (turtles, crocodiles)	Fieldwork, systematics, paleoecology
Maeve **Leakey**	1942–	England, Africa	Primates (hominids)	Anatomy
Maria **Leal-Bonde**		Brazil	Fish, amphibians, reptiles	Anatomy, evolution
Margaret **Lewis**		United States	Mammals (carnivores)	Science communication
Nieves **López Martínez**	1949–2010	Spain		Anatomy
Laura **MacLatchy**		United States	Mammals (primates)	Anatomy, biostratigraphy
Claudia Marsicano		Argentina	Mesozoic amphibians and reptiles (archosaurs)	Taxonomy, systematics, anatomy, biostratigraphy
Tiiu **Märss**	1943–	Estonia	Early vertebrates, fish	Paleoecology
Mary C. Maas		United States	Mammals	Evolution
Judy **Massare**		United States	Reptiles (marine)	Fieldwork
Amy McCune		United States	Fish	Anatomy (CT)
Ellen Miller			Mammals (primates)	Fieldwork
Angela **Milner**	1947–	England	Dinosaurs (theropods, birds)	Anatomy
Marisol **Montellano-Ballesteros**		Mexico	Vertebrates	
Jeanette Muirhead		Australia	Mammals (marsupials)	Systematics, paleobiogeography
Allison **Murray**		Canada	Fish	Anatomy
Virginia **Naples**		United States	Mammals	Systematics, fieldwork
Elizabeth **Nicholls**	1946–2004	United States, Canada	Mesozoic marine reptiles	Anatomy, systematics, evolution
Maureen **O'Leary**		United States	Mammals (marine, artiodactyls)	Systematics
Olga Otero		France	Fish	Fieldwork
Maria **Palombo**		Italy	Cenozoic mammals	Fossil preservation foundation
Maria **Paramo Fonseca**		Colombia	Reptiles (dinosaurs, marine)	Paleoecology
Brenda Pobiner		United States	Mammals (primates)	Paleoecology
Kaye Reed		United States	Mammals (faunal studies)	Anatomy, evolution
Karen Rosenberg		United States	Mammals (primates)	Anatomy
V. Louise Roth		United States	Mammals	Anatomy, fieldwork, systematics
Natalia **Rybczynski**		Canada, United States	Vertebrate faunas	Fieldwork, biostratigraphy
Julia **Sankey**		United States	Fish and reptiles (dinosaurs)	Evolution, systematics, paleoecology
Tamaki **Sato**	1972–	Japan		Fieldwork, biostratigraphy
Judy **Schiebout**	1946–	United States	Mammals	

(continued)

TABLE 5.1 (continued)

Name	Dates	Country	Research	Major contribution
Mary **Schweitzer**		United States	Reptiles (dinosaurs)	Paleohistology
Kathleen Scott		United States	Mammals (ungulates)	Anatomy
Karen **Sears**		United States	Mammals	Development
Gina **Semprebon**		United States	Mammals	Paleoecology, paleohistology
Nina **Shevyreva**	1931–1996	USSR	Mammals (rodents)	Fieldwork
Pat **Shipman**	1949–	United States	Mammals (primates)	Writer of popular paleo books
Mary **Silcox**		United States	Mammals (primates)	Anatomy, systematics
Felisa **Smith**		United States	Mammals	Macroevolution
Kathleen **Smith**		United States	Mammals	Development
Meredith **J. Smith**		Australia	Mammals (marsupials)	Anatomy
Moya **Meredith Smith**	1943–1998	England	Fish	Evolution of teeth
Maria **Soria Mayor**	1948–2004	Spain	Mammals (carnivores, ruminants)	Evolution of South American mammals
Barbara **Stahl**	1930–2004	United States	Fish	Evolution, paleohistology
Margaret **Stevens**		United States	Fish	Fieldwork
Nancy **Stevens**		United States	Mammals	Anatomy, paleoecology
Suzanne **Strait-Holman**		United States	Mammals	Anatomy
M. Elizabeth Strasser		United States	Mammals (primates)	Anatomy, evolution
Sun Ai-lin (Ai-ling)	193?–	China	Reptiles (dicynodonts)	Fieldwork, anatomy
Jessica **Theodor**		Canada	Mammals (ungulates)	Anatomy, paleoecology
Suyin **Ting**		China, United States		Biostratigraphy
Katherine **Trinajstic**		Australia	Paleozoic fish	Anatomy (CT), paleohistology
Linda Trueb	1942–	United States	Amphibians	Anatomy, systematics
Susan **Turner**	1946–	England, Australia	Early vertebrates, fish, tetrapods	Taxonomy, biostratigraphy, paleobiogeography
Blaire **Van Valkenburgh**	1952–	United States	Mammals (carnivores)	Anatomy, paleoecology
Pat **Vickers-Rich**	1944–	United States, Australia	Reptiles (birds, dinosaurs)	Fieldwork, research, evolution

			Early vertebrates	Paleohistology
Inessa Anatolyevna **Vislobokova**	1947–	Russia	Macroevolution	Fieldwork
Elisabeth **Vrba**	1942–	United States, Africa		Systematics, evolution
María Guiomar **Vucetich**		Argentina	Mammals (rodents)	
Mary **Wade**	1928–2005	Australia		Fieldwork, dinosaur footprints
Carol Ward	1941–	United States	Mammals (primates)	Anatomy, evolution
Anne **Warren**		Australia	Paleozoic–Mesozoic amphibians	Fieldwork, anatomy, systematics
Anne **Weil**		United States	Mammals	Anatomy, fieldwork
Joan **Wiffen**	1922–2009	New Zealand	Reptiles (dinosaurs)	Fieldwork
Alisa **Winkler**		United States	Mammals (rodents)	Systematics
Darla **Zelenitsky**		Canada	Reptiles (dinosaurs)	Paleohistology

Note: Where names are in bold, see more information in the text.

TABLE 5.2
Women VPs active during the 21st century (2000 to the present)

Name	Country	Research	Major contribution
Heather Ahren	United States	Mammals (carnivores)	Anatomy
Kate Andrzejewski	United States	Reptiles (dinosaurs)	Anatomy
Victoria **Arbour**	Canada	Reptiles (dinosaurs)	Paleobiology
Amy **Balanoff**	United States	Reptiles (dinosaurs)	Anatomy (CT)
Holly **Ballard**	United States	Reptiles (dinosaurs)	Paleohistology
Emily Bamforth	Canada	Reptiles (dinosaurs)	Fieldwork
Luciana **Barbosa de Carvalho**	Brazil	Reptiles (dinosaurs and crocodylomorphs)	Anatomy
Ewa **Barycka**	Poland	Mammals (carnivores)	Fieldwork, systematics
Allison Beck	United States	Reptiles (nonmammalian synapsids)	Anatomy
Karen **Black**	Australia	Mammals (marsupials)	Anatomy, development
Jessica **Blois**	United States	Mammals (faunal studies)	Paleoecology
Catherine **Boisvert**	Australia	Early vertebrates	Development and anatomy
Jennifer **Botha-Brink**	Africa	Stem mammals (therapsids)	Paleoecology
Virginie **Bouetel**	France	Mammals (whales)	Anatomy, evolution
Margaret **Bradshaw**	New Zealand	Fish	Fieldwork
Pip **Brewer**	England	Mesozoic mammals	History of science
Alyson Brink	United States	Cretaceous mammals	Anatomy
Kristen **Brink**	Canada	Vertebrates	Paleohistology of teeth
Constance Bronnert	France	Mammals (perissodactyls)	Evolution, phylogeny, paleobiogeography
Allison Bronson	United States	Fish	Anatomy (CT)
Lisa **Buckley**	Canada	Reptiles (dinosaurs, birds)	Trackways and trace fossils
Monica **Buono**	Argentina	Mammals (whales)	Anatomy, systematics
Sara **Burch**	United States	Reptiles (dinosaurs)	Anatomy, paleohistology
Carole **Burrow**	Australia	Paleozoic fish	Anatomy, evolution
Sanjukta **Chakravorti**	India	Mesozoic amphibians	Paleoecology, taphonomy
Belinda **Chávez**	Mexico	Reptiles (dinosaurs)	Anatomy
Donglei **Chen**	China, Sweden	Early vertebrates	Evolution, paleoecology
Amy **Chew**	United States	Mammals	Anatomy, evolution
Kerin **Claeson**	United States	Fish	Anatomy, evolution, systematics
Alice **Clement**	Australia	Fish	Anatomy, development
Lisa **Cooper**	United States	Mammals (bats, marine mammals)	

Name	Country	Taxa	Specialty
Gloria **Cuenos-Becos**	Spain	Reptiles (dinosaurs), Mesozoic mammals	Paleoecology, systematics
Elena Cuesta	Spain	Reptiles (dinosaurs)	Anatomy
Sue **Dawson**	Canada	Mammals	Anatomy, evolution
Paula Dentizen-Dias	Brazil	Vertebrates	Trace fossils
Vanesa de Pietri	New Zealand	Reptiles (birds)	Anatomy, paleoecology
Larisa **DeSantis**	United States	Mammals	Anatomy, systematics, biostratigraphy
Julia **Desojo**	Argentina	Reptiles (dinosaurs)	Paleoecology, taphonomy
Stephanie Drumheller	United States	Reptiles (dinosaurs)	Anatomy
Rachel Dunn	United States	Mammals	Anatomy (CT)
Catherine **Early**	Japan	Reptiles (birds)	Anatomy, systematics
Naoko **Egi**	United States	Mammals	Anatomy, trace fossils
Amanda Falk	Argentina	Reptiles (birds)	Life history
Mariela **Fernandez**	Argentina	Reptiles (dinosaurs)	Anatomy, systematics
Mercedes Grisel **Fernandez**	Argentina	Mammals	Anatomy, evolution
Gabriela **Fontararrosa**	United States	Reptiles (lizards)	Isotope geochemistry, paleoecology
Kena **Fox-Dobbs**	Canada	Mammals	Paleoecology
Danielle Fraser	United States	Cenozoic mammals	Anatomy
Elizabeth Freedman-Fowler	United States	Reptiles (dinosaurs)	Anatomy, biomechanics
Rachel **Frigot**	Germany	Reptiles (pterosaurs)	Evolution, development
Nadia **Fröbisch**	England	Amphibians	Evolution
M. Sarah Gibson	United States	Fish (actinopterygians)	Anatomy (CT)
Samantha **Giles**	Austria	Fish	Curator
Ursula Goehlich	United States	Reptiles (birds)	Author
Eugenia **Gold**	Africa	Reptiles	Museum administration
Elizabeth **Gomani-Chindebvu**		Reptiles (dinosaurs, crocodiles)	
Aida Gomez-Robles	England	Mammals (primates—hominins)	Evolution
Romala **Govender**	Africa	Mammals (marine)	Evolution
Carolina **Gutstein**	Chile	Mammals (marine)	Anatomy, systematics
Victoria Herridge	England	Mammals	Founder of *TrowelBlazers*
Anneke van **Heteren**	Germany	Reptiles (birds), mammals	Anatomy
Penny **Higgins**	United States	Mammals	Isotope geochemistry
Simone Hoffman	United States	Mammals	Anatomy
Samantha **Hopkins**	United States	Mammals (rodents)	Biostratigraphy, paleoecology
Alexandra **Houssaye**	France	Mammals (marine)	Paleohistology
Annie **Hsiou**	Brazil	Reptiles (lizards)	

(continued)

TABLE 5.2 (continued)

Name	Country	Research	Major contribution
Yu-zhi **Hu**	China	Early vertebrates, fish	Evolution, fieldwork
ReBecca **Hunt-Foster**	United States	Reptiles (dinosaurs)	Fieldwork, trackways
Helen Frances **James**	United States	Reptiles (birds)	Fieldwork, systematics
Katrina **Jones**	England, United States	Reptiles, mammals	Anatomy, development
Daniela Kalthof	Germany	Mammals (rodents, sloths)	Paleoecology; mammal curator
Sandy Kawano	United States	Stem tetrapods	Anatomy, evolution
Melissa Kemp	United States		Conservation paleontology
Yuri **Kimura**	Japan	Mammals (rodents)	Systematics, paleoecology
Emily **Lindsey**	United States	Mammals	Extinction
Juan Liu	United States	Vertebrates	Anatomy
Carolina **Loch**	Brazil, New Zealand	Mammals (marine)	Paleohistology
Adriana Lopez-Arbarello	Germany	Mesozoic fish	Evolution, Anatomy
Malena **Lorente**	Argentina	Mammals	Evolution, systematics
Jing **Lu**	China	Early vertebrates, fish	Anatomy
Kate **Lyons**	United States	Mammal faunas	Paleoclimatology
Susannah **Maidment**	England	Reptiles (dinosaurs)	Evolution
Fang-yuan **Mao**	China	Mammals	Fieldwork
Ryoko **Matsumoto**	Japan	Reptiles	Anatomy
Emma **Mbua**	Africa	Mammals (primates—hominids)	Anatomy, evolution
Jenny McGuire	United States		Conservation paleontology
Nadia McLaughlin	United States	Vertebrate faunas	Science education
Maria **McNamara**	Ireland	Vertebrates	Preservation (skin, color)
Julie **Meachen**	United States	Mammals	Paleoecology
Bolor **Minjin**	Mongolia, United States	Reptiles (dinosaurs), mammals	Fieldwork, outreach
Ruzan **Mkrtchyan**	Armenia	Vertebrates	Geology (depositional environments)
Julie Molnar	United States	Reptiles (dinosaurs)	Evolution, anatomy
Karen Moreno	Chile	Reptiles, mammals	Anatomy, biomechanics
Ashley **Morhardt**	United States	Reptiles (dinosaurs)	Anatomy
Jacqueline Nguyen	Australia, Vietnam	Reptiles (birds)	Anatomy, evolution
Sirpa **Nummela**	Finland	Mammals (marine)	Evolution
Haley **O'Brien**	United States	Mammals	Anatomy (CT)

Jingmai O'Connor	Reptiles (birds)	China, United States	Fieldwork, anatomy
Jennifer Olori	Early tetrapods	United States	Evolution, development
Elsa Panciroli	Mesozoic mammals	Scotland	Anatomy
Tonya Penkrot	Tertiary mammals	United States	Anatomy
Stephanie Pierce	Reptiles (early tetrapods)	England, United States	Anatomy
J. Kat Piper	Mammals	Australia	Anatomy, systematics
Kari Prassack	Mammals	United States	Taphonomy, paleoecology
Tuo Qiao	Early vertebrates, fish	China	Anatomy, evolution
Ana Quadros	Reptiles (snakes)	Brazil	Anatomy, evolution, systematics
Lovasoa Ranivoharimanana	Reptiles, mammals	Madagascar	Anatomy
Emily Rayfield	Reptiles (dinosaurs)	England	Anatomy, biomechanics
Taissa Rodrigues	Reptiles (pterosaurs)	Brazil	Fieldwork, systematics, anatomy
Gertrud Roessner	Mammals	Germany	Curator, fossil mammals
Irina Ruf	Early mammals	Germany	Anatomy
Lauren Sallen	Fish	United States	Evolution, macroevolution
Karen Samonds	Vertebrate faunas, Madagascar	United States	Fieldwork, anatomy
Daniela San Felice	Mammals (pinnipeds)	Brazil	Anatomy
Sanaa El-Sayed	Mesozoic faunas	Egypt	Biostratigraphy, paleontology
Ane Schroeder	Fish	Denmark	Anatomy, evolution
Claudia Serrano	Reptiles (dinosaurs)	Mexico	Paleoecology, taphonomy
Beth Shapiro	Mammals	United States	Ancient DNA, evolution
Michelle Spaulding	Mammals (carnivores)	United States	Systematics, evolution
Kathleen Springer	Vertebrates	United States	Paleoecology, geology (deposition environments
Mette Steeman	Mammals (marine)	Denmark	Anatomy, evolution, systematics
Juliana Sterli	Reptiles (turtles)	Argentina	Anatomy, evolution, systematics
Michelle Stocker	Reptiles	United States	Evolution
Denise Su	Mammals (primates)	United States	Paleoecology
Rebecca Terry	Mammals	United States	Paleoclimatology
Allison Tumarkin-Deratz	Reptiles (dinosaurs)	United States	Paleohistology
Ana Valenzuela-Toro	Mammals (marine)	Chile, United States	Anatomy, evolution, systematics, fieldwork
Evelyn Vallone	Fish	Argentina	Anatomy, evolution, systematics
Sara Varela	Mammal	Germany	Macroevolution (extinction)
Barbara Vera	Mammals (notoungulates)	Argentina	Anatomy

TABLE 5.2 (continued)

Name	Country	Research	Major contribution
Vera **Weisbecker**	Australia	Mammal	Anatomy (CT)
Gina **Wesley**	United States	Mammals (carnivores)	Anatomy
Abagael **West**	United States	Mammals	Anatomy, macroevolution
Robin **Whatley**	United States	Early mammals	Evolution, paleoecology
Cheryl Wilga	United States	Fish (sharks)	Anatomy, behavior
Galina **Zakharenko**	Russia	Early vertebrates, fish	Fieldwork, anatomy
Živilė **Žigaitė**	Lithuania, Sweden	Early vertebrates	Taxonomy, paleoclimatology
Virginia **Zurriaguz**	Argentina	Reptiles (dinosaurs)	Anatomy, evolution

Note: Where names are in bold, see more information in the text.

Africa. She is also known for her efforts to popularize science and won the Sub-Saharan Africa Regional Prize for the public understanding of science from the World Academy of Science in 2013. Among VP students Chinsamy-Turan has mentored is Romala Govender (see later in this chapter).

Gillian M. **King** (194?–) is an English vertebrate paleontologist who studied zoology at Oxford with Tom Kemp, then went to work at the South African Museum Earth Sciences Division, Cape Town. She conducted pioneering systematic and paleobiological work in the 1980s and 1990s on dicynodont reptiles in South Africa using a cladistic and biomechanical approach. She prepared the Anomodontia part of the *Handbuch der Paläoherpetologie* and wrote a classic book on dicynodonts in 1990 (e.g., King, 1990; Taylor, 1990). Returning to England in 1993, she went to work in the Faculty of Classics, Cambridge University, and turned to writing general books on vertebrates, including one on the history of herbivory (King, 1996).

Maeve **Leakey** (née Epps) (1942–) is an English-born Kenyan paleontologist who has conducted research focused on Plio-Pleistocene faunas and early hominid evolution in Africa. She married into the Leakey family, her husband, Richard, being the son of paleoanthropologists and archaeozoologists Louis and Mary Leakey (e.g., Morell, 1995; and see Chapter 4). Maeve initially studied zoology and marine zoology at the University of North Wales, and while working on her PhD (zoology, 1968) at the Tigoni Primate Research Center in Kenya, she met the Leakey family and went to work on Lake Turkana sites (Hemming, 1998). Maeve worked at the National Museums of Kenya from 1969 until 2001, where she was head of the division of paleontology for many years. She later received an honorary DSc in paleontology from UCL. She is currently a research professor and director at the Turkana Basin Institute (affiliated with Stony Brook University), where she has led numerous field excavations recently with her daughter Louise (see Chapter 6). She was elected a foreign member of the US National Academy of Sciences (2013). One of her key discoveries was made in 1999 when her research team found the skull of a 3.5-million-year-old hominoid, which she named *Kenyapithecus platyops* (Hemming, 1998; McCall, 2001).

Asia

China. **Sun** Ai-lin (Ai-ling) (193/194?–) was contemporary with Chang-Kang Hu and Yu-Ping Zhang (see Chapter 4) and a student of C.C. Young, and she was present at the founding of the IVPP in 1953. Her work on therapsids, particularly dicynodonts, is featured in the IVPP building on the "Nine Dragon"

TABLE 5.3
Selection of late 20th and early 21st-century Chinese women VPs

Family name	First name	Dates	Location	Research
Bi	Zhuzhan	194/5?–	Beijing	Mammals
Chang (Zhang)	Meemann	1936–Nanjing	Beijing, Stockholm	Early vertebrates, fish
Chang (Zhang)	Yu-Ping	1934–1985	Beijing	Mammals
Chen	Donglei		Uppsala	Fish, amphibians
Chen	Guanfang		Beijing	Mammals
Chen	Meng	197?–	Nanjing	Reptiles
Dong	Liping		Beijing	Reptiles
Fu	Qiaomei	1986–	Beijing, Mongolia, Jehol	Ancient DNA
Guan	Ying			Stratigraphy, paleoanthropology
Han	Defan		Beijing	Mammals
Hou	Sukuan		Beijing	Stratigraphy, biogeography
Hou	Yamei		Beijing	Mammals
Hu	Chang-Kang	1928–2011		Mammals
Hu	Han			Reptiles (birds, dinosaurs)
Hu	Yu-zhi "Daisy"	1992?–	Beijing, Canberra	Fish
Li	Jinling	195/6?–		Reptiles, mammals
Li	Qian			
Li	Yuqing			Mammals
Liu	Juan		Beijing, United States	Fish
Liu	Liping		Beijing, Stockholm	Mammals
Lu	Jing		Beijing, Canberra, Yunnan	Fish
Ma	Fengzhen	195/6?–	Beijing	Fish
Mao	Fang-yuan		Beijing, New York, Inner Mongolia	Mammals
O'Connor	Jingmai Kathleen	1983	Southern California, Beijing	Reptiles (birds, dinosaurs)
Pan	Lei			Mammals (primates)
Qi	Guoqin			Mammals
Qiao	Tuo	1979/82?–	Beijing, Yunnan	Early vertebrates, fish
Shang	Qinghua		Beijing	Reptiles
Shi	Qinqin		Beijing	Mammals
Sun	Ai-lin (Ai-ling)	193?–	Beijing (now retired)	Reptiles, mammals

Ting (Ding)				Reptiles (birds), mammals
Wang	Suyin	194?–	Beijing, then United States	Mammals
Wang	Banyue	198?–		Fish
Wu	Wei		Beijing	Mammals
Wu	Ruijin		Beijing, Shanxi	Mammals
Wu	Wenyu		Beijing	Mammals (hominids)
Wu	Xiujie		Beijing	Mammals
Wu	Yan		Beijing	Mammals
Xu	Yuxuan			Amphibians, reptiles (dinosaurs)
Yi	Hong-yu	1984– Leshan	Beijing, New York, Mongolia and Inner Mongolia, Scotland	Mammals (hominids)
Zhang	Xiaoling			Fish, reptiles
Zhao	Lijun		Zhejiang, Yunnan	Mammals (primates)
Zhao	Ling-Xia			
Zhou	Jianjian	194/5?–	Beijing, Gobi, Tarim	Fish

Sources: BFV, Women in Vertebrate Paleontology database; Lucas, 2001; and colleagues.
Notes: Those in Beijing were mainly members of the IVPP. Where names are in bold, see more information in the text.

wall filled with dicynodont skeletons. In 1989 she wrote the first VP book from China, *Before Dinosaurs: Land Vertebrates of China 200 Million Years Ago*. As scientific exchange became possible again after the Cultural Revolution, when she was assistant director of the IVPP in 1986, Sun visited Drumheller, Alberta, Canada, following which the Canada-China Project was established (Grady, 1993; Lucas, 2001; D. Spalding in litt.). Sun's work continued on into the late 20th century, and she worked with the first Chinese-Western expedition after the Cold War era—the British cooperative visit by Alan Charig (1927–1997), Angela Milner (see later in this chapter), and colleagues in 1981 (Lucas, 2001). One of her VP students was Li Jin-Ling, who also studied dicynodonts and Mesozoic reptiles, working in London with vertebrate paleontologist Barry Cox (1931–).

Japan. Tamaki **Sato** (1972–) is one of the first Asian women vertebrate paleontologists in Japan. She is on the faculty at Tokyo Gakugei University and studies evolution, paleoecology, and systematics of Mesozoic marine fish and reptiles, particularly plesiosaurs. Born in Takahashi, she received a dinosaur (*Diplodocus*) toy from her parents when she was in kindergarten, and this set her on her life's path. Sato gained a BS in geology from the University of Tokyo in 1995, after which she went to the University of Cincinnati to study for her MS with Glen Storrs. She completed her PhD in 2003, studying plesiosaurs and supervised by Elizabeth "Betsy" Nicholls (see discussion earlier in this chapter) at the Royal Tyrrell Museum and Russell Hall in the Department of Geology at the University of Calgary. After postdoctoral work at Hokkaido University, she worked at the National Museum of Nature and Science in 2008 until her appointment at Gakugei University. Working with Nicholls, she prospected for Triassic ichthyosaurs in the British Columbia mountains in 1998 and 1999, and later, with Darla Zelenitsky (see discussion later in this chapter), she discovered the first shelled eggs inside a mother dinosaur (Sato et al., 2015). In 2016 she won the prestigious Saruhashi Prize presented to Japanese women in science.

Australia and New Zealand

With a few exceptions, there were no trained women vertebrate paleontologists until the late 20th century in Australia and the 21st century in New Zealand (e.g., Rich, 1999). As Patricia Vickers-Rich and Neil Archbold (1991) note, this region did not "come of age" in vertebrate paleontology until the 1970s. Michael Archer and Georgina Clayton (1984) provide a summary of the progress in Australian vertebrate paleontological studies and give brief biographies of most of the active workers and their students at that time. Jeanette Partridge **Hope** (1942–) gained a BS honors at Sydney University in 1964. As an undergraduate in 1963 with an interest in caving, Hope explored the fossil bone deposits in the Wellington Caves and then found a Holocene thylacine skeleton on a caving expedition to the Nullarbor; these were the topics of her first papers. She earned a PhD at Monash University on the biogeography of Bass Strait island mammal faunas, supervised by Jim Warren. In the 1970s, Hope, as a technical officer in the Department of Prehistory at the Australian National University, Canberra, analyzed faunal remains from some key Quaternary archaeological sites, such as Willandra Lakes, western New South Wales, and others in New Guinea and the Pacific and also excavated Quaternary megafaunal paleontological sites in South Australia, Victoria, and New South Wales (Archer and Clayton, 1984). In the 1980s she moved into heritage management in the New South Wales National Parks and Wildlife Service.

As there have been few professionals, amateurs are indispensable and many have contributed by finding and researching important specimens, often fostered during the later decades by the formation of the Fossil Collectors' Association of Australasia, with its own bulletin, in 1980 (see also MacFadden et al., 2016, for the United States). One such amateur is Elizabeth "Liz" **Smith** (194?–), who has collected many Cretaceous fossil vertebrates and written a major book on the opal vertebrate discoveries of Australia (Smith and Smith, 1999).

Several VPs came to live in or returned to Australia at this time, and they formed the phalanx for training the next generation (see later in this chapter). By the time of a new series of symposia (CAVEPS, see Chapter 1) in 1987, there were over 70 people present, with 15 women presenting talks or posters. One of the founders of CAVEPS, Susan **Turner** (1946–) was born in Birmingham and grew up in Wolverhampton, United Kingdom. Early on she knew she wanted to be a bone hunter, having found Edwin Colbert's (1951) *Dinosaur Book* in the library. As an undergraduate at Reading University in 1965, she found her teacher, L. Beverly "Bev" Halstead (1933–1991). She stayed on in the Geology Department at Reading University as Halstead's research student and research curator, working on agnathan thelodonts and their utilization for biostratigraphy (e.g., Miles, 1971; Turner 1973). This topic was to influence her working life as she followed and discovered thelodont scales around the globe. In 1980 she immigrated to Australia and began to explore for Paleozoic fish microfossils around the continent and across Gondwana.

Her finds include some of the oldest Ordovician and Silurian thelodonts, the oldest Devonian sharks in Queensland and Canada, the first Carboniferous amphibian in the Southern Hemisphere, and the youngest (albeit controversial) dicynodont (the former with Anne Warren [see later in this chapter], and tetrapods with her husband, R.A. "Tony" Thulborn). From 1991 to 1996 she co-led an international UNESCO–International Union of Geological Sciences IGCP project to consider Paleozoic microvertebrates worldwide (Blieck and Turner, 2000). Subsequent work allowed her and colleague Alain Blieck to define a major gap in the Late Ordovician, which they named Talimaa's Gap after Karatajūtė-Talimaa (see Chapter 4). In the late 2000s they and colleagues refuted claims that conodonts were truly vertebrates (e.g., Turner et al., 2010b). Turner has worked on fossil fish across all continents and has mentored many students, including M. Ginter, J. Vergoossen, Carole Burrow (see later in this chapter), Mags Duncan, V. Hairapetian, M. Yazdi, and H. Ferron.

Patricia "Pat" Arlene **Vickers-Rich** (1944–; Plate 25) is an American Australian geologist and vertebrate paleontologist with research interests in the origin and evolution of Australian vertebrates and their environments. Specialist in fossil birds and dinosaurs, her work on Southern Hemisphere polar biotas has completely revised our understanding of Mesozoic life at high latitudes. She was born in California on a farm and was the first in her family to go to college. She received a BA in paleontology in 1966 from UCB and in the same year married fellow student Tom Rich. Vickers-Rich was inspired by fossil bird expert Hildegarde Howard (Chapter 4) to go further, and Ruben Arthur Stirton and Alden Miller were mentors at the University of California. She went on to receive an MA in 1969 and a PhD in fossil flightless birds in 1973 from the Geology Department at Columbia University, New York. Walter Bock and Malcolm McKenna mentored her in New York. Vickers-Rich made her first visit to Australia in 1971 as part of Dick Tedford's (1929–2011) AMNH team to examine Miocene sites around Lake Frome (Tedford, 1991b). She continued working as a paleontologist at the AMNH in 1972. She held a faculty appointment at Texas Tech University from 1973 to 1976.

In 1976 Vickers-Rich immigrated to Australia and held a succession of appointments (ranging from lecturer to personal chair of paleontology and founding director, Monash Science Centre, Monash University, Melbourne, Victoria). After 35 years, she is still working at Monash as emerita professor and since 2017 has lectured at Swinburne University. During that time, Pat and Tom have led field expeditions to Dinosaur Cove in southern Australia (Vickers-Rich and Rich, 1993–2019), and she has worked in Patagonia, in search of Cretaceous polar dinosaurs and fossil birds (Turner et al., 2010a). Apart from numerous research papers, she has published several award-wining popular science books (e.g., Murray and Vickers-Rich, 2004) and contributed to education at all levels, with early childhood science her focus. She was also one of the first VPs to visit and research in China in 1979 and 1982, with a major project leading to an English-Chinese dictionary of paleontological terms. Since 2003 her focus has been on the Precambrian, with work in Australia, Russia, and Namibia, and all her fieldwork has been funded by the income from exhibitions or else self-funded. Her students include J. Long, N. Schroeder, D. Seegets-Villiers, and Corrie Williams, who have worked on a range of fossil vertebrates.

Mary **Wade** (1928–2005) was an Australian paleontologist born in Adelaide, South Australia. Her family moved when she was young to Thistle Island in Spencer Gulf off the southern coast, where she first became interested in geology (Cook, 2005). Despite desultory schooling, she gained a place at the University of Adelaide and graduated with a BS with honors in geology in 1954 and received a PhD for her dissertation on Tertiary foraminifera in 1958. Although she might be better known for her work on the Precambrian Ediacaran fauna in Australia (e.g., Turner and Vickers-Rich, 2007), she made several notable contributions to vertebrate paleontology. Moving to become curator of geology at the Queensland Museum in 1970, she and Michael Archer (University of New South Wales, Sydney [UNSW]) explored for Tertiary vertebrates in northern Australia. With vertebrate paleontologist Tony Thulborn of the University of Queensland, Wade organized and supervised the excavation during 1976–1977 of more than 3,000 dinosaur footprints in the Tully Ranges in central-west Queensland. Based on their work, the site near Winton that became known as Lark Quarry or the Dinosaur Stampede (the inspiration for the *Jurassic Park* movie) has been made an Australian Natural Heritage site (Poropat, 2017). The result of predatory theropods causing a stampede of a range of smaller dinosaurs, the trackways clearly show the movement of the animals in the mud, sliding and slipping as they ran (e.g., Thulborn and Wade, 1984; Thulborn, 1990; see Romillo et al., 2013, for an alternate interpretation and Falkingham et al., 2016, for rebuttal). In 1987 Wade recovered the second skull of the Queensland ornithopod *Muttaburrasaurus*. She also excavated specimens of the giant Cretaceous pliosaur *Kronosaurus queenslandicus* and new material of the rare Jurassic sauropod *Rhoetosaurus* at the holotype site, thought to be lost in the 1920s (Rich and Vickers-Rich, 2003). Wade made a major study of the giant ichthyosaur

Platypterygius australis and was studying the most complete Early Cretaceous plesiosaur fossil at Hughenden, a specimen that she had excavated
(see Turner et al., 2010a).

Anne **Warren** (née Howie) (1941–) is an Australian vertebrate paleontologist specializing in the study of Paleozoic and Mesozoic amphibians with a
Gondwanan focus. She received a BS (1963) from the University of Sydney
and a PhD in 1967 from Cambridge, mentored by F.R. Parrington (1905–1981),
with a dissertation on capitosaurid labyrinthodonts from East Africa. Warren returned to Australia in the late 1960s, working first in the Sydney basin
and then exploring the Early Triassic of Queensland with Australian Museum and Queensland Museum colleagues, which has yielded numerous
new temnospondyl amphibian taxa. She was first affiliated with Melbourne
University and then the Department of Zoology at La Trobe University,
where she is now an emerita scholar. Renewed collecting in Queensland
brought forth a Jurassic temnospondyl and plesiosaurs and later an early Carboniferous tetrapod fauna (see also the foregoing discussion of Turner).
Throughout her career, Warren's research has been supported by the Australian Research Council, enabling fieldwork with museum visits to institutions in Argentina, India, Japan, Russia, South Africa, Scandinavia, France,
Germany, and North America. Among VP students whom she has mentored
are A. Yates, L. Davey, J. Garvey, C. Northfield, K. Parker, and K. Pawley.

Joan **Wiffen** (1922–2009) was a homegrown and self-trained amateur vertebrate paleontologist in New Zealand. She pioneered dinosaur hunting and
added considerably to the record of Mesozoic reptiles. She grew up in Hawkes
Bay, North Island. Reportedly, her father did not believe in education for
girls, so it was not until after performing war work during World War II that
she went to school and became a teacher. Her dinosaur hunting began later
in life when she and her husband, Pont, were in their 50s. They tracked down
a map of North Island from a petroleum company that noted "reptilian bones"
in the Te Hoe valley and began to explore (Cox and Wiffen, 2002). Wiffen has
the distinction of having found the first dinosaur fossils in New Zealand. She
became known as the "Dinosaur Lady"; she and her husband collected numerous Late Cretaceous dinosaurs, including nodosaur, carnosaur, and sauropod, the first major finds from New Zealand, in addition to various mosasaurs, turtles, bony fish, and many other finds (see Farlow and Brett-Surman,
1997). American vertebrate paleontologist Ralph Molnar, then at the
Queensland Museum, mentored her, and he and Wiffen published numerous
papers about her discoveries. In recognition of her significant fossil discov-

eries, she received an honorary PhD from Massey University in 1994, the Commander of the British Empire in 1995, and the SVP's Skinner Award in 2004 (Turner et al., 2010a).

Early Vertebrates and Jawed Fish. Carole **Burrow** (Queensland Museum, Brisbane, Australia) pursues studies on a diverse range of Australian Paleozoic fish and received a PhD (2001) from the University of Queensland under the mentorship of Susan Turner and Tony Thulborn. She specializes in Paleozoic stem gnathostomes, "sharks," and acanthodians and is a noted paleohistologist. Katherine "Kate" **Trinajstic** (Curtin University) uses synchrotron CT imagery to study the exceptionally preserved Devonian Gogo placoderms (jawed fish found in the Gogo Formation of northwestern Australia), which led to the discovery of soft tissues, including muscles and nerves, and the fact that these fish were the oldest vertebrates to give birth to live young, 200 million years earlier than previously thought. Her current research involves how vertebrates evolved a skeleton. She received a BS from Murdoch University and a PhD (2000) on Australian Devonian micro- and macrovertebrates from the University of Western Australia, mentored by John Long. In recognition of her research, she was awarded the Prime Minister's Science Prize in 2010.

Geologist Margaret **Bradshaw** (Canterbury Museum, Christchurch, South Island, New Zealand) has been contributing to New Zealand vertebrate paleontology since the 1970s. She has collected Devonian fish in Antarctica since the 1990s and helped find the country's first early vertebrates near Nelson, New Zealand. She was the first female recipient of the United Kingdom's Polar Medal in recognition of her research in polar areas.

Mammals. Suzanne J. **Hand** (UNSW) studies the paleontology, phylogeny, and biogeography of Australian Cretaceous and Tertiary monotremes, marsupials, and especially bats. She received a BS (1980) from the UNSW and a PhD (1986) from Macquarie University, Sydney, and has conducted many field trips to the northern Australian Riversleigh site (e.g., Archer and Clayton, 1984). Hand has written many books on Australian vertebrates, such as *The Antipodean Ark* (Hand and Archer, 1987), and she has mentored many VP students: R. Beck, P. Fabian, E. Hall, S. Hocknull, T. King, A. Klinkhamer, C. Lopez Aguirre, Jacqueline Nguyen, C. Palmer, and K. Travouillon.

Europe

Denmark. Ella **Hoch** (1940–) is a Danish vertebrate paleontologist and historian of geoscience, as well as a strong supporter of the EAVP, now retired from the Geological Museum of Copenhagen University and the Gram Museum of Paleontology, Haderslev, Denmark. She has conducted research over four decades, often with European colleagues, on a diverse range of Cretaceous to Holocene vertebrates, including sharks, birds, macropods, and whales in Denmark, Greenland, and Papua New Guinea (e.g., Tedford, 1991a).

Estonia. Tiiu **Märss** (1943–) is now a senior researcher at Tallinn University of Technology, Estonia. She graduated in 1983 as a candidate of geological and mineralogical sciences (corresponding to a PhD) from the Institute of Geology, Academy of Sciences of the Estonian SSR, mentored by Elga Mark-Kurik, working on early vertebrates from the Paleobaltic Sea of Estonia (see Chapter 4; Schultze et al., 2009). Following perestroika and her involvement as a leader of UNESCO–International Union of Geological Sciences International Geological Correlation Programme 328 (1991–1996) and afterward International Geological Correlation Programme 406, her research interests expanded to involve the taxonomy, phylogeny, morphology, and biostratigraphic distribution of Silurian-Devonian early vertebrates, especially in the Northern Hemisphere (e.g., Märss and Ritchie, 1998). She coauthored the classic handbook on thelodonts (Märss et al., 2007), and has worked on anaspids, heterostracans, and stem chondrichthyans from Europe and Canada.

Finland. Ann-Marie **Forstén** (1939–2002) was a Finnish vertebrate paleontologist who worked at the Finnish Museum of Natural History in Helsinki and conducted research on fossil horses. A student of Björn Kurtén, and the first female VP in her country, she studied *Hipparion* for her MS work (1963) and her PhD thesis, "Revision of the Palearctic *Hipparion*" (1968), still considered a classic, applying quantitative methodology. In the 1970s to 1980s, Forstén wrote on American *Mesohippus*, Pliocene *Hipparion* from as far afield as Tunisia, and, later, European and Chinese *Equus* (Gayet and Babin, 2007; Donner, 2014).

France. France de **Lapparent de Broin** (1938–) (officially retired in 2003 and honorary research associate at the MNHN) is an expert on the anatomy and phylogenetic and depositional environment of Mesozoic and Cenozoic crocodiles and turtles. She has made them known to the paleoherpetological and wider community over a 50-year career (Rage, 2015). Born into a famous French family of geologists and paleontologists (e.g., Albert F. de Lapparent [1839–1908] was her uncle), she "inherited a tradition of scientific research" and became a well-rounded scientist with broad interests and expertise in both geology and biology (Laurin and Bardet, 2015:437). She completed a thesis in 1965 for research on Cretaceous crocodiles from the Sahara and then took a position at the Centre national de la recherche scientifique and turned her attention to turtles working across the globe (e.g. Taquet, 1998), receiving a *thesis d'état* (equivalent to a PhD) under the mentorship of vertebrate paleontologist Robert Hoffstetter (1908–1999) at the MNHN Paleontological Lab (Rage, 2015).

Claire **Derycke** (also Derycke-Khatir) is a French paleoichthyologist and histologist from Arras, northern France, now a Centre national de la recherche scientifique professor who works at the Université Sciences et Technologies de Lille at Villeneuve d'Ascq and Liège University, Belgium. She has studied Paleozoic fish from Europe, Turkey, Mexico, and Africa, discovering many new species. She gained her MS in geology in 1987 and her doctorate in 1994 working on Late Devonian shark and bony fish microremains from the Manche to the Rhine, mentored by vertebrate paleontologist Alain Blieck. Since the early 1990s, her focus has been on biostratigraphy and paleogeography, using fish to gain paleoenvironmental data to understand the geology of the Carnic Alps. In 2000 she received the Prix Gosselet from the Société des Arts, des Lettres et de l'Agriculture de Lille.

Poland. Maria Magdalena **Borsuk-Białynicka** (1940–) is a vertebrate paleontologist at the Institute of Paleobiology, Polish Academy of Sciences, Warsaw. Born in Warsaw, she received her MS (1961) and PhD (1971) at Warsaw University as a student of Roman Kozlowski, working on Pleistocene mammals. She received a DSc (1982) from Jagiellonian University in Kraków. She has held teaching appointments at Warsaw University and Adam Mickiewicz University in Poznan while also supervising graduate students at Silesian University, Opava, Czech Republic, rising to head of VP at the institute where she is now an emerita professor. Her research expertise is focused on the functional anatomy and phylogenetics of fossil amphibians and reptiles, including dinosaurs (Triassic diapsids and Mesozoic lizards, especially from the Late Cretaceous of Mongolia), and she often works in cooperation with Susan Evans (see discussion later in this chapter). She took part in the first (1964) Polish-Mongolian dinosaur expedition (see Farlow and Brett-Surman, 1997, fig. 4.6).

Spain. Emerita SVP member Maria Teresa **Alberdi** (1949–) of the Departamento Palaeobiologia, Museo Nacional de Ciencias Naturales, and the Spanish National Research Council, Madrid, studies the evolution and paleoecology of Cenozoic mammals (especially horses) and faunas in her home country, elsewhere in Europe (e.g., with Maria Palombo, see discussion below, and Vera Eisenmann), and in South America.

Nieves **López Martínez** (1949–2010) was part of a group of women re-sponsible for the modernization of paleontology in Spain, promoting inter-disciplinary research and science communication in paleontology. Born in Burgos, Spain, she studied at the Complutense University of Madrid, gradu-ating with a degree in biological sciences in 1970. She received two PhDs from the University of Montpellier, France, under the mentorship of French paleontologist Louis Thaler (1930–2002), and at the University of Madrid, working with Spanish paleontologist Emiliano Aguirre (1925–). Her re-search was focused on mammalian paleontology, particularly lagomorphs (rabbits and their kin), and at the Miocene Somosaguas fossil site. López Martínez also studied faunal changes in the Pyrenees during the Cretaceous extinction event. She promoted what she called "social paleontology" and was a pioneer of the internet "palaeoblogosphere" (see Fernández and Gó-mez Cano, 2015).

María Dolores "Loli" **Soria Mayor** (1948–2004) was a Spanish vertebrate paleontologist with research interests in Cenozoic carnivores, extinct rumi-nants, and the evolution of mammals in South America. She received a BS in biological sciences in 1971. She was hired as a Spanish National Research Council collaborator in 1974, and in 1981 she obtained the position of senior scientist at the Museo Nacional de Ciencias Naturales, where she was head of the Department of Paleobiology from 1996 to 2001.

United Kingdom. Jennifer Alice **Clack** (née Agnew) (1947–2020; Plate 26) was an English vertebrate paleontologist who was raised on the outskirts of Manchester, United Kingdom. She received a BS in zoology from the Univer-sity of Newcastle upon Tyne in 1970. In her final year at the university, she took a course in vertebrate paleontology taught by Alec L. Panchen (1930–2013), furthering a childhood interest. Since no jobs were available, she pur-sued a career in museums, obtaining a graduate certificate in museum stud-ies from Leicester University. She spent three years as a display technician (exhibits) at the Birmingham Museum and Art Gallery before moving to the Education Department where she was in charge of natural history education. While there, she was encouraged by her head of department, Ann Meredith, to do some part-time research. Panchen suggested that she study the last un-described stem tetrapod (embolomere) specimen in the United Kingdom in his lab. She prepared the specimen and serendipitously discovered its previ-ously covered braincase. With grant support and encouragement from her then boyfriend (later husband) Rob Clack in 1978, she began a PhD on *Pholi-dagaster* back at Newcastle upon Tyne, gaining it in 1984. Clack became curator of vertebrate paleontology at Cambridge University. She joined the University Museum of Zoology in Cambridge as an assistant curator in 1981. She was promoted to senior assistant curator in 1995 and in 2005 to curator in VP at the museum. In recognition of her work, in 2000 Clack was awarded an honorary DSc and in 2006 a personal chair by the University of Cambridge, after which she took the title professor of vertebrate paleontology. She retired as professor emerita in 2015, and her student Per Ahlberg organized a Fest about the Early Tetrapod World in her honor at the SVPCA in 2017. Other VP students that she has mentored include M. Coates, Stephanie Pierce (see later in this chapter), M.S. Lee, M. Friedman, and B. Swartz.

Clack is best known for her work on the fish-to-tetrapod transition—the origin, evolutionary development, and radiation of early tetrapods and their lobe-finned fish relatives—as revealed in her book *Gaining Ground: The Origin and Evolution of Tetrapods* (Clack, 2002, second edition 2012).The discovery of fossils of *Acanthostega* and *Ichthyostega* from classic localities in the Greenland mountains in 1987, led by Clack and her Danish and British colleagues, confirmed that they were stem tetrapods. These amphibians were "aquatic animals that rarely, if ever, made forays onto dry land," a challenge to the previous hypothesis that they were fish linked to invasion of the land (Shubin, 2003:766). With her discoveries of new fossils from Scotland that exemplify the founding of the tetrapod crown group (the Tetrapod World: Early Evolution and Diversity, known as the TWeed Project), she has helped close the so-called Romer's Gap at the base of the Carboniferous.

As noted by Koppes and Allen (2013), her research "profoundly changed the understanding of the origin of terrestrial vertebrate life." She was awarded the Daniel Giraud Elliot Medal from the US National Academy of Sciences in 2008. Clack is the first vertebrate paleontologist since G.G. Simpson (1958) to have been awarded the medal, only the second female (the other being Libby Hyman in 1951), and a rare British awardee—the first woman—joining D'Arcy Thompson and Scot Robert Broom (1946). In 2013 she also received the T. Neville George Medal from the Geological Society of Glasgow. She achieved Britain's highest honor, becoming a fellow of the Royal Society in 2009, and in 2018 she won the Lapworth Medal, the Palaeontological Association's most prestigious award, based on her significant contributions.

Like Clack, Angela C. **Milner** (née Girven) (1947–) is an English vertebrate paleontologist who received a BS (1969) and a PhD (1978) from the University of Newcastle upon Tyne under the mentorship of Panchen. She then joined the NHM in London, where she (now emerita) studies the anatomy, faunas, and paleobiogeography of theropod dinosaurs and birds. She has used CT scanning and three-dimensional reconstructions to investigate the structure of the inner ear and braincase of various birds and reptiles. She has conducted fieldwork in China (with Sun Ai-lin; see earlier in this chapter), Africa (southern Niger; Whybrow, 2000), Scotland, and England, and worked on Antarctic dinosaurs. She has tracked and men and women who signed Lady Maud Smith Woodward's famous tablecloth (Milner, 2016). Among VP students whom she has mentored are P.M. Barrett and P. Upchurch. She was elected an honorary member of the SVP in 2018.

Early Vertebrates and Jawed Fish. Sarah E. **Gabbott** (University of Leicester, United Kingdom) is a British geologist and paleontologist specializing in the taphonomy of early vertebrates and fish using experimental geochemical and microstructural analysis techniques to explore soft-tissue fossilization in enigmatic fossils such as *Jamoytius*. Zerina **Johanson** (NHM) studies the embryological development in early vertebrates, combining paleontological research with developmental studies of living vertebrates (see Appendix 1). She received a BS (1989) and MS (1992) from the University of Alberta and a PhD from Macquarie University, Sydney, moving to England in the late 1990s. She has mentored VP students including H. Beckett, C. Dobson, J. Keating, R. Smith, and A. Thierry. Johanson's colleague Moya Meredith **Smith** (formerly of Guy's Dental Hospital and King's College, now at NHM) is a histologist who has studied the evolution of teeth, especially in lungfish and sharks, putting forward the hypothesis that such originated from branchial scales rather than the traditional skin denticles (e.g., M.M. Smith and Coates, 1998). Lithuanian Živilė **Žigaitė** (former Swedish Research Council fellow, Department of Organism Biology, Evolutionary Biology Center, Uppsala University) studies early vertebrates, especially thelodonts from the Paleozoic of Asia, following the work of her mentor Karatajūtė-Talimaa. She is particularly interested in paleodiversity and paleogeography and uses geochemistry and paleohistology techniques.

Amphibians and Reptiles. Susan "Sue" E. **Evans** (UCL) conducts research on the functional anatomy, paleoecology, and systematics of fossil amphibians and reptiles, especially lizards. She received a BS (1974) from Bedford College and a PhD (1977) from UCL. She worked in Saudi Arabia and then Birkbeck College, and beginning in 2001 she conducted fieldwork on Cretaceous vertebrates in Madagascar. She is a fellow of the Linnaean Society and the Zoological Society and has mentored many students, including Ryoko

Matsumoto (see discussion later in this chapter). Nathalie **Bardet** (MNHN, Paris) studies the anatomy and evolution of Mesozoic marine reptiles. She received a BA (1983) from the Université de Nice and a PhD (1992) from the Université de Paris. Her work brought clarity to plesiosaur evolution, and in 1994 she identified two major extinction events affecting marine reptiles. Her French students and colleagues include Peggy Vincent and Delphine Angst, and she has worked with them to rediscover Mary Anning's specimens, naming one for her (Vincent et al., 2014).

Dinosaurs. Emily **Rayfield**'s (University of Bristol) research is focused on the functional anatomy of extinct reptiles, especially dinosaurs, using computational methods including finite element analysis and geometric morphometrics (see Appendix 1). Current research projects focus on nonavian dinosaurs and birds, the water-to-land transition, and the origin of mammals. She received a BA (1996) from Oxford University and a PhD (2001) from Cambridge University. She has mentored the following VP students: A. Taylor, M. Morales Garcia, Stephanie Pierce, S. Jasinoski, L. Porro, and M. Young. Rayfield became the 7th female and the youngest president of the SVP in 2019. The same year, she was awarded the Bigsby Medal by the GSL, and she was recently awarded the President's Medal from the Palaeontological Association.

Mammals. Anjali **Goswami** (NHM and UCL) studies vertebrate evolution and development. She uses three-dimensional morphometric methods to incorporate data from embryos to fossils to test genetic and developmental hypotheses of modularity and morphological diversity and reconstruct macroevolutionary patterns through deep time (see Appendix 1). She received a BS (1998) from the University of Michigan and a PhD (2005) from the University of Chicago, mentored by John Flynn. She has conducted fieldwork in Madagascar, Chile, Peru, and India. She is a fellow of the Linnaean Society and was awarded the Bicentenary Medal of the Linnaean Society in 2016. Currently, she serves on the SVP Executive Committee as member at large. She has mentored the following VP students: M. Bell, S. Bennett, R. Benson, A.-C. Fabre, R. Felice, A. Gomez-Robles, T. Halliday, Katrina Jones (see later in this chapter), G. Price, M. Randau, C. Verity Bennett, D. San-Felice, J. Smaers, A. Watanabe, and Vera Weisbecker. Maria Rita **Palombo** (Sapienza Università di Roma, Italy) conducts field research on biochronology, paleoecology, and evolution of Cenozoic mammals from Eurasia. She received a PhD (1970) from the Università di Roma. She conducted pioneering fieldwork in Inner Mongolia in 1999–2000. More recently, her research has focused on the origin and evolutionary patterns of insular mammals (e.g., in the Mediterranean islands) and the influence of climate change on mammal dispersal (including hominins).

Middle East

Israel. Yael **Chalifa** (née Ben-Shem) (1940–2006) (affiliated with Avshalom Institute in Tel Aviv) was born in Tel Aviv and was one of the few Israeli female vertebrate paleontologists. She specialized in and described new taxa of Cretaceous bony fish from her country. She received an MS and then PhD in 1986 for her work on enchodontiform fish at the Hebrew University of Jerusalem, supervised by Eitan Tchernov (1935–2002), and went on to a postdoc with Mark Wilson at the University of Alberta. She was unable to secure a VP position, so she became a schoolteacher, although she continued to pursue research and publish papers into the 1990s (M. Wilson and Mienis, 2008). Naomi F. **Goldsmith** (Department of Evolution, Systematics and Ecology, Hebrew University, Jerusalem, and Ben Gurion University, Negev, now retired), worked with Eitan Tchernov and others on ctenodactylid rodents in the Miocene Negev faunas.

United States, Canada, and Mexico

Anna "Kay" **Behrensmeyer** (1946–; Plate 27), while doing fieldwork in Cameroon, nonchalantly remarked that a green mamba, a highly poisonous snake, had been sighted near her field site all week. Given this, you might expect her to walk slowly and warily approach outcrops looking for fossils, but, as colleague Scott Wing recounted to interviewer Ellen Currano (2014), Behrensmeyer moved quickly and purposefully through the grass. It was clear that she could not wait to explore the next outcrop and was not going to let venomous snakes stand in her way of finding fossils.

Behrensmeyer is recognized as a pioneer in the subfield of taphonomy, the study of processes that affect animal and plant remains as they become fossilized. She considers one of her proudest achievements having led the Amboseli (Kenya) Bone Taphonomy Project for more than 40 years. She edited the definitive text *Fossils in the Making* (Behrensmeyer and Hill, 1980). As she stated in an interview (Currano, 2014), "My long term goal has been to touch down in as many time periods as possible, in the terrestrial record, to see how vertebrate (and some plant!) taphonomy and paleoecology plays out over the Phanerozoic up through the Recent."

Behrensmeyer received a BA from Washington University in Saint Louis, Missouri, and a PhD from Harvard University in 1973 under the mentorship of paleomammalogist Bryan Patterson (1909–1979). After postdoctoral fellowships at UCB and Yale and a brief teaching stint at the University of California, Santa Cruz, she began her career at the NMNH in 1981 as curator of vertebrate paleontology. At the museum, she is codirector of the Evolution of Terrestrial Ecosystems Program and an associate of the Human Origins Program. Behrensmeyer is also compiling a taphonomic reference collection of bones and fossils at the NMNH.

She received the SVP Romer-Simpson Medal in 2018, the fourth woman to be so honored. Also in 2018, she was recognized by the PS as the PS Medalist and awarded the Raymond C. Moore Medal of the Society for Sedimentary Geology. That same year, she received the US National Academy of Sciences G.K. Warren Prize for her contributions to our understanding of fluvial geology. Among the VP students that she has mentored are R. Bobe, A. Villaseñor, and C. Colleary.

Jane **Colwell-Danis** (1941–) is the first formally trained female vertebrate paleontologist in Canada (Tanke, 2019). She was born in Fort Stockton, Texas, into a family that nurtured her interest in nature, rocks, and fossils. A visit to the University of California Museum of Paleontology encouraged by a high school teacher aware of her interest in fossils resulted in her receiving an MS in paleontology from UCB in 1965 on South American Miocene notoungulate mammals. She presented one of two female-authored presentations at the GSA meeting in Fresno that same year. Paleoherpetologist Wann Langston Jr. (1921–2013), while on sabbatical and visiting UCB, encouraged her to seek work outside the United States. On the strength of Langston's letter, she was hired as Richard Fox's assistant at the University of Alberta in Edmonton, Canada. She prepared fossils, curated the paleontology collections, and began conducting fieldwork at mammal-bearing localities in southern Alberta and Saskatchewan. In 1968 she met Dale Russell, of the Canadian Museum of Nature, in Dinosaur Provincial Park (now a World Heritage Site). She married Gilles Danis, Russell's technician and, because of a rule against the employment of married couples, they moved to Ottawa. She was hired as a curatorial assistant to invertebrate paleontologist J.A. Jeltsky of the Geological Survey of Canada. From 1971 to 1977 she continued fieldwork and curatorial and preparation work, as well as authoring a popular book, *The Age of Dinosaurs in Canada* (Danis, 1973). She moved with her family to Edmonton, where she worked briefly as a tour guide in the Education Department at the Provincial Museum of Alberta. Director David Spalding found a way to hire the couple in different capacities, getting around the marriage rule. Later, she packed and moved specimens to the newly constructed (now Royal) Tyrrell Museum of Paleontology, where she worked from 1979 to 1991, but her position was eliminated as a cost-cutting measure. She returned to the Tyrrell as a volunteer in 2008, sorting Cretaceous microvertebrate matrix (Tanke, 2017).

Margery Chalifoux **Coombs** (1945–) is an American vertebrate paleontologist and emerita professor at the University of Massachusetts, Amherst (see Appendix 2). She received a BA in biology from Oberlin College in 1967 and a PhD in biological sciences from Columbia University in 1973. She married fellow vertebrate paleontologist Walter Coombs. She taught vertebrate paleontology, comparative vertebrate anatomy, and evolution at the University of Massachusetts from 1972 until her retirement in 2012. Coombs is internationally known for her work on fossil perissodactyls, especially chalicotheres, pursuing research on skeletal anatomy, paleogeography, taphonomy, paleoecology, biostratigraphy, phalangeal fusion, dental microwear, and the comparative morphology of large mammalian, clawed herbivores living and extinct. She recalls in an interview with us, "My first SVP meeting was in New York City in 1969. I was one of only two women presenting a talk that year. We were both placed at the beginning of the program, and by some quirk of fortune I was first. Thus, I gave my first SVP talk without ever seeing a previous presentation. My first words at SVP were 'I hope you can see me over the top of the podium.'" Over the next 46 years, she served the SVP in many capacities, such as on the executive committee (member at large), as associate editor of the journal, on honorary membership and ethics committees, and on two different student prize committees, rarely missing a meeting. Coombs was recognized as an SVP honorary member in 2015. Among VP students whom she has mentored are D. DeBlieux, Virginia Naples (see later in this chapter), A. Pratt, Gina Semprebon (see later in this chapter), B. Wall, L. Holbrook, and E. Dewar.

Elizabeth (Betsy) L. **Nicholls** (1946–2004) was an American Canadian vertebrate paleontologist, curator of marine reptiles at the Royal Tyrrell Museum in Alberta, Canada. Nicholls was born in Oakland, California. At the age of 10, she moved with her family to Melbourne, Australia. She later returned to the United States, where she completed her BS in paleontology at UCB. In 1969 she moved with her husband, a geologist, to Calgary, Alberta, and the University of Calgary, where she received an MS, and then a PhD in 1989. She became interested in paleontology when her father, a professor at Berkeley, took her to visit Sam Welles. Nicholls looked at the fossils around his room and is reported to have said, "I want to be a paleontologist" (D. Brinkman, pers. comm., 2009). A few years later, at the age of 12, she wrote to Roy Chapman Andrews asking whether a girl could be a paleontologist, and she kept Andrews's answer of encouragement as a treasured memento. Her research focused on Mesozoic marine reptiles (e.g., plesiosaurs, mosasaurs, turtles, thalattosaurs, and ichthyosaurs), and as Robert Carroll recollects, "she was clearly fascinated by the entire assemblage—their functional anatomy, biogeography, ways of life, and relationships" (Carroll in Korth and Massare, 2006:107). For painstakingly excavating the largest ichthyosaur skeleton ever recovered over a three-year period (which she later named *Shonisaurus sikanniensis*), under extremely challenging field conditions, she was awarded the Rolex Award for Enterprise in 2000 (Turner et al., 2010a). As Robert Carroll (in Korth and Massare, 2006:108) writes, "Betsy's research provides a wonderful model for the unbiased investigation of marine reptiles or any other group of organisms. Go out in the field and collect all the fossils that can be found, no matter how difficult the terrain or the excavation of specimens. Prepare, describe, and illustrate with great attention to the details and significance of anatomical features. Compare with other specimens of the same and related taxa. Evaluate the nature of their adaptation, and their evolutionary history. Cooperate with colleagues, where ever they may be, or whatever groups they may be particularly attached to, have fun, and spread the joy throughout your discipline."

Kathy **Lehtola** (née Mavis) (1945–) worked in the 1960s on her MS in geology at Michigan State University on the then oldest fossil jawless fish "ostracoderms" from the Ordovician. She went on to redescribe *Astraspis desiderata* Walcott based on new northern Michigan remains. Her work represents some of the rare works on Paleozoic fish by women in North America (Lehtola, 1973, 1983).

Barbara Jaffe **Stahl** (1930–2004) was a paleoichthyologist born in Brooklyn who credited her interest in paleontology to visits to the AMNH as a child. She received a BA from Wellesley College in 1952, an MS from Radcliffe in 1953, and a PhD from Harvard University in 1965, the last graduate student of Alfred Romer. She began teaching in 1954 at Saint Anselm College in Manchester, New Hampshire, and was the first woman on that faculty. She wrote the classic text *Vertebrate History: Problems in Evolution* in 1974, which remained in print for 20 years. Much of her research focused on sharks, especially Paleozoic holocephalans, and she wrote volume 4, *Chondrichthyes III*, of the *Handbook of Paleoichthyology* on that topic in 1999. She was a longtime member of the SVP, having been nominated by Tilly Edinger (Bruner, 2004; Turnbull and Zangerl, 2004).

German-born Elisabeth **Vrba** (1942–) is an American paleontologist who grew up in South Africa. She earned her PhD in zoology and paleontology at the University of Cape Town in 1974, studying zoology and mathematical statistics. Her research interests, based on two decades of fieldwork, focused on African faunas especially bovids—their taxonomy, paleobiogeography, phylogeny, and paleoecology. She attended the important 1976 Wenner-Gren African VP symposium at Burg Wartstein with Behrensmeyer and Judy Harris (van Couvering). Since the late 1980s Vrba has been emerita professor at Yale University, where she taught in the Department of Geology and Geophysics. In her important theoretical work on macroevolutionary processes, she developed the turnover-pulse hypothesis, which suggests that major climatic changes often result in a period of rapid extinction followed by speciation (Vrba, 1980, 1993). She and paleontologist Stephen Jay Gould coined the term *exaptation* to describe a change in the function of a trait during evolution (Gould and Vrba, 1982). One of her many students, Faysal Bibi, works at the National Museum of Ethiopia.

Several other US women made significant contributions to VP, mostly during the late 20th century. Judith A. **Schiebout** (1946–) (retired from Louisiana State University, Baton Rouge) conducted field expeditions and studied biostratigraphy and mammals from a number of Tertiary sites in the southeastern United States, including Big Bend, Texas, and Louisiana, as well as in China (see Appendix 2). She received a BA (1968), MA (1970), and PhD (1973) from the University of Texas, Austin. Among VP students whom she mentored are B. Albright, B. Standhardt, A. Dooley, Julia Sankey, Suyin Ting, and W. Lancaster. She received Outstanding Educator Awards from the Association for Women Geoscientists Foundation (2001) and the Gulf Coast Association of Geological Societies (2002).

When Margaret **Stevens** (née Skeels) (retired from Lamar University, Texas) first graduated from the University of Michigan, she discovered and collected Devonian placoderms and holocephalians from Ohio. After her marriage during the 1970s, she studied the stratigraphy and paleontology of various mammals (i.e., rodents, mammoths, and mastodons). She collected many fossil vertebrates for the Texas Memorial Museum at the University of Texas and was honored as the 2nd recipient of the SVP Skinner Award in 1991.

Early Vertebrates and Fish. Gloria **Arratia** (University of Kansas, Lawrence) is a Chilean-born paleontologist who conducts research on the morphology, development, and systematics of teleost fish, especially ray-finned fish. She has received numerous awards, including the Humboldt Prize, and she has been named a member of the Academy of Sciences of Chile and an honorary member of the American Society of Ichthyologists and Herpetologists. She has mentored a number of VP students, including P. Brito, P. Lambers, A. de la Pena, F. Poyato-Ariza, V. Parmar, D. Rubilar, G. Giordano, S. Gouric-Cavalli, C. Hernandez, E. Ordonez, C. Shen, Lynne Bean, J. Casciotta, J. Zhang, and Adriana Lopez-Arbarello. Allison M. **Murray** (University of Alberta, Canada) conducts research on the osteology, phylogenetic relationships, and paleobiogeography of fish and their extinction patterns across the Cretaceous–Tertiary boundary. She received a BS (1987) from the University of Victoria, an MS (1994) from the University of Alberta, and a PhD (2001) from McGill University. She has conducted paleontological fieldwork and museum studies in East Africa (Kenya, Tanzania, Ethiopia), North Africa (Morocco, Egypt), and Asia (China, Indonesia).

Birds. Julia **Clarke** (Plate 29) (University of Texas, Austin) studies the origin and evolution of early birds. She received a BA from Brown University (1995) and a PhD from Yale University (2002). She has conducted fieldwork in Antarctica and China (see Appendix 1). She served as SVP member at large from 2009 to 2010. She has mentored the following VP students: C. Boyd, D. Kespa, Z. Li, and N.A. Smith. Helen Frances **James** (NMNH) studies fossil and extant birds from the Hawaiian Islands. She received a BA (1977) from the University of Arkansas and a PhD (2000) from the University of Oxford.

Dinosaurs. Brenda **Chinnery** (Austin Community College, Texas) studies the paleobiology of dinosaurs. She received a BA (1995) from the University of Colorado and a PhD (2002) from Johns Hopkins University. She has conducted paleontological fieldwork in Colorado, New Mexico, South Dakota, and Wyoming. Catherine A. **Forster** (George Washington University, Washington, DC) conducts research on the systematics and paleontology of dinosaurs, including the iconic *Triceratops*, as well as bird origins and eucynodonts. She received a BA and BS from the University of Minnesota and a MS (1985) and PhD (1990) from the University of Pennsylvania and did postdoctoral work at the University of Chicago (1990–1994). She has conducted fieldwork in many countries, including Argentina, Madagascar, South Africa, and China. She has also been a champion for conserving Madagascar's paleontological heritage (Krause et al., 2006). She served on the SVP Executive Committee as member at large (2000–2003) and president (2013–2015). Among VP students she has mentored is Kristi **Curry Rogers** (Macalester

College, Saint Paul, Minnesota) (see Appendix 1). She received a BS (1996) from Montana State University and an MS (1999) and a PhD (2001) in anatomical sciences from Stony Brook University; she is an active field researcher. She served on the SVP Executive Committee as member at large from 2010 to 2012. Jessica A. **Harrison** studied ungulate systematics and evolution, especially camels. She received a PhD from the University of Kansas and worked at the NMNH designing the Hall of Fossil Reptiles until the mid 1980s. Currently, she is a psychiatrist working in Tucson, Arizona. Frances "Frankie" **Jackson** (Montana State University, retired) pursues research on the fossil eggs associated with dinosaurs and birds and is interested in parental care behavior. She discovered paleontology late and received a BA (1992) and PhD (2007) from Montana State University. Another scientist studying dinosaur growth patterns and molecular paleontology is Mary Higby **Schweitzer** (North Carolina State University Raleigh). She earned a BS in 1977 from Utah State University and a PhD from Montana State University in 1995 under the mentorship of dinosaur paleontologist Jack Horner. She is best known for her claim to have discovered the remains of blood vessels in dinosaur fossils and to have identified a pregnant *Tyrannosaurus rex*. Darla **Zelenitsky** (University of Calgary, Alberta, Canada) became the first professionally trained dinosaur woman in Canada. She studied geology at the University of Manitoba, earning a BS in 1991, then gained her MS (1995) and PhD (2004) in Calgary, mentored by Philip Currie. She studies dinosaur eggshell structure and the evolution of sensory and reproductive systems in dinosaurs (see Farlow and Brett-Surman, 1997) and did pioneering work on dinosaur nesting sites in China, Mexico, and Korea (Zelenitsky, 2000).

Reptiles (Nondinosaurs). Judy A. **Massare** (retired from the College at Brockport, State University of New York) studies the evolution and paleoecology of Mesozoic fish and marine reptiles, especially ichthyosaurs. Part of her work has been on iconic Mary Anning specimens (e.g., Massare and Lomax, 2014). She received a BS from the University of Rochester and an MA and PhD from Johns Hopkins University supervised by Robert "Bob" Bakker. James (formerly Jane) **Robinson** received a PhD in biology from UCLA and later taught at UC Berkeley and studied plesiosaur anatomy and locomotion. Currently a musician, his songs focus on scientific themes, especially paleontology.

Mammals. Debra K. **Bennett** studied the anatomy and evolution of fossil and living horses. She received a PhD from the University of Kansas and worked at the NMNH before founding the Equine Studies Institute in Bul-verde, Texas. She received SVP's Romer Prize in 1980. Annalisa **Berta** (pro-fessor emerita, San Diego State University, California) began in the 1970s studying fossil carnivores and went on to study and write several books about marine mammal evolution, anatomy, and systematics, especially concerning whales and pinnipeds (e.g., Berta, 2017). Her current research is focused on the anatomy and evolution of feeding, including the transition from teeth to baleen and raptorial to filter feeding in baleen whales. She received a BA from the University of Washington (1974) and a PhD from UCB (1979) under the mentorship of W.A. "Bill" Clemens. She served on the SVP Executive Com-mittee as member at large (1994–1996) and as president (2005–2006). She was the first female co-senior editor of the *JVP* and associate editor of *Marine Mammal Science*. She was elected a fellow of the American Association of Science in 2015. She has mentored the following VP students: P. Adam, M. Churchill, M. Colbert, Lisa Cooper (see later in this chapter), E. Ekdale, T. Harper, S. Kienle, Agnese Lanzetti, J. Martin, M. McGowen, Rachel Rac-icot (see Chapter 6), and M. Voss. Christine **Janis** (professor emerita, Brown University, Providence, Rhode Island) studies the anatomy of ungulates, es-pecially feeding and locomotor adaptations (see Plate 5). She received a BA from the University of Cambridge (1973) and PhD from Harvard University (1979). Janis cowrote the classic (2002) textbook *Vertebrate Life*, now in its 10th edition. She was awarded the George Gaylord Simpson Prize for Pale-ontology in 1985 for her work on the evolution of horns in ungulates. She was elected a fellow of the PS in 2007 and an SVP honorary member in 2019. Janis has mentored the following VP students: M. Carrano, F. Borja Figuei-rido Castillo, Elsa Panciroli, and Jessica Theodor (see later in this chapter). Irina **Koretsky** (Howard University, Washington, DC) works on the anat-omy, evolution, and systematics of pinnipeds, especially phocid seals. She received a PhD from Howard University in 1999. Margaret **Lewis** (Stockton University, Galloway, New Jersey) specializes in the evolution of carnivo-rous mammals. She received a BA from Rice University (1988) and an MA and PhD from Stony Brook University (1988–1995). She served as SVP member at large in 2002–2003. Marisol **Montellano-Ballesteros** (Uni-versidad Nacional Autónoma de México, Mexico City) conducts research on fossil vertebrates from Mexico (see Appendix 2). She received a PhD (1986) from UCB under the mentorship of W.A. Clemens. She has conducted fieldwork on fossil mammals from Cretaceous–Tertiary deposits in Baja California and mainland Mexico. She served as SVP member at large from

2004 to 2005. Virginia **Naples** (Northern Illinois University, DeKalb) is a functional morphologist who studies the anatomy, evolution, and systematics of fossil and living sloths, carnivorans, and cetaceans. She received a BS (1972), MS (1975), and PhD (1980) from the University of Massachusetts. Nancy A. **Neff** studied the systematics and evolution of carnivores, especially cats before becoming a systems and software engineer. She received a BS (1974) from the University of Michigan and a PhD (1982) from the City University of New York and American Museum of Natural History. She is the first recipient of SVP's Romer Prize (1979). Maureen **O'Leary** (Stony Brook University, Stony Brook, New York) conducts research on the evolution, anatomy, and systematics of whales. She is perhaps best known for development of the online database MorphoBank, which enables scientists to study the Tree of Life. She also conducts fieldwork in the Cretaceous and Tertiary of West Africa (Mali). She received a PhD from Johns Hopkins University (1997) under the mentorship of Ken Rose. V. Natalia **Rybczynski** (retired from the Canadian Museum of Nature, Ottawa, Ontario) has conducted research on the paleobiology and paleoenvironment of the Cenozoic High Arctic, including the land-to-sea transition in pinnipeds (see Appendix 2). She received a BS from Carleton College (1994), an MS from the University of Toronto (1996), and a PhD from Duke University (2003). She has conducted fieldwork in Cretaceous and Tertiary deposits of Canada (e.g., Ellesmere Island, Devon Island, Old Crow), Cretaceous of Montana (Milk River) and Utah (Grand Staircase), Eocene (Fayum) of Egypt, Miocene of Bolivia, and Pliocene of Idaho (Hagerman). She has mentored the following VP students: T. Cullen, Danielle Fraser (see later in this chapter), and R.S. Patterson. Gina **Semprebon** (Bay Path University, Longmeadow, Massachusetts) studies the ecomorphology (tooth microwear) of herbivorous mammals. She received a BA (1978) from American University and an MS (1996) and PhD (2002) from the University of Massachusetts. Nancy **Simmons** (AMNH) conducts research on the anatomy and evolution of fossil and extant bats. She received a BA (1981) from Pomona College and PhD (1989) from UCB under the mentorship of W.A. Clemens. Among students that she has mentored are VPs J. Geisler and Karen Samonds (see later in this chapter). She has conducted fieldwork (biology and paleontology) at various sites in the United States (e.g., Arizona, California, Colorado, Montana, Nevada), Tobago, Belize, French Guiana, Malaysia, and Thailand. She received SVP's Romer Prize in 1988. Suzanne **Strait-Holman** (Marshall University, Huntington, West Virginia) studies feeding and dietary adaptations in various mammals. She received a

BA from Hampshire College (1984) and a PhD from Stony Brook University (1991). She received SVP's Romer Prize in 1990. Jessica M. **Theodor** (University of Calgary, Alberta, Canada) studies ungulate diversity through time, focusing on morphologic evolution and environmental changes. Research in her lab is focused on the diets of extinct herbivores, which have been related to changing environments. She received a BA (1989) from the University of Toronto and a PhD (1996) from UCB under the mentorship of W.A. Clemens. She is president-elect of the SVP (2020–2022). Among the VP students that she has mentored are Danielle Fraser, C. Barrén-Ortiz, and C. Zurowski. Su-yin **Ting** (Ding) (Louisiana State University, Baton Rouge) began her work at the IVPP and moved to the United States in the 1990s, where she has worked with Judy Schiebout. She studies the biostratigraphy of Tertiary mammals from the United States (Louisiana) and China. Blaire **Van Valken-burgh**'s (UCLA) research is on the paleobiology and paleoecology of carnivores, notably the quantification of guild structure in fossil carnivore communities and the evolution of feeding adaptations (see Appendix 1). She received a BS from Stockton State College and an MA (1979) and PhD (1984) from Johns Hopkins University. She served on the SVP Executive Committee as member at large from 1997 to 1999 and as president from 2008 to 2010. She was elected a fellow of the PS in 2013. She has mentored the following VP students: P.J. Adam, W. Anyonge, M.F.A. Balisi, W.J. Binder, D. Bird, C.A.-C. Brown, A.A. Curtis, T. Deméré, G.L. Duckler, A.R. Friscia, F. Hertel, Julie A. Meachen (see later in this chapter), B. Pang, T. Sacco, J.X. Samuels, E. Scott, and G.J. Slater. Anne **Weil**'s (Oklahoma State University, Stillwater) research interests include early mammal evolution, phylogeny, and biogeography. She received a BA (1988) from Harvard University, an MA (1992) from the University of Texas, Austin, and a PhD from UCB. She has conducted fieldwork in the Late Cretaceous of various sites in North America. Alisa J. **Winkler** (Southern Methodist University, Dallas, Texas) conducts research on the systematics and paleobiogeography of fossil mammals, especially rodents. She received an MS from the University of Texas, Austin, and a PhD from Southern Methodist University.

Primates and Human Evolution. Laura **MacLatchy** (University of Michigan) conducts research on hominoid origins and primate locomotion. She received a PhD from Harvard University in 1995. Pat **Shipman** (1949–) (retired from Pennsylvania State University) conducted field-based research that involved reconstructing the ecology of ancient environments (focused on taphonomy; e.g., Shipman, 1981). She has also written numerous popular paleontology books about fossil birds, including *Archaeopteryx*, primates, and hominoids, some with her late husband, the paleoanthropologist Alan Walker. She received a BA (1970) from Smith College and an MA (1973) and PhD (1977) from New York University. Mary **Silcox** (University of Toronto, Ontario, Canada) studies the anatomy and systematics of fossil primates, especially plesiadapiformes. She received a PhD (2001) from Johns Hopkins University under the mentorship of Ken Rose. She has mentored the following VPs: O. Bertrand, S. Lopez-Torres, and K.A. Prufrock.

Faunal Studies (Including Biostratigraphy and Paleoecology). Catherine **Badgley** (University of Michigan) conducts research on the structure, ecology, and biogeography of modern and fossil mammalian faunas (see Appendix 1). The common link of her various research projects is the ecological and evolutionary response of biodiversity to global change. She received a BA (1972) from Radcliffe College (Harvard University) and an MS of forest science (1974) and a PhD (1982) from Yale University. She has served on the SVP Executive Committee as member at large (1997–1998), secretary (1999–2002), and president (2007–2008). In 2012 she received the SVP's Joseph T. Gregory Award and she was elected a fellow of the PS. She has mentored the following VP students: J. El Adli, M. Balisi, J. Bloch, D.L. Fox, G. Gunnell, L. Roe, K.T.M. Smiley, M. Smith, and Bian Wang. Rachel **Benton** (Big Badlands National Park, South Dakota) studies the geology and fauna of the White River Badlands, South Dakota. She received a BS from Denison University, an MS from the South Dakota School of Mines, and a PhD from the University of Iowa. Laurie J. **Bryant** (retired, Bureau of Land Management, Utah) was responsible for issuing permits to collect dinosaur and other fossils on public lands in Utah. She conducted field-based research on the Cretaceous–Tertiary extinction event. She received a PhD from UCB under the mentorship of W.A. Clemens. Karen **Chin** (University of Colorado and Colorado Museum of Natural History, Boulder) studies Mesozoic ecosystems, especially trace and body fossils, including coprolites, and has discovered examples of remarkable preservation such as undigested muscle tissue (see Brett-Surman et al., 2012). She received a PhD from the University of California, Santa Barbara (1996), for her work on the paleobiological implications of herbivorous dinosaur coprolites. Chin has popularized dinosaurs through her specialty, writing a general text that intrigues younger audiences (Chin and Holmes, 2005). She has mentored many students, including H. McLean. Jaelyn J. **Eberle** (Plate 30) (University of Colorado, Boulder) pursues research on the evolution of mammals across the Cretaceous–Tertiary boundary and Arctic vertebrate evolution, paleoclimate, and paleobiogeography. She received a BS from the University of Saskatchewan (1991) and a PhD from the University of Wyoming (1996). She has conducted fieldwork in the late Cretaceous and Paleogene of Alaska and Wyoming, the Paleogene of the Canadian High Arctic, the Eocene of British Columbia, and the Late Cretaceous and Paleogene of Colorado and was awarded the Romer Prize in 1995. She has served on the SVP Executive Committee as member at large (2004–2006) and as secretary (2016–2018). Elizabeth **Hadly** (Stanford University, Stanford, California) studies vertebrate diversity through time and space in terms of environmental gradients, using tools such as population genetics, morphology, phylogeography, isotopes, and geographic information system and remote sensing data. She received a BA (1981) from the University of Colorado, Boulder, an MS (1990) from Northern Arizona University, and a PhD (1995) from UCB. She served on the SVP Executive Committee as member at large from 2013 to 2015. Among the VP students and postdocs whom she has mentored are the following: J. Blois, R. Feranac, M. Kemp, S. Newsome, and R. Terry (see later in this chapter). Julia T. **Sankey** (California State University, Stanislaus) studies faunal diversity through time in the context of depositional environments. She received a BS (1987) from Albertson College of Idaho, an MS (1991) from Northern Arizona University, and a PhD (1988) from Louisiana State University under the mentorship of Judy Schiebout. She has led paleontological field expeditions in Big Bend National Park, Texas; North Dakota; and Raton Basin, New Mexico, and coedited a book, *Vertebrate Microfossil Assemblages* (Sankey and Baszio, 2008). Nancy **Stevens** (Ohio University) conducts research on the anatomy (locomotory and feeding adaptations) and paleoecology of various mammals, including primates and carnivores, in Africa and Vietnam. She conducts fieldwork in Madagascar and on the Arabian Peninsula. She received a BS (1992) from Michigan State University, an MPhil (1994) from the University of Cambridge, and an MS (1999) and PhD (2003) from Stony Brook University. Currently she serves on the SVP Executive Committee as member at large (2018–2020). Among VP students whom she has mentored are E. Gorsac and Haley O'Brien (see later in this chapter).

Developmental Biology and Evolutionary Developmental Biology (Evo-Devo). Emily **Buchholtz** (Wellesley College, Wellesley, Massachusetts) studies morphological diversity and development in marine reptiles and mammals (see Appendix 1). She received a BA (1969) from the College of Wooster (1971), an MS from the University of Wisconsin, Madison, and a PhD

(1974) from George Washington University. Ann C. **Burke** (Wesleyan University, Middletown, Connecticut) is an evolutionary morphologist who pursues research on the role of developmental processes in vertebrate evolution. She received a BA from New York University and an MA and PhD from Harvard University. Her VP career began when she was a preparator at the AMNH. After graduate school, she became interested in evo-devo and focused her study on the evolution of vertebrate body plans, especially the turtle. Karen **Sears** (UCLA) conducts research on developmental variation in mammals, especially sensory system development and limb development, and major mammalian evolutionary transitions. She received a BS (1998) from the University of Wisconsin, Madison, and a PhD (2003) from the University of Chicago, mentored by John Flynn. She received SVP's Romer Prize in 2002. Among the students and postdocs whom she has mentored are N. Anthwal, Lisa Cooper, A. Howenstine, and A. Sadier. Kathleen **Smith** (Duke University, Durham, North Carolina) studies skull development in vertebrates, especially mammals. She received a BA (1973) from the University of California, Santa Cruz, and a PhD (1980) from Harvard University. Among the VP students whom she has mentored are Natalia Rybczynski and M. Sanchez-Villagra.

Macroevolution. Felisa **Smith** (University of New Mexico, Albuquerque) pursues research on the evolutionary and ecological consequences of body size in various mammals. She received a BA from the University of California, San Diego (1980), and a PhD (1991) from the University of California, Irvine. Among the students she has mentored are VPs M. Pardi and Meghan Balk.

South America

Ana María **Báez** is an Argentine vertebrate paleontologist in the Department of Geological Sciences, University of Buenos Aires (now retired), who specialized in paleoherpetology. Báez completed her PhD on pipid frogs from the Upper Cretaceous of Salta in 1975 at the University of La Plata, under the mentorship of Rosendo Pascual. Her research highlights herpetofaunas of the Late Cretaceous and the Cenozoic of South and North America. Báez mentored students (e.g., Claudia Marsicano) in the biology and paleontology of fossil and living anurans (Tonni, 2005). She served as SVP member at large from 2007 to 2009.

Zulma Nélida Brandoni de **Gasparini** (1944–) is an Argentine vertebrate paleontologist and senior researcher at CONICET, where she has been since

1972. Gasparini was born and educated in La Plata, Argentina, graduating with a BS in zoology in 1967 and a DSc in 1977 from the University of La Plata. During the early years of her career, she was mentored by vertebrate paleontologists Pascual and José Bonaparte, who reportedly supported her work but also recommended that she not continue given various issues, including the difficulty of undertaking such work as a woman. Thanks to the encouragement of her parents and husband, she persisted, and she is now internationally recognized for her more than 40 years of study of fossil marine reptiles from the Mesozoic of South America. She has collected marine reptiles (i.e., crocodiles, mosasaurs, plesiosaurs, and pterosaurs) from various places around the world ranging from Patagonia to Antarctica. Among the honors she has received was her election as a member of the Argentine National Academy of Sciences (Tonni, 2005).

Reptiles. Andrea B. **Arcucci** is an Argentine vertebrate paleontologist who was a student of José F. Bonaparte (1928–2020) at CONICET and the Facultad de Ciencias Naturales, Universidad Nacional de Tucumán, Buenos Aires. She works at the Universidad Nacional de San Luis, Argentina. With Paul Sereno she described the Middle Triassic dinosaur-like ornithodiran *Marasuchus lilloensis* in 1994. She has also worked on other reptiles, including Late Triassic turtles from South America, thecodonts, and a proterochampsid, working often with compatriot tetrapod specialist Claudia Marsicano, as in their research on early archosaurs (in Weishampel et al., 2004). Marta **Fernandez** is an Argentine vertebrate paleontologist at the Universidad Nacional de La Plata who studies crocodyliforms, ichthyosaurs, and fish. Maria Eduarda de Castro **Leal-Bonde** is a Brazilian vertebrate paleontologist who formerly worked at the Federal University of Ceará, Fortaleza, Brazil, and has now moved to Copenhagen. She conducts research on fish, amphibians, and reptiles. Penelope **Cruzado-Caballero** is an Argentine researcher at CONICET and a professor at the Paleobiology and Geology Research Institute at the Universidad Nacional de Río Negro. She conducts research on the evolution and paleobiology of dinosaurs, especially hadrosaurs. Maria **Paramo Fonseca** is a Colombian paleontologist who, since 2006, has studied various reptile groups, including dinosaurs, ichthyosaurs, and plesiosaurs, at the Universidad Nacional de Colombia in Bogotá. She also helped establish a foundation for the preservation and rescue of fossils in Colombia.

María Guiomar **Vucetich** is an Argentine vertebrate paleontologist, research paleontologist at CONICET since 1971, and professor of vertebrate paleontology at the Universidad Nacional de La Plata, Argentina. She is a recognized specialist on the evolution of caviomorph rodents. Her contributions include work on the phylogeny of these mammals, as well as the climatic, biogeographic, and biostratigraphic implications derived from their study (Tonni, 2005).

USSR (Later Russia)

Olga B. **Afanas'ieva** (1960–) is now chief of the Fossil Fish Section, PIN. Her research interests include the morphology (histology) of early vertebrates (agnathan osteostracans) from the Silurian and Devonian of Russia and the Baltic states. Afanas'ieva, a specialist in paleohistology, investigates the evolution and morphology of the exoskeleton and has described many new taxa. She coedited the fish volume of the 2004 *Fossil Vertebrates of Russia* and gained H. Rausing prizes. Swedish philanthropist Hans Rausing (1926–2019) for many years supported the PIN in its scientific activities, including monograph publications, field activities, personal awards, and stipends, especially in the 1990s (Oleg Lebedev, pers. comm. 14 September 2019).

Margarita A. **Erbajeva** (aka Erbaeva, Erbajaeva) (1938– ; Plate 28) is a principal scientific researcher of the Siberian branch of the Russian Academy of Sciences Institute of Geology at Ulan-Ude and an honorary member both of the SVP (2009) and of the Russian Paleontological Society (2016). Erbajeva first studied in the Department of Biology at Irkutsk State University (1955–1960) and became a PhD student (1962–1965) at the Zoological Institute, Russian Academy of Sciences, Leningrad. Her teacher was Igor Mikhailovich Gromov (1913–2003), son of Vera Gromova (see Chapter 3). She gained her DS in biology at the PIN in 1991. In her long scientific career, she has taken part in many joint expeditions, such as the Raleigh International Expeditions to Siberia in 1993 and 1994 as a scientific adviser, the 1994–1996 joint Russian-French project "Environmental Dynamics during Pliocene-Pleistocene in Transbaikalia," and the joint Kazakh-American project (1995–1998) on Zaisan basin paleontology. More recently she has been the Russian principal investigator on the joint Chinese-Russian project (2008–2010) "Evolution, Biodiversity and Dispersal Events of Ochotonids and Arvicolids during the Pliocene in Northern China and Baikalian Region in the Context of Global Change" and participated in the joint Austrian-Mongolian project on the geology and paleontology of Mongolia. Her scientific interests encompass the systematics, phylogeny, and evolution of small mammals of East Siberia, including extant and extinct ochotonids and leporids and their implications for paleoenvironmental and climatic reconstruction.

Another student of I. M. Gromov, Nina Semenova **Shevyreva** (1931–1996), was born in Moscow. After completing school, she started as a preparator at the PIN. After gaining her MS on recent rodents at the Biology Faculty of MSU through evening courses after the war, she got a research position at the PIN. She studied mainly fossil rodents for 30 years (1961–1991), the subject of her PhD, including squirrels from the Paleogene of Asia—Mongolia, central Asia, and Kazakhstan. Shevyreva discovered at least 50 new localities, as well as mixodontes, lagomorphs, and the first marsupial fossil of Asia (Gabunia et al., 1986; Agadjanian et al., 1998).

Inessa [Innessa] Anatolyevna **Vislobokova** (aka Visloberkova) (1947–) graduated from the Geological Faculty of MSU in 1970 and attended graduate school at the Siberian branch of the Russian Academy of Sciences Institute of Geology, Novosibirsk (1973), gaining her candidate thesis in in 1974. She moved to the PIN in 1973 and earned her doctorate of biological sciences in 1990 on the fossil deer of Eurasia, including Kazakhstan and Mongolia. Vislobokova studies the evolution and ecology of artiodactyls and horses and has been awarded two H. Rausing prizes (1998, 2002).

21st Century (2001 to the Present)

Women more recently active in VP (see Table 5.2) are emerging leaders in our field. The future of paleontology and use of the fossil record is multifaceted, and there is growing recognition of the wide range of research questions in many disciplines that can benefit by incorporating data from the fossil record (Jablonski and Shubin, 2015; Reisz and Sues, 2015). In the 21st century, developmental studies, and evo-devo in particular, is one growing area. Sánchez (2012), for instance, recounts the history and recent work of several women (e.g., Cécile Poplin, Moya Smith, q.v.). Another new area is conservation paleontology, which fits well into the unfolding UNESCO Global Geoparks program and World Heritage, where women work in sites such as the Jurassic Coast, Dinosaur Provincial Park, and Messel Pit, Germany.

Here, we list many women of the 21st century, providing more detail on a few; all are in the early years of their VP careers. Information came mainly from personal knowledge or communications and faculty profiles and other webpages. Given space considerations, we cannot list all those women who are now PhD students or postdoctoral researchers in VP, although many are listed in the online database (online appendix S1).

Africa

Jennifer **Botha-Brink** (National Museum, Bloemfontein, South Africa) studies the paleoecology and extinction events of Permian–Triassic vertebrates, especially therapsids, using histology and stable isotope analyses. Elizabeth **Gomani-Chindebvu** (director of culture, Malawi) received an MS and PhD at Southern Methodist University in Dallas, where she studied reptiles (dinosaurs and crocodiles), mentored by Louis Jacobs. Sanaa **El-Sayed** and other Egyptian students (Mansoura University Vertebrate Paleontology Center, Egypt) conduct research on the stratigraphy, sedimentology, and paleontology of Mesozoic vertebrate faunas, mainly fish and dinosaurs from Egypt. El-Sayed received an MS and PhD from Mansoura University, Mansoura, Egypt. She is currently director of the Mansoura University Vertebrate Paleontology Center and a lecturer at the university. El-Sayed was a member of an Egyptian team of scientists funded by the National Geographic Society that collected, prepared, and described an exceptionally preserved marine catfish from Wadi Al-Hitan, also known as the Valley of Whales, in Egypt. She is the first woman VP from the Middle East to be senior author on an internationally published paper. El-Sayed received the SVP Patterson Award in 2018. Romala **Govender** (Iziko South African Museum, Cape Town, South Africa) works on the anatomy and evolution of marine mammals and South African faunas (see Appendix 2). Emma **Mbua** (National Museums of Kenya, Mount Kenya University) studies hominid anatomy and evolution. Lovasoa **Ranivoharimanana** (University of Antananarivo, Madagascar) studies the functional morphology of various Triassic reptiles and mammals.

Asia

Armenia. Ruzan **Mkrtchyan** (Yerevan State University) conducts research in paleontology and archaeology. She has conducted fieldwork in Plio-Pleistocene deposits of Armenia. She was the 2011 recipient of the SVP Program for Scientists Award.

China. Among the many new young women vertebrate paleontologists in China, Jing **Lu**, Yu-zhi **Hu** (now at the Australian National University, Canberra), and Tuo **Qiao** are in Zhu Min's paleoichthyology lab and Donglei **Chen** is currently in Per Ahlberg's Uppsala University lab. All are studying the new early vertebrates and stem gnathostomes being found in Yunnan, Xinjiang, and elsewhere in China. **Mao** Fang-yuan is studying Jurassic mammals from the Yanliao Biota, Inner Mongolia, and she has a team exploring the Upper Cretaceous at Bayanmandehu.

India. Only students with an MS in geology can now study paleontology via the Indian Statistical Institute. Sanjukta **Chakravorti** (Indian Statistical Institute) studies the morphology and evolution of Mesozoic amphibians in her country.

Japan. Naoko **Egi** (Kyoto University) studies the functional morphology and systematics of various mammals (primates, carnivores, ungulates). She received a PhD from Johns Hopkins University under the mentorship of Ken Rose. Yuri **Kimura** (National Museum of Science and Nature, Tokyo) received her PhD from Harvard University in 2013. She studies the systematics and paleoecology of rodents and is currently working on a new town mascot, based on the aquatic fossil desmostylid *Paleoparadoxia*. Ryoko **Matsumoto** (Kanagawa Prefectural Museum of Natural History) was mentored by and often works in cooperation with Susan Evans. She mainly works on champsosaurs. She received the SVP Estes Grant in 2011.

Mongolia. Bolortsetseg "Bolor" **Minjin** (Ulaanbaatar and New York Institute of Technology) conducts research on the paleobiology of Cretaceous dinosaurs and early mammals of Mongolia (see Chapter 6 and Appendix 1).

Russia. Galina V. **Zakharenko** (PIN) studies paleoecology and vertebrate assemblages, especially placoderm fish from the Devonian of Russia.

Australia and New Zealand

The 21st century has seen a burgeoning of young researchers particularly in Australia (Turner and Berta, 2019); here we highlight a few. Karen **Black** (UNSW) specializes in Riversleigh's mammal-rich Cenozoic faunas in northwestern Queensland. Her interests lie in ontogeny and biocorrelation, especially of vombatomorph marsupials (including koalas and diprotodontoids). French Canadian Catherine Anne **Boisvert** (Curtin University, Perth) studies the early evolution of jawed vertebrates and shark evo-devo. Alice **Clement** (Flinders University, Adelaide) is studying the anatomy, evolution, and interrelationships of Devonian lungfish and stem tetrapods using three-dimensional tomography. Katarzyna J. "Kat" **Piper** (Monash University, Melbourne) studies the anatomy and systematics of Australian Pleistocene mammals, especially kangaroos. Vera **Weisbecker** (University of Queensland, Saint Lucia) uses data-collection methods, including CT scanning, dissection, morphometrics, histology, and chemical staining, to study Australian mammal brain evolution and development.

Carolina **Loch** S. Silva (University of Otago, Dunedin, New Zealand) is a Brazilian-born scientist working on the anatomy and evolution of the teeth of modern and fossil marine mammals (see Appendix 2).

Europe

Denmark. Ane Elise **Schroeder** (Danish Natural History Museum, Copenhagen) is studying exquisitely preserved bony fish fossils from the Early Eocene Fur Formation in Denmark, an untapped asset, to understand the diversification of teleosts. Mette E. **Steeman** (Museum of Southern Jutland, Denmark) studies the anatomy, evolution, and phylogenetics of fossil and extant whales and currently holds an administrative position at the museum.

Finland. Sirpa **Nummela** (Helsinki University, Finland) studies sensory evolution in various fossil and extant mammals, including pinnipeds and whales.

France. Virginie **Bouetel** (MNHN) studies the anatomy and evolution of whales. Alexandra **Houssaye** (MNHN) studies mammal bone histology, biomechanics, and three-dimensional morphometrics. She received a PhD (2009) from the MNHN and did postdoctoral research at Bonn University, Germany (Humboldt Fellowship), and the European Radiation Facilities in Grenoble, France. Her current research, funded by the European Research Council, is focused on how bone adapts to heavy weight (graviportality).

Germany. Julia M. **Fahlke** (formerly Museum für Naturkunde, Berlin, Germany) has studied whale anatomy and evolution using three-dimensional technology. Nadia **Fröbisch** (Museum für Naturkunde, Berlin, Germany) studies the evolution and ontogeny of fossil and extant amphibians. She received SVP's Romer Prize in 2006. Anneke H. van **Heteren** (Zoologische Staatssammlung München, Munich, Germany) studies the functional morphology of birds and mammals. Gertrud **Roessner** (Ludwig Maximilians Universität, Munich) is curator of fossil mammals and conducts research on fossils artiodactyls. Irina **Ruf** (Senckenberg Research Institute, Frankfurt am Main, Germany) studies the ear anatomy and evolution of early mammals.

Poland. Ewa **Barycka** (1979–2008) (Polish Academy of Sciences) had realized her childhood dream and was studying the evolution and systematics of feliform carnivores before her untimely death (Wolsan, 2008).

Spain. Gloria **Cuenos-Bescos** (University of Zaragoza, Spain) studies the paleoecology and systematics of dinosaurs and Mesozoic mammals from Mexico, Spain, and Portugal.

United Kingdom. Pip **Brewer** (NHM) studies Australian and South American fossil mammals, Mesozoic mammals, and the history of science. Rachel **Frigot** (Keele University) studies the biomechanics of pterosaurs. Samantha "Sam" **Giles** (University of Birmingham) conducts research on the brains and cranial anatomy of extant and fossil ray-finned fish using CT scans. Her research helped to debunk the commonly held assumption that sharks were primitive fish. Giles found that our common ancestors are not like sharks but instead like bony fish (Osteichthyans) originating around the end of the Silurian. Comparing the brains of fossil and modern actinopterygian (ray-finned) fish, she and colleagues claim that this group, which includes the most vertebrate species that have ever been, are some 40 million years younger than previously known, thus questioning the earlier evolutionary history of this group (Giles et al., 2017). She received a BS from Bristol University (2011) and a PhD from the University of Oxford (2015), mentored by Matt Friedman. She was awarded a L'Oréal-UNESCO fellowship in 2016 and recognized as a "rising star" by this organization in the following year. She received the GSL's Lyell Fund in 2018, which is awarded for outstanding published research. Susannah **Maidment** (NHM) studies the evolution and paleobiology (especially locomotion and the evolution of quadrupedality) of ornithischian dinosaurs. She has worked extensively on stegosaurs and is recognized as a world expert on this group. She is also pursuing research using geology to better understand spatial sampling biases in the fossil record. She received an MS (2003) from Imperial College London and a PhD (2007) from the University of Cambridge, mentored by Paul Upchurch and David Norman. She received the Hodson Award of the Palaeontological Association in 2016 and the GSL's Edmund Johnson Garwood Fund in 2018. Maria E. **McNamara** (University College, Cork, Ireland) studies the paleobiology of exceptionally preserved fossils (e.g., pterosaur integument, fossil vertebrate skin colors). Elsa **Panciroli** (National Museum of Scotland and University of Edinburgh, Scotland) conducts research on the anatomy, ecology, and locomotion of Mesozoic mammals and the origins of modern ecosystems. She employs CT and other digital imaging methods. She also writes a blog about science, *Giant Science Lady: Ramblings of a Scottish Palaeontologist.*

South America

Argentina. Monica R. **Buono** (CONICET) studies the anatomy, systematics, and paleobiology of whales. Julia Brenda **Desojo** (CONICET) works on the anatomy, phylogeny, and biostratigraphy of Triassic archosauriformes, especially pseudosuchians. Mercedes Grisel **Fernandez** (Universidad Nacional de Buenos Aires) studies the anatomy and systematics of South American mammals, such as notoungulates. Mariela Soledad **Fernandez** (CONICET) studies dinosaur reproduction. Gabriela **Fontararrosa** (CONICET) studies the anatomy and evolution of lizards. Malena **Lorente** (Museo de La Plata) studies the evolution and systematics of various South American mammals. Juliana **Sterli** (CONICET) studies turtle anatomy and evolution. Her research has shown that heterochronic changes (paedomorphosis) have been important in the early evolution of turtles. She received a PhD in 2004 from the Museo de Historia Natural de San Rafael, Mendoza, Argentina. Her research has been supported by the National Geographic Society. She received the Estes Grant in 2006. Evelyn **Vallone** (CONICET) studies the anatomy, systematics, and evolution of Pleistocene bony fish, including the South American lungfish. Barbara **Vera** (CONICET) studies South American mammals (notoungulates). Virginia **Zurriaguz** (Universidad Nacional de Río Negro) studies the evolution and anatomy of dinosaurs.

Brazil. Luciana **Barbosa de Carvalho** (Brazilian National Museum) studies crocodylomorphs and dinosaurs. Annie S. **Hsiou** (Universidade de São Paulo) studies fossil lizards. Ana B. **Quadros** (Museo de Zoologia, Universidade de São Paulo) studies the anatomy, evolution, and systematics of snakes. Taissa **Rodrigues** (Universidade Federal do Espírito Santo) studies pterosaur diversity, anatomy, and systematics. She received a BS (2004) and MS (2007) from the Universidade Federal de Minas Gerais and a PhD (2011) from the Museu Nacional/Universidade Federal do Rio de Janeiro. She co-organized a workshop on women in paleontology at the 2019 SVP meeting in Brisbane, Australia. Daniela **San Felice** (Fundação Zoobotânica do Rio Grande do Sul) conducts research on fossil and extant pinniped morphometrics.

Chile. Carolina S. **Gutstein** (Universidad de Chile) studies the functional morphology and phylogeny of toothed whales. Ana **Valenzuela-Toro** (currently a PhD student at the University of California, Santa Cruz) conducts research on the evolution of various marine mammals, especially pinnipeds and whales (see Appendix 1).

United States, Canada, and Mexico
Victoria M. **Arbour** (Royal British Columbia Museum) studies the paleobiology of armored dinosaurs. She received a BS (Hons) from Dalhousie University, Halifax, in 2006. She completed an MS (2009) and PhD (2014) at the University of Alberta, both under the mentorship of Philip Currie. Arbour

has studied biomechanical analyses of ankylosaur tail clubs. Currie credits Arbour for involving the paleontology discipline in the University of Alberta's Women in Scholarship, Engineering, Science and Technology program, making the study of dinosaurs more appealing to women. Amy M. **Balanoff** (Johns Hopkins University) studies the phylogeny of maniraptorian dinosaurs using CT imagery (see Chapter 6). Holly Woodward **Ballard** (Oklahoma State University, Stillwater) studies dinosaur bone histology. Jessica **Blois** (University of California, Merced) conducts research on Quaternary ecology, biogeography, and phylogeography. Kristen **Brink** (University of British Columbia, Vancouver) conducts research on the geochemistry of teeth and paleoecology. Lisa G. **Buckley** (Peace Region Palaeontology Research Centre, Tumbler Ridge, British Columbia) conducts research on the tracks and traces of dinosaurs and birds. She was mentored by Philip Currie. Sara **Burch** (State University of New York at Geneseo) studies forelimb evolution and morphometrics and muscle reconstruction in theropod dinosaurs. Amy E. **Chew** (Brown University, Providence, Rhode Island) studies mammal evolution and paleoecology. She received a PhD from Johns Hopkins University under the mentorship of Ken Rose. Kerin **Claeson** (Philadelphia College of Osteopathic Medicine) researches the anatomy and evolution of Tertiary sharks and rays, especially from the Eocene Monte Bolca Lagerstätte. She received a PhD from the University of Texas, Austin, mentored by Christopher Bell. Lisa N. **Cooper** (Northeast Ohio Medical University, Rootstown) studies the development, evolution, and skeletal biology of bats. She received a PhD from Northeast Ohio Medical University, mentored by J.G.M. Thewissen. Larisa R.G. **DeSantis** (Vanderbilt University, Nashville, Tennessee) received a PhD from the University of Florida, mentored by Bruce MacFadden. She studies carnivores and paleoecology. She currently serves as SVP member at large (2019–2022) (see Appendix 1). Catherine **Early** (University of Florida, Gainesville, and the Dave Blackburn Lab at the Florida Museum of Natural History) studies the comparative cranial anatomy of fossil and modern birds using CT imagery (see Appendix 1). Kena **Fox-Dobbs** (University of Puget Sound, Tacoma, Washington) studies the paleoecology of mammals using biogeochemical approaches (e.g., isotopes). Katrina **Gobetz** (James Madison University) studies fossil burrows including those of lungfish and mammals. She received SVP's Romer Prize in 1999. Eugenia **Gold** (Suffolk University, Boston, Massachusetts) studies theropod neurobiology and has increased the visibility of women in paleontology with her children's book *She Found Fossils* (see Bibliographic Sources and Further

Reading). Penny **Higgins**'s (University of Rochester, New York) research focus is the isotopes and geochemistry of fossil mammal teeth and their paleoenvironments (see Plate 40). Funded by the National Science Foundation (NSF), Samantha **Hopkins** (University of Oregon, Eugene) conducts field-based research on the paleoecology of the Oligo-Miocene of Oregon using rodent biostratigraphy. She has a major research project examining the evolutionary relationships among diet, body size, and natural history in living mammals. She received a BS (1999) from the University of Tennessee and a PhD (2005) from UCB and did a postdoctoral fellowship at the National Evolutionary Synthesis Center in North Carolina. Among VP students whom she has mentored are the following: W. McLaughlin, D. Reuter, P. Barrett, and J. Orcutt. She currently holds an SVP Executive Committee position as secretary (2018–2021). ReBecca **Hunt-Foster** (Dinosaur National Monument, Utah and Colorado) studies dinosaurs and their tracks from North America (see Appendix 1). Katrina Elizabeth **Jones** (Harvard University) is a functional morphologist who studies the functional, phylogenetic, and developmental factors that drive evolution. She is currently a member of the Stephanie Pierce Lab, investigating the thoracolumbar anatomy of synapsids. She received a PhD from Johns Hopkins University under the mentorship of Ken Rose and she was awarded SVP's Romer Prize in 2014. Emily **Lindsey** (George C. Page Museum, La Brea Tar Pits) pursues research on Pleistocene and Quaternary mammal extinctions in North and South America. Kate **Lyons** (University of Nebraska, Lincoln) studies how global climate change affects fossil and modern mammalian diversity patterns. Julie **Meachen** (Des Moines University, Iowa) is a functional morphologist and paleoecologist with research interests focused on Pleistocene megafauna. She received a PhD (2008) from UCLA, mentored by Blaire Van Valkenburgh. Following a postdoctoral fellowship at the National Evolutionary Synthesis Center in Durham, North Carolina, she now leads an NSF-funded investigation of Natural Trap Cave in Wyoming, where she is excavating to determine the role that climate change plays in both the morphology and the genetics of the Ice Age mammals from the site. Her other projects relate to Pleistocene mammals, including dire wolves and saber-toothed cats of Rancho La Brea and a frozen Beringian wolf puppy from the Yukon, Canada. Ashley **Morhardt** (Washington University, Saint Louis, Missouri) studies dinosaur brains and vertebrate soft tissues using iodine-based contrast-enhanced computed tomography. Haley **O'Brien** (Oklahoma State University, Stillwater) studies the impact of thermoregulatory cranial vasculature in large-bodied mammals using CT imagery. Her macroevolutionary projects use techniques such as mathematical algorithms and ecological niche modeling to examine how an organism's anatomy has influenced its response to climate change. She received a BS (2009) from the College of Charleston, South Carolina, and a PhD (2016) from Ohio University, Athens, mentored by Larry Witmer and Nancy Stevens. She was awarded SVP's Romer Prize in 2016. Jingmai **O'Connor** (Plate 31) (IVPP) is a Chinese American vertebrate paleontologist and a professor at IVPP. She studies the origin and early evolution of birds, including feathers, flight, development, systematics, taxonomy, and paleohistology. She was mentored by Donald Prothero as an undergraduate student at Occidental College (BA, 2004) and received a PhD (2009) from the University of Southern California, mentored by Luis Chiappe. O'Connor has conducted fieldwork in six countries and is recognized as one of the world's leading experts on Mesozoic birds (see Appendix 1). Jennifer C. **Olori** (State University of New York at Oswego) studies the evolution of development in early tetrapods. She received a BS from Cornell University and a PhD from the University of Texas, Austin, mentored by Christopher Bell. She received the SVP's Estes Memorial Grant (2007) and the Romer Prize (2010) for her research on growth and development in salamander-like extinct microsaurs. Tonya **Penkrot** (Arizona State University, Tempe) studies the functional anatomy of Paleocene and Eocene mammals. Stephanie E. **Pierce** (Harvard University) is a functional morphologist reconstructing the three-dimensional anatomy and locomotion potential of early tetrapods to test hypotheses of limbed movement across the water-land transition. She received a BS and MS from the University of Alberta and a PhD from the University of Bristol. She has conducted fieldwork collecting early tetrapods in West Virginia in collaboration with Cambridge University and the TWeed Project (see the discussion of Jenny Clack). She has mentored the following VP students and postdocs: K. Jones, T. Simões, B. Dickson, P. Lai, Z. Morris, and M. Wright. Kari **Prassack** (Hagerman Fossil Beds, Idaho) studies carnivore taphonomy and ecology and human evolution. Karen **Samonds** (Northern Illinois University, DeKalb) studies the origin and evolution of Madagascar's vertebrate fauna, including life history variation in extinct primates. With Marisol Montellano-Ballesteros (see earlier discussion), Claudia **Serrano** (Institute of Geology, Universidad Nacional Autónoma de México, Mexico City) and Belinda Espinosa **Chávez** (Paleontology Museum, Benemérita Escuela Normal de Coahuila) are working on the paleobiology, paleoecology, and taphonomy of Cretaceous dinosaurs of Mexico, with special emphasis on the biological and geological processes that control their preservation. Beth **Shapiro** (University of California, Santa Cruz) conducts research on how populations and species change through time. Her VP interests are focused on the use of ancient DNA (e.g., of mammoths) to understand environmental change, as well as the evolution of hominins. Michelle **Spaulding** (Purdue University North Central, Westville, Indiana) studies the phylogeny and evolution of carnivores. She received a PhD from Columbia University, mentored by John Flynn. Michelle **Stocker** (Virginia Tech University, Blacksburg) studies the evolutionary history of reptiles. She received a PhD from the University of Texas, Austin, mentored by Christopher Bell. Rebecca **Terry** (Oregon State University, Corvallis) investigates the structure and dynamics of modern to ancient terrestrial populations and communities in response to climate change and anthropogenic land use. She was awarded SVP's Romer Prize in 2007. Allison R. **Tumarkin-Deratz** (Temple University, Philadelphia) studies bone microstructure and growth patterns in fossil and modern amniotes, principally archosaurs. Gina **Wesley** (Montgomery College, Montgomery County, Maryland) studies the paleobiology of carnivores. She received a PhD from the University of Chicago under the mentorship of John Flynn. The research interests of Abagael **West** (University of Pittsburgh, Pennsylvania) include mammalian morphology and macroevolutionary patterns, the timing of major placental radiations and innovations, and biogeography and vertebrate evolution on Gondwanan landmasses. She received a PhD from Columbia University–AMNH, also mentored by Flynn. Robin **Whatley** (Columbia College, Chicago) studies the evolution and paleoecology of early mammals and has helped to promote fellow women VPs in their endeavors, along with Alison Beck and Karen Sears (Whatley and Beck, 2016).

Summary

By the late 20th and early 21st centuries, women VPs had increased significantly in number and were pursuing research on fossil vertebrates around the world. During this time, VP expanded from its geologic roots, becoming more biological with the integration of other subdisciplines, including development, functional anatomy and biomechanics, genomics, and conservation paleontology, showcasing new techniques (e.g., morphometrics, computer modeling, and stable isotopes). Women VPs today occupy various roles as researchers and teachers, documenting the patterns and processes of evolution through study of the fossil record and investigating species, clades, and ecosystems. The rise of women VPs in Asia and Africa is notable. Highlights include the research of Anna "Kay" Behrensmeyer, who is widely known for her field-based taphonomic research, and the late Jenny Clack, who is recognized for her landmark research on the origin of tetrapods and the transition from water to land. Kate Trinajstic and Pat Vickers-Rich have made pioneering field discoveries of Australian fossil vertebrates, especially fish (placoderms), dinosaurs and birds.

Plate 1. SVPCA meeting group photo at the Queen's University of Belfast, Northern Ireland, 1986. Boldface names identify the women members of the group. *Front row from left:* Maggie **Rowlands**, A. Kitchener, Denise **Sigogneau-Russell**, Pat M. **Ferguson-Lees**, Jenny **Clack**, Dave S. Brown, Stan Wood, R.H. Reid, Sheila Mahala **Andrews**, Bob Savage, Alec Panchen, Angela **Milner**, Alan Charig, Percy Butler, Brian Gardiner, Arthur Cruickshank, John Martin, Sally V.T **Young**, Susan **Gay**. *Second row from left:* Sue **Evans**, A.D. Wright, Don Russell, Frances **Mussett**, R.P.S. Jeffries, Rob Clack, Alan Gentry, P. Dullenmeij, Bill Clemens, Phil Doughty, N. Monaghan, Eric Buffetaut, Colin Patterson, L. Beverly Halstead, Tom Kemp, Gillian M. **King**, Michael Benton, S. **Banerjee**, Dave Norman, A. **McCabe**, T. **Gale**, B. **Noddle**. *Back row from left:* E.F. Freeman, Douglas Palmer, Jim Hopson, J.-J. Jaeger, Haiyan **Tong** (Mme Buffetaut), J.-M. Mazin, Andrew Milner, Nick Fraser, Henry Gee, Dave Unwin, Per Ahlberg, G. Dickson, P. Spencer, N. Court, T. Suzuki. *Source:* Courtesy of SVPCA Richard Forrest and Michael J. Benton.

Plate 2. First VP meeting in Australia (Gondwana Symposium, held in Canberra 1973 under the auspices of the International Union of Geological Sciences), with the names of women VPs and wives in bold. *Back row from left:* Alan Thorne, Eileen **Finch**, Neville Pledge, Jeanette **Hope**, Alan Charig, Michael Plane, Al S. Romer, Bruce Erickson. *Front row from left:* Anne **Warren**, Gavin Young, Ruth **Romer**, Geoff Hope, Robert Jones, Alex Ritchie. *Source*: Courtesy of Alex and Shona Ritchie.

Plate 3. CAVEPS meeting in Sydney, Australia, March 1989, with group photo at the Australian Geographic headquarters. Boldface names identify the women members of the group. *Front row from left:* Colin Groves, Jeanette **Muirhead**, Rhys Walker, unknown, Corrie **Williams**, Dietland **Knuth**, Coral **Gilkeson**, Sue **Creagh**, Anne **Warren**, Sue **Hand**. *Back row from left:* Tony Thulborn, Michael Loy, Zhang Gue-Rui, Gavin Young, Robert Jones, Tim Hamley, Susan **Bergdolt**, Bernie Cooke, Brian Mackness, Miranda **Gott**, Mike Durant, John Barry, Walter Boles, Neville Pledge, Julie **Barry**, Arthur White, Alex Ritchie, Jim Laverack, Mary **White**, Sue **Laverack**, Pat **Vickers-Rich**, unknown, Tom Rich, unknown, John Long, Henk Godthelp, Peter Murray, Paul Willis, Michael Archer, John Scanlon. *Source:* Photo from Vickers-Rich and Archbold (1991, fig. 23), courtesy of Patricia Vickers-Rich and *Australian Geographic*.

Plate 4. Mary Anning and hammer (from Turner et al., 2010a). Print modified from the portrait in the Natural History Museum Library. *Source*: Photo © The Trustees of the Natural History Museum.

Plate 5. Christine **Janis** taking on the persona of Mary Anning, at the SVPCA in 2018. *Source*: Courtesy of Christine Janis and photographer Michael Streng.

Plate 6. Etheldred Benett, silhouette, made ca. 1837 (taken from Torrens, 1985). *Source*: Reproduced courtesy of the Geological Society of London.

Plate 7. Barbara Rawdon Hastings, 1828, portrait at the age of about 17, by Thomas Anthony Dean (1801–1860), after Emma Eleanora Kendrick. *Source*: Photo © National Portrait Gallery, London. CC BY-NC-ND 3.0.

Plate 8. Mary Buckland, photograph by unknown, ca. 1850. *Source*: Buckland Family archive held by Phyllis D.E. Cursham (née Buckland), from Burgess, 1900.

Plate 9. *Eugnathus philpotae* Ag. (Agassiz, 1833–1844, vol. 2, plate 52; taken from Kölbl-Ebert, 2002). *Source*: Creative Commons.

Plate 10. Hester Lucy Stanhope, portrait by Robert Jacob Hamerton, printed by Charles Joseph Hullmandel, published by Richard Bentley, lithograph, ca. 1830s. *Source*: Wikipedia, © National Portrait Gallery, London.

Plate 11. Dorothea Bate in the field with a workman in the pit at Bethlehem, ca. 1937. A fossil elephant tooth can be seen just behind the man's right arm. *Source*: Copyright Natural History Museum, London, used with kind permission of the NHM Library & Archives. With thanks to Karolyn Shindler for providing the digital image.

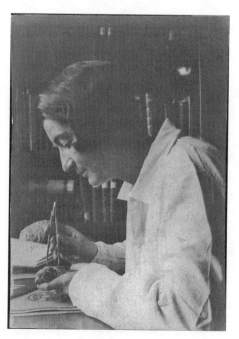

Plate 12. Tilly Edinger in the Senckenberg Museum, Frankfurt, October 1926. *Source*: From the Archives of the Museum of Comparative Zoology, Ernst Mayr Library, Harvard University.

Plate 13. Antje Schreuder, Dutch paleomammalogist, photo by Cor van Weele (from Det and Hoek-Ostende, 2002). *Source*: © Naturalis Biodiversity Center Archives, Leiden

Plate 14. Annie Alexander in the field, photographed by Louise Kellogg. *Source*: Copyright University of California Museum of Paleontology and the Regents of the University of California. Reproduced with permission.

Plate 15. Fanny Rysam Mulford Hitchcock. *Source*: Ewing, 2017. From the Collections of the University of Pennsylvania Archives.

Plate 16. Mignon Talbot around the time she found her dinosaur specimen. Photo taken by Asa Kinney in 1911 or earlier. *Source*: Mount Holyoke College Archives and Special Collections.

Plate 17. Vera Isaacovna Gromova, from Nalivkin, 1979. *Source*: Courtesy of Nauka, Russian Academy of Sciences.

Plate 18. SVP charter meeting photo at lecture room in the Biological Laboratories Harvard University on December 28, 1940 (see Wilson, 1990). *Left to right:* E.H. Colbert, unknown, G.G. Simpson, C.W. Gilmore, unknown, W. Granger, C.H. Falkenburgh, R.V. Witter, A.S. Romer, R.G. Chafee, Tilly **Edinger**, R.M. White, R.H. Denison, G.L. Jepsen, H.E. Wood, B. Schaeffer, H.J. Sawin, unknown (probably Anne **Roe Simpson**), W.E. Moran, T.E. White, J.L. Kay, C.B. Schultz, F.C. Whitmore. *Source*: Photographer probably Nelda **Wright**. Courtesy of SVP Archives.

Plate 19. SVP Charter member Carrie Barbour preparing an *Apatosaurus* dinosaur femur from the Morrison Formation of Wyoming in 1901 (photo by Erwin Barbour). *Source*: Courtesy Archives and Special Collections, University of Nebraska–Lincoln Libraries and University of Nebraska State Museum.

Plate 20. SVP Charter member Hildegarde Howard measuring bird bones at LACM. *Source*: © The National History Museum of Los Angeles County Archives.

Plate 21. SVP Charter member Grace Russell with her husband, Loris Shano Russell, in 1986 at the Albertosaurus quarry near Drumheller, Alberta, Canada. Photo modified from Maurice Stefanuk. *Source*: Courtesy of the Royal Ontario Museum.

Plate 22. Mary Leakey in the field in Olduvai Gorge, Tanzania, Africa. Photo by Des Bartlett. *Source*: Courtesy of the late Andrew P. Hill.

Plate 23. SVP foreign member Dorothy Rayner. *Source*: Photographer unknown. Courtesy of the School of Earth and Environment Photo Archive, University of Leeds.

Plate 24. Estonian Elga Mark-Kurik in the field at the entrance to the Arukula caves, Latvia, in 1950 (photo from the late Elga Mark-Kurik). *Source*: Photographer unknown. Courtesy of Oleg Lebedev.

Plate 25. Patricia Vickers-Rich in the field in Patagonia, Argentina. *Source*: Photo by Lesley Kool for Patricia Vickers-Rich.

Plate 26. The late Jenny Clack at Willie's Hole (Paleozoic) excavation, Scotland, 2015. *Source*: Courtesy Andy Ross. Permission of National Museums Scotland.

Plate 27. Anna "Kay" Behrens-meyer at the late Pleistocene Nyokie fossil site in the southern Kenya rift valley, Africa, 2016. *Source*: Photograph by Jennifer Clark.

Plate 28. Margarita Erbajeva (*third from right*) with her colleague Alexandr Shchet-nikov and diploma students in the field in Mongolia, July 2018. *Source*: Courtesy of Margarita Erbajaeva.

Plate 29. Julia Clarke on Vega Island, James Ross Archipelago, Antarctica, 2011. *Source*: Photo by Dr. Jin Meng.

Plate 31. Jingmai O'Connor in the field at the Jurassic Elliott Formation in South Africa. *Source*: Courtesy of Wenje Zheng.

Plate 30. Jaelyn Eberle at the Strathcona Fiord fossil forest on central Ellesmere Island in Canada's High Arctic, July 2010. *Source*: © Jaelyn Eberle.

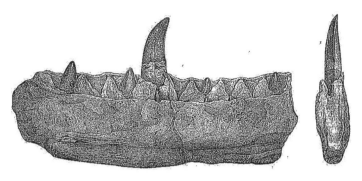

Plate 32. Theropod dinosaur *Megalosaurus* lower jaw drawn by Mary Buckland. *Source*: Buckland, 1824.

Plate 33. Orra White Hitchcock, ca. 1860, by John L. Lovell. *Source*: Courtesy Archives and Special Collections at Amherst College, MA.

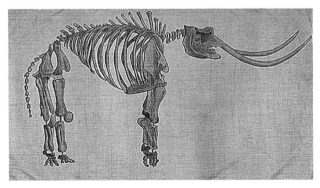

Plate 34. Pen-and-ink and watercolor of *Mastodon maximus* Cuvier, 1812, by Orra Hitchcock, original 25.5 in. × 37 in. *Source*: Courtesy of Amherst College Archives & Special Collections.

Plate 35. Cretaceous frog (*Beelzebufo ampinga*) with largest extant Malagasy frog (*Mantidactylus guttulatus*). *Source*: Luci Betti-Nash, scientific illustrator, Department of Anatomical Sciences, Stony Brook University.

Plate 36. Margaret Matthew Colbert in her studio in Arizona in 1993. *Source*: Courtesy of Matthew Colbert.

Plate 37. Joanne Wilkinson in the field in 2016, excavating a Pliocene macropod limb, Lion Hill Locality, Chinchilla Rifle Range, western Darling Downs, South East Queensland, Australia. *Source*: Courtesy of Joanne Wilkinson and photograph by Kristen Spring.

Plate 38. Maria Louk'yanova during the 1958 PIN excavation for Upper
Permian vertebrates, Ocher, Cisurals, Russia.
Source: Courtesy of Orlov Paleontological Institute, Russian Academy of
Sciences, Moscow; modified from Farlow and Brett-Surman (1997, fig. 4.3).

Plate 39. Bolor Minjin and team excavating a dinosaur in Mongolia. *Source*: Courtesy of Jonathan Geisler.

Plate 40. "Bearded Lady" portrait of Pennilynn "Penny" Higgins, University of Rochester, at Hanna Basin, Wyoming. *Source*: Dr. Penny Higgins, Vertebrate Paleontologist and Geochemist, University of Rochester. © Kelsey Vance.

CHAPTER 6

Artists, Preparators, Technicians, Collections Managers, and Outreach Educators

> Great art, as Plato stated, need not falsify reality or make nature
> artificial. . . . Great art improves our understanding of nature's reality.
> Stephen J. Gould (2001:viii)

This chapter considers in more detail those women who have contributed essential work in VP, often behind the scenes and, until more recent times, often unacknowledged. In the 18th to 19th centuries, women regularly took on the paid, or more often unpaid, roles of artist, research assistant or preparator, and educator about VP. Here we include more on artists (Table 6.1), preparators, CT and other imaging specialists, museum curators, collection managers, and outreach educators.

We remembered Carrie Barbour in Chapter 4 for her pioneering role in VP in America. Here we recognize her similar importance as a museum professional and a preparator of fossils in a career at University of Nebraska State Museum that lasted 25 years. She was one of the many women who assisted their colleagues in the supportive roles we explore here.

Paleoart

Paleoart is broadly defined as the scientific or naturalistic rendering of paleontological subject matter, in this case pertaining to vertebrate fossils. It is one of the most important vehicles for visually communicating discoveries and data among paleontologists and to the public and has been used by museums since the earliest times (Rudwick, 1976, 1992, 2005; Sellers, 1980; Witton, 2018). The scientific tradition of drawing in ink or graphite originated as a visual medium in the early 1800s in England. Older works of possible "proto-paleoart," suggestive of ancient fossil discoveries, originate from "prehistoric" Australia (e.g., Mihirung [flightless birds], *Thylacoleo* [marsupial lion], *Diprotodon* [giant wombat]), although the relationship of these works to known fossil material is often speculative (see Turner, 1986; Vickers-Rich and Archbold, 1991, fig. 1; Murray and Vickers-Rich, 2004; Mayor,

TABLE 6.1

Notable women VP pioneers and early illustrators, 19th–20th centuries

Name	Dates	Publication date	Place/publisher	Reference
Cecile Braun Agassiz	1799/1802?–1848	1833–1844	Neuchâtel, Switzerland	Drawings, color plates for Agassiz, 1833–1844
Mary Ann **Anning**	1799–1847	e.g., 1839	*Annals & Magazine of Natural History*	Self-taught? Drawings, reconstructions of her own fossils—fish, reptiles
Cecilia **Beaux**	1855–1942	1875	US Geological Survey, Washington, DC	For Cope in Hayden, 1875, The Vertebrata of the Cretaceous Formations of the West; Aldrich, 1982
Mary Morland **Buckland**	1797–1857	1820s–1830s	England	Illustrated bones for Cuvier and Buckland; Kölbl-Ebert, 1997a; Turner et al., 2010a; Turner and Berta, 2020
Margaret Matthew **Colbert**	1911–2007	1930s–1990s	AMNH; New Mexico Museum of Natural History	For Colbert, e.g., 1968, 1984, 1983; and see Elliott, 2000; Turner and Berta, in press
Lois MacIntyre **Darling**	1917–1989	1955–1977	William Morrow	Colbert, 1955, 1968, 1984; own books *Before and after Dinosaurs* (1959); *Sixty Million Years of Horses* (1960)
Ina von **Grumbkow**	1872–1942	1915		Tendaguru site paintings; Turner et al., 2010a
Helen **Haywood**	1908–1995	1965, 1970	England	Children's books
Orra White **Hitchcock**	1796–1863	1841	Amherst, Massachussets	Dinosaur footprints, fish fossils for Edward Hitchcock; Aldrich, 1982
Margaret **Hobson**	179?/181?–post-1848	ca. 1845	Australia	Illustrated *Diprotodon* bones—see Hobson, 1849, plate
Harriet **Huntsman**	1850?–1938?	1898	United States for Kansas Geological Survey	Drawings dinosaur bones for Williston; Aldrich, 1982
Hope **Johnson**	1916–2010	1970s–1990s	Alberta, Canada	1974, with J.E. Storer, *A Guide to Alberta Vertebrate Fossils from the Age of Dinosaurs* 2009 Guide edition

Name	Dates	Active	Location	Notes
Ely **Kish**	1924–2014		National Museums of Canada	Colbert, 1983; Canadian Museum of Nature; Kish and Russell, 1977; children's books
Graceanna **Lewis**	1821–1912	1876	Centennial Exposition, Philadelphia	Pelycosaur *Dimetrodon*; Aldrich, 1982
Milda **Liepina**	1902–1991	1950s–1960s	Scientific papers	Stensiö Lab, Naturhistoriska riksmuseet, Stockholm
Mary Woodhouse **Mantell** and daughter Ellen Maria **Mantell**	1795–1869; 1818–1892	1822–1844	England	Mantell, 1822, 1844; Turner et al., 2010a
Emma **Parkinson**	1788–1867	1841	England	Colorwork for Parkinson, 1811; plates reprinted in Mantell, 1850; Lewis, 2017
Erna **Pinner**	1890–1987	1933–1950	Jonathon Cape, London	Children's books, e.g., Swinton and Pinner, 1948
Phoebe **Traquair**	1852–1936	1880s	Scotland, *Geological Magazine*	Drawings for R.H. Traquair papers; Ewan et al., 2006
Alice Bolingbroke **Woodward**	1862–1951	Numerous	BMNH, England	NHM drawings, book illustrations, postcards; e.g., Turner et al., 2010a; Turner and Berta, in press
Gertrude Mary **Woodward**	1854–1939	Numerous	BMNH, England	Drawings in works including Woodward, 1889–1901; see Turner and Long, 2016

Sources: Lambrecht et al., 1938; Sarjeant, 1978–1987; Cleevely, 1983, Burek and Higgs, 2007; modified from Turner et al., 2010a; *SVP New Bulletin*; internet.
Note: Where names are in bold, see more information in the text.

2011). Other artworks from the late Middle Ages of Europe, typically wood-blocks, portraying sometimes fossils but often mythical creatures are inter-preted now as more plausibly inspired by fossils of prehistoric large mam-mals and reptiles that were seen and misinterpreted (Mayor, 2011). Some of the earliest medieval representations of actual fossils depict vertebrates, such as paleoniscid fish and shark teeth (e.g., Münster, 1544; Gesner, 1565; Adams, 1938; Davidson, 2000). Beginning in the 19th century a new wave of paleo-art portrays fossil vertebrates in colorful detail in natural settings.

Artists: Illustrators, Sculptors, and Video and Digital Animators

Women have been involved in all forms of paleoart (i.e., drawings, recon-struction, restoration) since the early 19th century, although few women have been mentioned in compilations (e.g., Davidson, 2008; Lescaze, 2017; Witton, 2018). Until recent times most institutions—universities and major museums—invariably had a team of artists and designers in house; some re-main major employers of paleoartists. Here we focus on some of the key women artists past and present; we explore these women further elsewhere (Turner and Berta, in press).

Early to Middle 19th Century

AUSTRALIA

In the 19th century, women were often the artists for their husbands, fathers, and brothers. London-born Margaret **Hobson** (née Adamson) (181?–unknown; married 1837) was a natural history artist for her husband, medically trained Australian-born naturalist Edmund C. Hobson (McCallum, 1966). They came to Australia in 1839. Her only recorded work is sketches of a fos-sil jaw and teeth of the marsupial *Diprotodon* found on Mount Macedon, Vic-toria, in 1845. The drawings were lithographed by G.A. Gilbert for the *Tas-manian Journal of Natural Science, Agriculture, Statistics &c.* (Hobson, 1849; Vickers-Rich and Archbold, 1991).

EUROPE

Germany. Another important early artist was Cecile **Braun** (ca. 1799/1802–1848) of Karlsruhe, from an academic family and sister of Alexander, a stu-dent friend of Louis Agassiz. Agassiz and Braun met around 1816 when he was 19 or 20 at her family home, and Agassiz already realized her artistic talent. They married in 1822 and went to Neuchâtel, where he had taken his first job in the museum. Braun made many drawings for his great work

Recherches sur les Poissons Fossiles (Agassiz, 1833–1844) before her tragic death (Lurie, 1960).

United Kingdom. As noted in Chapter 2, Mary Anning was one of the few independent scientific collectors and researchers, and she drew her own fossils and sometimes made restorations (Taylor and Torrens, 1986, 1995; Turner et al. 2010a).

Before her marriage, Mary Buckland (née Morland; see Chapter 2) had kept her own fossil collection and produced models of fossils. She created several illustrations for Georges Cuvier's *Animaux fossiles* (Buckland, 1823:177; Barber, 1980; Torrens, 2004). The book ostensibly brought Mary and William Buckland together on a coach journey through Dorset in 1824. William had a letter of introduction to "Miss Morland," and he noticed the young woman opposite him reading Cuvier. To his surprise, it was Mary herself; they married the next year. While this perhaps is an apocryphal family story (see Gordon, 1894; McGowan, 2001), it was Mary's artistic prowess that brought them together. She drew cave hyaena for William's 1823 *Reliquiae Diluvianae* (Buckland, 1823, plate 4). Alongside Mary's drawings in the 1823 book is the work of either Frances (1803–1881) or Louisa **Duncombe** (1807–1852) (of Duncombe Park, Yorkshire), one of whom William met when he was exploring Kirkdale Cave, who similarly created pencil drawings of cave fossils (Buckland, 1823, plate 13). Mary also drew and produced lithographs for Buckland's first description of the British dinosaur, *Megalosaurus*, in 1824 (Plate 32) and of Pleistocene mammalian teeth and bones found in the famous Kents Cavern, Torquay, Devon. Some of Benjamin Waterhouse Hawkins's restorations of extinct vertebrates (most famously the first dinosaurs depicted) for the 1851 Royal Crystal Palace Exhibition in Hyde Park, London, were modeled after her drawings for William's 1836 Bridgewater Treatise (Gordon, 1894:198).

Mary Mantell (née Woodhouse; see Chapter 2) notably drew *Iguanodon* teeth, the first dinosaur ever recognized, as well as illustrating for her husband, Gideon, including 364 engravings for his 1822 book on Sussex geology (e.g., Edmonds, 1979). Their daughter, Ellen Maria **Mantell** (1818–1892), took up the mantle (pun intended) and illustrated his *Medals of Creation* (1844) with drawings—100 in 1,000 pages—the first "modern" synthesis of paleontology in English. Ellen became an illustrator for publisher John Parker, whom she later married (Critchley, 2010).

Emma Rook **Parkinson** (1788–1867) was illustrator for her father, James, executing hand-colored (presumed) watercolors of vertebrates for his classic three-volume *Organic Remains* in 1811; some of these were reprinted by Gideon Mantell in his 1850 *Pictorial Atlas* (Lewis, 2017).

UNITED STATES

One of the first American female scientific illustrators, Orra White **Hitchcock** (1796–1863; Plates 33 and 34) was a gifted artist who learned to draw as a child from private tutors and at school and went on to be a teacher of drawing and science at Deerfield and Amherst Academies. She illustrated for her husband, pioneer vertebrate tracks worker Edward Hitchcock, at Amherst College, Massachusetts, making giant visual aids on canvas, and accompanied him on geological excursions throughout the Connecticut Valley, drawing landscapes for his reports. Orra produced one of the first restorations of an American mastodon and made some of the first drawings of dinosaur footprints and fossil fish (Aldrich, 1982, 2001; Turner et al., 2010a). Her banner at Amherst, "Five Lines of *Orthichnites* Fossil Footprints," showed the five kinds of "birdlike" traces, actually fossil dinosaur tracks, and was one of 61 drawings she did for use in Edward Hitchcock's classes on geology and natural history.

Harriet "Hattie" **Huntsman** (fl. 1880s) was an American illustrator who drew dinosaur bones for Samuel W. Williston's *Paleontology* (1898) for the Kansas Geological Survey (Aldrich, 1982, fig. 7; Turner et al., 2010a). She may be Hattie Greenough Huntsman (1850–1938), of Tankersley, Yorkshire, who immigrated to Ohio around 1870 (Kathy Lehtola, pers. comm., August 2019).

Graceanna **Lewis** (1821–1912), an American illustrator and Quaker, was a noted naturalist in her own right. Her botanical paintings and a sketch of the pelycosaur *Dimetrodon* drawn for a chart were displayed at the 1876 Centennial Exposition in Philadelphia and received praise from Thomas Huxley (Aldrich, 1982). Her interest in botany and zoology, especially birds, resulted from early years spent at the Kimberton Boarding School, which was run by botanist and social reformer Abigail Kimber (1804–1871). Given the limited opportunities for women in the 1870s, Lewis was unable to secure a permanent college teaching position even though she was well known and widely respected by her peers (Lukens, 2013).

Later 19th to Early 20th Century

United Kingdom. By far the most prolific of scientific writers of VP papers in the second half of the 19th century is Ramsay Heatley Traquair (1840–1912). He had the good fortune in Dublin in the early 1870s to meet a young Irish woman named Phoebe Moss, a student training at the Royal Dublin School of Design, who was assigned to illustrate his research papers. She studied between 1869 and 1872, and after their marriage in June 1873 they moved to Edinburgh, where Ramsay was appointed in 1874 to the Museum of Science and Art and in 1879 as keeper of natural history at the Royal Scottish Museum. Phoebe Anna **Traquair** (née Moss) (1852–1936) became one of Scotland's most noted artists, contributing to the Arts and Crafts movement with enamel work, illuminated manuscripts, embroideries, paintings, murals, and furniture decoration. Her work of late Pre-Raphaelite style graces several churches in England and Scotland and the poetry of Dante Gabriel Rossetti, Alfred Lord Tennyson, and the Brownings (Ewan et al., 2006). Her fine black-and-white drawings, mostly of fossil fish, illustrate many of Ramsay Traquair's nearly 130 papers, including his major Palaeontographical Society Monograph on paleoniscoid fish, which he continued to publish until his retirement in 1906.

By the late 19th century, two young women, children of Henry Woodward, keeper of geology and paleontologist at the BMNH, became known for their illustrations of books and scientific papers. Like many Victorian middle- and upper-class children, they were educated at home by governesses and encouraged to draw from a young age. By her late teens, Alice Bolingbroke **Woodward** (1862–1951) was skilled enough to illustrate for her father's lectures and his colleagues' papers. Zoë Lescaze (2017) has suggested that Alice was inspired by American illustrator Charles Knight's paleoart, but this seems unlikely as she was already drawing in her teens. Instead, she and her siblings were probably brought up on contemporary British books. Many were used in or created for H.R. Knipe's *Nebula to Man* (1905) and *Evolution in the Past* (1912) and then shared with the Rev. H.N. Hutchinson for his *Extinct Monsters and Creatures of Other Days* (1910), which no doubt brought her additional income and acclaim. Further drawings (e.g., Hutchinson, 1910, fig. 105) were drawn especially for the book and are based on her own admirable restorations (Hutchinson, 1910, plates XLIV–XLVI). Susan Turner et al. (2010a) published some of her vertebrate restorations from the NHM archives for the first time. Alice's prehistoric restorations, from fish to dinosaurs to mammals, and scenes became famous thanks to the *Illustrated News*. One of her later works (Anon, 1925), was entitled "Larger than any known land animal: a fossil-hunter's vision of the enormous Gigantosaurus" and subtitled "As it probably appeared in life in Tanganyika: the huge dinosaur, as visualized by a fossil hunter dreaming by his desert fire, beside an unearthed humerus twice as long as that of the 80 ft *Diplodocus* in the Natural History Museum." This reconstruction drawing was based on scientific data from bones discovered in Tanzania (see discussion of Viktorine Grumbkow in Chapter 3). In the following decades, up to the 1960s, the NHM turned her dinosaur paleoart into collectible souvenir postcards (see Turner and Berta, in press).

Alice's elder sister, Gertrude **Woodward** (1861–1939), was similarly brought up in Chelsea and introduced to paleontology at an early age. She became an excellent color-wash illustrator and gained employment at the newly built BMNH. Gertrude worked primarily with her father's successor, Arthur Smith Woodward, until his retirement in 1924 (Turner and Long, 2016) but also others, and her drawings and lithographs appeared in scientific journals of the time, such as the *Geological Magazine*. She worked on restorations of fish, as well as fine drawings of specimens, for Arthur Smith Woodward's (1889–1901) multivolume BMNH *Catalogue of Fossil Fishes*, which became the standard work worldwide well into the 20th century. She even drew the infamous *Eoanthropus* (Piltdown man) remains for him (Turner et al., 2010a).

United States. Accomplished American artist Eliza Cecilia "Leilie" **Beaux** (1855–1942) was born in Philadelphia. When Beaux was fourteen years old, she entered the Misses Lymans' School, with lessons in natural history, zoology, and geology, and later learned lithography from a distant relative, artist Catherine Ann Drinker. After classes in 1871 at the Van Der Wielen School, Beaux became a commercial artist, and on the strength of her drawings of bones and a recommendation from lithographer Thomas Sinclair (1805–1881), Beaux was hired to make drawings of fossils for E. Drinker Cope (Davidson, 2008; Turner et al., 2010a). One plate of his *Vertebrata of the Cretaceous Formations of the West* contained lithographs of carefully drawn specimens of the hadrosaur *Cionondon arctatus*, and these illustrated Cope's contribution to the US Geological Survey report on the territories (Hayden, 1875; Aldrich, 1982). Once she had proved herself, she was faced with the challenge of her next assignment, the head of an extinct ass that was Cope's pride and joy (Tappert, 1994). In the end, Beaux did not find scientific illustration suitable as a career and spent the rest of her life painting portraits and landscapes. Notably, she was the first woman to have a teaching position at the Pennsylvania Academy of Fine Arts. Beaux considered herself a "New Woman," a 19th-century woman who explored educational and career opportunities that had generally been denied to women (e.g., Mathews, 2014).

She was considered by the American painter William Merritt Chase "not only the greatest living woman painter, but the greatest who has ever lived" (Parker, 2008).

20th Century

EUROPE

Sweden. One of the most celebrated illustrators in the later 20th century was Latvian Milda **Liepina** (née Freidenberg) (1902–1991), who drew immaculate illustrations for the Swedish fossil fish workers in Stockholm, working primarily with Tor Ørvig (1916–1994) and Erik Stensiö, whose lab at the Naturhistoriska Riksmuseet, Stockholm, dominated much of fossil fish work in the early to mid-20th century. She also retouched photographs, a not always recognized or approved practice, for his parts of the classic *Traité de Paléontologie* (e.g., Janvier, 1996).

United Kingdom. Jennifer "Jenny" **Halstead** (née Middleton) illustrated for her future husband, vertebrate paleontologist L. Beverly Halstead, first when he was associated with the Royal Dental Hospital in London in the 1960s, where they met. She studied art and design at Sutton and Cheam School of Art, then won guest scholarship to the Royal Academy Schools while training in medical art at the Central Middlesex Hospital, London. Middleton joined in writing a book on *Bare Bones* with Halstead (Halstead and Middleton, 1972). They married in 1973 and went on to write and illustrate many books on fossil vertebrates together, especially on the life history of dinosaurs, aimed at children, with project work in China, India and West Africa (e.g., Halstead and Halstead, 1981, 1982, 1983, 1984, 1985; Halstead, 1993; Sarjeant, 1995). In addition, she contributed to several editions of *Gray's Anatomy*.

Helen **Haywood** (1908–1995) was an English writer and illustrator of children's books, including *Exploring Fossils* (1965) and *My Book of Prehistoric Creatures* (1970). She was born in London in 1908 but was taken as a child to Chile, where her father, an engineer, worked on the trans-Andean railway. She remained in Chile until she was about 15 years old. Haywood was a keen student of science and an amateur naturalist and anthropologist. Her art was underlain by science, and she did pioneering research into the skin colors of reptiles to enhance her dinosaur illustrations (Connelly, 2009).

Margaret "Maggie" **Lambert Newman** (194?–) was the chosen illustrator in the late 1960s for several mammal workers, especially Björn Kúrten (*Cave Bear Story, Pleistocene Mammals*), with Elaine Anderson and Anthony

"Tony" Sutcliffe, fossil mammal curator at the NHM (UK mammals, e.g., *First Land Mammals*, by Robin Place), at least one of which was made into an NHM postcard. Lambert was a regular on the British VP scene at that time, attending the SVPCA and events in London. In the 1970s she married Barney Newman, former preparator at the NMH, and they moved to live in Africa, where, in recent years, she has illustrated, for example, the key Karoo vertebrate *Lystrosaurus* (see Botha-Brink, 2016).

Erna Wilhelmine **Pinner** (1890–1987) was an exile from Nazi Germany and, like Tilly Edinger (Chapters 3 and 4), she was born into a well-off Frankfurt Jewish family. At the age of 16 she began her artistic education at the Städel Art Institute in her hometown, after which she studied in Berlin and Paris. Her adventurous and interesting career was cut short in 1935 when she was expelled from the Reich Chamber of Fine Arts and, for safety, immigrated with her mother to England (Phillips, 2001). There, at the age of 45, she retrained, learning biology and becoming a writer on natural history and illustrator. She worked with vertebrate paleontologist and science popularizer W.E. Swinton (1900–1994) on the 1948 children's book *The Corridor of Life*, which contained rather cartoonish reconstructions. Pinner continued to write for the *Natural Science Review* in Stuttgart until she was 90.

UNITED STATES AND CANADA

Carol **Abraczinskas** (1969–) is an American VP illustrator at the University of Michigan Museum of Paleontology, Ann Arbor. Previously, she was a scientific illustrator at the University of Chicago. A BFA graduate of the School of the Art Institute of Chicago, she began her career in 1989 documenting Egyptian and Nubian artifacts. Later that year she joined Paul Sereno at the University of Chicago on collecting trips as principal scientific illustrator. Her award-winning drawings have been featured in exhibits at the FMNH and the Museum of Science and Industry in Chicago, as well as in national magazines and scientific journals, including on the covers of the *JVP*, *Cladistics*, and *Evolution*. She was awarded the SVP Lanzendorf–National Geographic PaleoArt Prize for Scientific Illustration in 2000, 2005, and 2008.

Luci **Betti-Nash** (195?–) is an American illustrator in the Anatomical Sciences Department at Stony Brook University, Stony Brook, New York. She grew up near New York City and was influenced by her father, a graphic designer. She studied art, printmaking, and design at Stony Brook in the late 1970s. Her art and interest in science from her early years resulted in her becoming a scientific illustrator. She is known for her illustrations of Mesozoic mammals and amphibians from Madagascar (Plate 35) and hominid fossils from Africa.

Joy **Buba** (née Margaret Joy Flinsch) (1904–1998) was an American illustrator, sculptor, and author. Her work is held at various museums, including the National Portrait Gallery, the Vatican Library, and the National Statuary Hall. Her paleoart illustrations were published in H.F. Osborn's *Proboscidea* (1936 and 1942 volumes). She studied at the Städel Kunst Institute in Frankfurt and at the Academy of Fine Arts in Munich, Germany.

Born to an unconventional, free-spirited artist mother and the renowned vertebrate paleontologist William Diller Matthew, Margaret **Colbert** (née Matthew) (1911–2007; Plate 36) chose a career as an artist, eventually specializing in restorations of extinct animals (e.g., Colbert, 1965). Colbert graduated in art and sculpture from the California School of Arts and Crafts in 1931 and began her six-decade career at the AMNH drawing fossil bones. There she met her future husband, noted paleontologist Edwin "Ned" Colbert, on an earlier visit; they married in 1933 and began a partnership that brought vertebrate paleontology worldwide. A respected restoration artist, illustrator, and sculptor, Colbert's paintings and drawings graced many of her husband's books and she made dioramas and sculptures for several museums, such as those at the Petrified Forest National Park Visitor Center and the New Mexico Museum of Natural History (e.g., Allmon, 2006). She helped to create the SVP logo with Ruth Romer, as well as a noted bust of her father, turning more to fine art and working well into her 80s (Elliot, 2000). In 1935, at the age 24, she was awarded the Daniel Giraud Elliot Medal from the National Academy of Sciences in recognition of her paleontological illustrations.

Sylvia Massey **Czerkas** (1943?–) is an American illustrator, sculptor, editor, and writer living in Utah who is known for her work on dinosaurs and encouraging other women illustrators, even experienced artists such as Margaret Colbert (Elliot, 2000). She published a book (with Donald Glut), *Dinosaurs, Mammoths and Cavemen* (1982), showcasing the paintings of Charles R. Knight and wrote an entry on reconstruction and restoration in the *Encyclopedia of Dinosaurs*. She edited and illustrated over 30 dinosaur books (e.g., Czerkas and Olson, 1987; Czerkas, 1988; Czerkas and Czerkas, 1989; Czerkas and Czerkas, 1990).

Bonnie **Dalzell** is a graphic artist who has illustrated fossil and modern vertebrates. She holds BA and MA degrees in Paleontology from UCB. Lois MacIntyre **Darling** (1917–1989) was an American artist and writer. Darling and her husband, Louis, wrote and illustrated nine books, including several on paleontology, such as *Before and after Dinosaurs* (1959) and *Sixty Million Years of Horses* (1960). She illustrated Colbert's (1955) *Evolution of the Vertebrates* with simple but effective black-and-white restorations.

Martha **Erikson** (194?–) was an American illustrator at the YPM, working with several paleontologists on projects such as figures illustrating ceratopsian dinosaur teeth (Ostrom, 1966).

Pamela **Gaskill** (1929?–2006), born in Jamaica, was "the most talented artist and preparator" (Carroll, 2004). Gaskill worked with Bob Carroll at the McGill Museum of Natural History, Redpath, Canada, and others as a scientific illustrator and researcher on tetrapod amphibians until her untimely death. She worked with him on the more than 1,700 illustrations for the major textbook *Vertebrate Paleontology and Evolution*, with at least 500 new drawings. In 1978 they worked together on the American Philosophical Society Memoir on the Microsauria (Carroll and Gaskill, 1978; Carroll, 1988).

Louise **Germann** (née Waller) (1896–1957) was an illustrator and sculptor at the AMNH, working with Edwin Colbert. Her illustrations appear in Colbert's *The Dinosaur Book* (1951) and *Evolution of the Vertebrates* (2001).

Hope **Johnson** (1916–2010) was an amateur paleontologist who collected, prepared, and drew vertebrate fossils. She was born in Vancouver, British Columbia, and as a child was interested in nature. She was homeschooled and briefly attended Victoria College. During World War II, she was serving in the Canadian Women's Army Corps when she met and married Olafur "Ollie" Johnson (1909–1987), an Icelandic civilian meteorologist who supported her interests. She joined a local art club and in the 1940s and 1950s taught herself about geology and paleontology. She made drawings for her own use, and museum staff recognized the value of these illustrations and worked with her to publish them. Johnson's first published illustrations were in *Dinosaurs in the Southwest Prairie Area* in 1968. Dale A. Russell, then at the National Museum of Canada, Ottawa, helped her get a job as "fossil custodian" at Dinosaur Provincial Park, where she worked in 1969–1970 and 1972–1973. She then developed hand-painted exhibits at the park and made many drawings of dinosaur bones for the Provincial Museum of Alberta. Johnson collaborated with vertebrate paleontologist John E. Storer (formerly of the Saskatchewan Museum of Natural History) on a handbook of Alberta fossil vertebrates (Johnson and Storer, 1974, 2009). She was awarded an honorary LLD in 1981 from the University of Lethbridge for her work in the public education of paleontology. She continued illustrating and collecting fossils for the rest of her life (Tanke, 2019).

Eleanor "Ely" M. **Kish** (b. Kiss) (1924–2014) was an American illustrator who became a paleoartist later in life. She was born in Newark, New Jersey, and took to art at an early age. She remembered, "I was the only girl in carpentry class" (quoted in Kingsmill, 1990:22). At school, befriended by the school principal, she visited her first art gallery at age 14. On the principal's advice, Kish took four years of night school to learn anatomy, composition, color, and perspective. Graduating from a technical high school for girls, she traveled across the United States to California and then to Mexico, where she painted billboards and designed greeting cards, furniture, cars, and houses for 15 years. In the late 1950s she settled in Canada, becoming a Canadian citizen and working into the 1990s. As Kish noted in a *Washington Post* interview (Welzenbach, 1990), "I was introduced to painting dinosaurs by Dale A. Russell . . . in 1974. . . . They sent me down to Russell to do a slide projection show for children on dinosaurs. And I knew nothing about dinosaurs—but I thought everyone knew what dinosaurs looked like. And when I got there he said, I just can't tell you what they looked like. It has to be worked out. And from that point he gave me illustrations of Charles Knight's and other artists' concepts. I worked out my slide pictures from them. He liked those so much he hired me to illustrate *[A] Vanished World: The Dinosaurs of Western Canada*, . . . in 1977." As she exclaimed, "It was the first time I had a chance to paint for my living" (Kingsmill, 1990:24). At the age of 53, she was a recognized top paleoartist just three years after she had drawn her first dinosaur. She illustrated another book with Russell in 1989, *Odyssey in Time: Dinosaurs of North America*. She is especially known for a series of images that portray dinosaurs and landscapes during the Cretaceous mass extinction events (Spears, 2014). One of her last paintings was the large mural formerly displayed in the Life in the Ancient Seas hall at the NMNH. The mural colorfully depicted, with anatomical precision, animal life over more than 500 years of earth history.

Tess **Kissinger** is an American paleoartist and writer working in Philadelphia who trained at Carnegie Mellon University. She is an author and illustrator with her partner, Bob Walters, of the book *Discovering Dinosaurs* (2014), and they have written about restoring dinosaurs (see Scotchmoor et al., 2002). The pair have illustrated dinosaur murals for various museums, including the Carnegie Museum of Natural History and Dinosaur National Monument in Vernal, Utah and in 2007 won the SVP Lazendorf PaleoArt Prize. Among her current work is a series of fine art paintings, including ones of ceratopsian dinosaur skulls.

Jaime Patricia **Lufkin** (1921–) is an American illustrator who worked at UCB's Museum of Paleontology, producing drawings of fossil vertebrates for numerous graduate students and faculty from 1964 to 1991 before she became freelance. She was born in Seattle, Washington, and studied at the University Washington (1941), the College of Sequoias in Visalia, California (1952), and the Schaefer-Simmern Institute of Art Education in Berkeley (1956–1958).

Kalliopi **Monoyios** (1978–) is an independent American scientific illustrator. She worked in Neil Shubin's lab for a decade until 2015, where she discovered the close relationship between science and art and the importance of making visual communication available to the public. For Shubin she helped document the exciting discovery of early tetrapod *Tiktaalik roseae* and made many original illustrations for his popular book (Shubin, 2007). She is currently focused on raising the profile of scientific art through her *Scientific American* blog *Symbiartic*.

Pat **Ortega** began to focus on paleoart in the mid-1980s. As a child in California, she often visited zoos and museums and drew the animals that she observed. Later visits to the La Brea Tar Pits and LACM to view the dioramas of fossil animals depicted by paleoartist Charles Knight inspired her art. Ortega's attention to detail has made her an award-winning illustrator whose art can be found in more than 60 books and other publications. She was the 2004 recipient of the Lanzendorf–National Geographic Award for Scientific Illustration.

Diane **Scott** is a Canadian preparator and illustrator of fish and reptilian fossils. She works as research assistant and graphic artist in the Biology Department at the University of Toronto Mississauga and is described as a "preparator and illustrator extraordinaire" by Robert Reisz, her lab director (Dilkes and Reisz, 1989:9).

Mary **Parrish** (196?–), now retired, began working at the NMNH in 1979 and became staff scientific illustrator in the Department of Paleobiology in 1983. Her illustrations have been published in many scientific journals and books, as well as in several NMNH exhibits, including the large mural currently on display in *The Last American Dinosaur* exhibit. She began researching and managing the care of paleobiology's historical illustration collection in 1995.

Other paleoartists include Elaine **Kasmer** (Johns Hopkins University), who worked for Ken Rose on Indian mammalian fossils (Anon, 2014; Rose et al., 2014), Ivy **Rutzky** (retired, AMNH), who created over 40 fossil fish restorations for the AMNH book *Discovering Fossil Fishes* (Maisey, 1996; Davidson, 2008), and Maidi **Wiebe Leibhardt** (1922–1998), German-born scientific illustrator and naturalist (FMNH) and illustrator of childrens' books.

21st Century

Artwork in the new century has been taken to another level, mostly with the means of computer animation, and "virtual" paleontology (e.g., Sutton et al., 2014) is now available. Websites such as Deviant Art and blogs allow anyone to present a restoration of a fossil vertebrate. Again, we note some of the new illustrators.

ASIA

China. Aijuan **Shi** (IVPP) is a Chinese paleoartist who creates digital art of fossil dinosaurs and birds.

EUROPE

France. Elisabeth **Daynès** (1960–) is a French sculptor specializing in hominid portraits. Her artwork is present in museums worldwide, including the Musée des Merveilles, Tende, France; the FMNH, Chicago; the Transvaal Museum, Pretoria, South Africa; the Sangiran Museum, Java, Indonesia; the Naturhistoriska Riksmuseet, Stockholm; and the Museum of Human Evolution, Burgos, Spain. One of her most notable works, at the Krapina Neanderthal Museum, Croatia, reconstructs an entire 17-member Neanderthal family. She was awarded the Lanzendorf–National Geographic PaleoArt Prize in the three-dimensional art category in 2010.

United Kingdom. Rebecca **Groom** is a self-taught UK artist perhaps best known for her scientifically accurate soft dinosaur toys and other paleontological soft sculptures.

UNITED STATES AND CANADA

Stephanie **Abramowicz** is an illustrator and photographer at the Dinosaur Institute, LACM. She has worked with the Dinosaur Institute since 2006, creating images for museum exhibits as well as scientific and popular publications. She has a background in fine arts and a BFA from the University of Southern California.

Raven **Amos** is a digital artist living in Alaska who is known for the bold visual styles and color schemes in her stylized depictions of dinosaurs and their surrounding landscapes. She studied animation art and graphic design for two years at the Art Institute of Seattle from 2000 to 2002.

Karen **Carr** is a wildlife and natural history artist based in Silver City, New Mexico, who works in both traditional and digital media. Her work has been displayed in publications, zoos, museums, and parks across the United States, Japan, and Europe. Recently she has done illustrations for the NMNH and authored or illustrated several books for young readers, including *Dinosaur Hunt* (2002). Born in Fort Worth, Texas, Carr is the daughter of an artist father and scientist mother, and these two shared influences and careers have shaped her life and work. While at the University of Texas at Austin, Carr studied natural sciences and art, and she gained a BFA from North Texas State University.

Danielle **DuFault** is a Canadian scientific illustrator at the ROM in Toronto. She creates both traditional and digital art, illustrating dinosaurs and other paleo subjects. She was the 2016 recipient of the Lanzendorf–National Geographic Award for Scientific Illustration.

Scientific illustrator and visualization expert April Isch **Neander** has worked in Zhe-Xi Luo's lab at the University of Chicago since 2012, specializing in mammalian paleontology, though she fills the additional roles of lab manager, CT data specialist, and University of Chicago PaleoCT Lab technician. Her illustrations range from technical pieces for scientific journals to popular art for the public. Neander earned an MS in biomedical visualization at the University of Illinois at Chicago. She was awarded the 2006 Lanzendorf–National Geographic Digital Modeling and Animation Award.

Jennifer **Hall** is an American paleoartist working at the Carter County Museum in Ekalaka, Montana, as both an artist and a science communicator. She was a printmaking major at the Pennsylvania Academy of Fine Arts in Philadelphia and then studied geology and dinosaurs, earning her degree at the University of Pennsylvania. Her work as a paleoartist began with her employment as a paid artist for dinosaur artists Bob Walters and Tess Kissinger. She talks of being a full-time pastry artist by day and moonlighting as a science illustrator at the beginning of her career (Diep, 2017). More recently she has become interested in taxidermy and creating sculptures of feathered dinosaurs.

Emily **Willoughby** (1986–) is an American paleoartist, illustrator, writer, and researcher living in Minneapolis, Minnesota. She is best known for her illustrations of feathered maniraptorian dinosaurs and her artwork and writings have appeared in the Shanghai Museum of Natural History in China,

Nature Genetics, and books including Paul Barrett and Darren Naish's (2016) *Dinosaurs: How They Lived and Evolved*, Steve White's (2017) *Dinosaur Art II*, and her own coauthored book (2016), *God's Word or Human Reason?*

Recognition for Paleoartists

Since 2000 the SVP has awarded the Lanzendorf–National Geographic PaleoArt Prize, created by American collector of dinosaur-related artwork John J. Lanzendorf (1941–), a noted patron of paleoartistry, to recognize outstanding achievement in paleontological scientific illustration and naturalistic art. Awards are made in the following categories:

- **Scientific Illustration:** works that are regarded as text-figures within the body of a scientific paper or as online supplementary material
- **Two-Dimensional Art:** works that typically would *not* be regarded as text-figures within the body of a scientific paper or as online supplementary material
- **Three-Dimensional Art:** works that exist in the three-dimensional, physical (not digital) realm (e.g., sculpture)
- **National Geographic Digital Modeling and Animation Award:** works, including animations, created in three-dimensional digital programs

Among the 68 art awards given since 2000, only 10 (6.8%) have been given to women (online appendix S3).

Preparators

Vertebrate paleontology requires the skilled preparation of fossils for research and education. Many VPs have to prepare their own fossils, but institutions such as museums may employ trained preparators or fossil conservators, who are individuals who clean, restore, and repair fossils. Museum preparators bring to the job an uncommon degree of adaptability and patience, diverse skill sets, and backgrounds in art and science, such as chemistry, paleontology, or archaeology. Peter Whybrow (1985) and Patrick Leiggi and Peter May (1994) have explored the history and development of techniques and uses, although no women are mentioned among the originators or developers. Willis and Muirhead (1987) and Whitmore and Kool (1991, 1996 [two editions]) provide more information, particularly on terrestrial vertebrates. Again, we focus on some of the key women who have been or are skilled in the preparation of vertebrate fossils, including with advanced technologies.

In the 19th century, preparation depended mainly on manual labor and blunt instruments (hammers, chisels), followed sometimes with the use of microscopes. As many early scientists were also medics, medical and dental tools were found to allow for finer preparation. Mary Anning (and her mother, Molly) excelled in the excavation of fossils (see her hammer in Turner et al., 2010a), but she did have help from strong men on occasion. As Whybrow (1985) notes, in fact most VP papers in the early days tell us little of how fossils were extracted or prepared.

In the 20th century, more chemical techniques were used, such as acids for preparing phosphatic vertebrate and other fossils (e.g., Whybrow, 1985; Blum et al., 1989; Passaglia and McCarroll, 1996; Jeppsson et al., 1999). Also during this time, British zoologist and paleontologist Igerna Sollas, along with her father, pioneered a serial sectioning methodology that enabled study of the internal anatomy of fossils, illustrating three-dimensional structure for the first time, which they then applied to their research of Devonian fish (Chapter 3).

By the late 20th century, scanning electron microscopes, ultrasound, and microwaves were in use. In addition, new, exciting, noninvasive techniques including CT scanning, surface laser scanning, and computer digitization enabled detailed anatomical studies.

Since 1999 the SVP has supported a professional vertebrate preparator grant, the Marvin and Beth Hix Preparators' Grant. We highlight here a few preparators with diverse employment experiences, including some of the grantees.

Africa

Egypt. Sanaa El-Sayed, working with Mesozoic fish and dinosaurs at the Mansoura University Vertebrate Paleontology Center, Egypt, was the SVP's 2016 Hix Preparator Grant recipient (see Chapter 5).

Kenya. Rose **Nyaboke** (National Museums of Kenya) is responsible for the vertebrate fossil collections at the National Museums of Kenya in Nairobi. In 2018 she received the SVP Marvin and Beth Hix Preparators' Grant. In 2018 she visited the University of Michigan Museum of Paleontology and learned about the conservation, organization, and databasing of collections and about the various preparation, molding, and casting processes that are involved in the study and exhibit of fossils.

Australia

At the start of the 20th century, with the growth of universities and museums, Australian vertebrate fossils were less likely to be sent overseas (e.g., Vickers-Rich and Archbold, 1991; Rich and Vickers-Rich, 2003). There were still few women in vertebrate paleontology, but one, Irene Maud **Longman** (née Bayley) (1877–1963), assisted with the preparation of some of Queensland's most iconic fossils, including newly found dinosaurs and the giant *Kronosaurus queenslandicus*, well known from the restored skeleton sent to the MCZ (Turner, 1986). Longman helped Heber Longman, her husband and director of the Queensland Museum, Brisbane, to prepare specimens. In the late 1920s she went on to a more illustrious but brief career as the first woman in the Queensland parliament (Mather, 1986; Fallon, 2003).

Lesley **Kool** (1950–) is a fossil preparator and honorary research associate in the Department of Paleontology, Museums Victoria, Melbourne, where she began as a volunteer after emigrating from England. In 1996 Patricia Vickers-Rich offered her a research assistant job at Monash University in Clayton, Victoria, where she worked for the next 20 years. She learned to acid-etch Miocene fossils from Bullock Creek in the Northern Territory. With over 30 years of experience working for Tom Rich and Vickers-Rich at Dinosaur Cove and elsewhere in Victoria, especially on the Cretaceous dinosaur material (e.g., Vickers-Rich and Rich, 1993–2019), she wrote an important introduction to techniques (see earlier). Kool is currently running a preparation group from her home, teaching the next generation of fossil preparators.

Joanne **Wilkinson** (1962–; Plate 37) is the senior fossil preparator and geoscience volunteer coordinator for the Queensland Museum Geosciences Program. She has more than 30 years of experience at the Queensland Museum, with skills in acid and mechanical preparation. In 2005 she was awarded the Queensland Museum Scholarship to learn fossil preparation techniques in Chicago, in New York, and at the Royal Tyrrell Museum. She has participated in numerous fossil excavation and collecting trips in Queensland and works particularly on dinosaur bones and Pliocene to Pleistocene marsupials. As Wilkinson recounted in her interview with us, "I remember like yesterday the day I walked the valleys of an area near Lethbridge, Alberta. Dinosaur bones were literally falling out of the sides of the gullies and dinosaur teeth were scattered on the ground. When I touched the bones they fragmented and there was no way I could retrieve them without them breaking. It sparked an interest that has led to a career in the technical side of paleontology, fossil preparation."

Europe

Norway. Lily **Monsen** (193/194?–) was a technical assistant at the Palaeontological Museum of the University of Oslo in the 1960s. She took part in the August expedition to Spitsbergen, Svalbard, in 1961, under the leadership of Natascha Heintz (see Chapter 4), to study the geology and to photograph and cast Cretaceous iguanodont footprints from a vertical cliff (Heintz, 1962). These were then the most northerly dinosaur remains known (Colbert, 1968; Thulborn, 1990).

Russia. Maria **Louk'yanova** (192/193?–unknown) was a technician at the PIN in Moscow in Soviet times, well known for her participation in the Soviet expeditions to collect vertebrate fossils. She took part in the 1946–1949 Russian expedition to Mongolia and in the 1958 PIN excavation to collect Upper Permian vertebrates at Ocher in the Cisurals (Plate 38; correct transliteration of name and update from Turner et al., 2010a).

Sweden. Erik Stensiö, in his lab in Stockholm (see earlier discussion), employed his cousin Agda **Brasch** (1902–1990) in the 1930s and Britta **Arnell** (dates unknown), the daughter of the dentist who helped Stensiö develop his mechanical mallet to prepare the fossil fish that were the basis for his acclaimed work (Miles, 2012; Turner and Miles, in press). Aina Sofia **Laurell** (1890–1984), who became his wife in 1920, had worked as both a preparator and an illustrator and photographer until she left the museum in about 1935.

United Kingdom. Sarah **Finney** (now Wallace-Johnson) is an English preparator who worked in the late Jenny Clack's Lab at the Museum of Zoology of the University of Cambridge from 1989 and is now conservator at the Sedgwick Museum. Using classic dental tools with tungsten carbide rods under the microscope, she has revealed important specimens such as *Acanthostega*, the Devonian amphibian. In 1998 Finney joined Clack (see Chapter 5) and two of her graduate students on a woman-only expedition to East Greenland to collect more Devonian tetrapod fossils.

Vicen **Carrió**, from Spain, is the conservator and preparator of geological and paleontological material at the National Museums of Scotland. She studied biology at the University of Valencia before moving to Edinburgh in 1992. Her role has included the care, protection, and conservation of the TWeed Project specimens (Early Carboniferous fish and tetrapods from the Scottish Borders), led by Clack.

United States and Canada

Karen **Alf** (1954–2000) worked on a fossil exhibit titled *Prehistoric Journey* at the Denver Museum of Natural History. Among her accomplishments are the mounts of the smaller of the two *Coelophysis* skeletons and the skeleton of *Sphenocoelus unitensis*. Alf was also involved in the discovery of a dinosaur egg clutch (Carpenter and Alf in Carpenter et al., 1994; Carpenter, 1999) in the Morrison Formation at Garden Park, Colorado. Alf began preparation of fossils at the Black Hills Institute in South Dakota, then went to the Denver Museum of Natural History in 1978 as a volunteer, and continued later as a staff preparator, running a fossil preparation class (Carpenter, 2001; Turner et al., 2010a).

Carrie **Ancell** is a senior preparator and part of Patrick Leiggi and Amy **Atwater's** team at the Museum of the Rockies, Bozeman, Montana. She did noted work on the first ornithischian *Hesperosaurus* from that state (Maidment et al., 2018).

Ana **Balcarcel** is a senior preparator at the AMNH. She received a BA from Harvard University. She has collected fossils in the Wyoming badlands, the Gobi Desert of Mongolia, the Peruvian Amazon, and the Caribbean.

Shelley M. **Cox** (George C. Page Museum, Los Angeles) is the laboratory preparator and volunteer supervisor at the George C. Page Museum, assisting volunteers in the cleaning and preparation of fossils recovered from the Rancho La Brea Tar Pits. After gaining a BA in history in 1972, she began her association with the museum more than 40 years ago when she started as a volunteer herself. Over her career, she has supervised more than 4,000 volunteers (Rosen, 2014).

Amy **Davidson** has been a preparator at the AMNH fossil preparation lab since 1993, where she has worked primarily on mammal and dinosaur fossils from the Gobi expeditions and Chilean Andes. She has an extensive background in sculpture and modelmaking, combined with a love of anatomy and natural history. She received her training as an apprentice to a master micro-preparator at the MCZ.

Mariana **Di Giacomo** is a paleontologist specializing in fossil preservation. She graduated in 2012 with an MS in zoology from the Universidad de la República, Montevideo, Uruguay. In 2006–2007 she was tutored by fossil preparators at the Museo de La Plata, La Plata, Argentina, and so began her interest in the conservation of fossil bone. In 2014 she began a PhD in the Preservation Studies program at the University of Delaware, Newark. She is also a fellow at the NMNH, where she is working on her dissertation.

Di Giacomo writes, "If you had told the 7-year-old me she would be here now, she would have answered, 'I know, I told you I wanted to be a paleontologist!'" (Di Giacomo, pers. comm.).

Donna **Engard** (1948–2005) was a preparator and later collections manager and curator at several museums. She began as an exhibit designer at the Cranbrook Institute Museum in Bloomfield Hills, Michigan. She collected fossils as a child and later donated her collection to the institute, where it serves as an example of a well-documented collection. She later moved to Colorado and took a certification course in paleontology at the Denver Museum of Nature and Science in 1991. She formed the nonprofit Garden Park Paleontology Society, which established the Dinosaur Depot Museum in Cañon City, Colorado. Engard was instrumental in developing a long-term loan agreement between the depot and the NMNH to ensure that dinosaur specimens, many collected by the crews of Othniel Charles Marsh (1831–1899) in the late 1800s, were properly prepared (Monaco, 2006).

Marilyn **Fox** is chief preparator (now retired) in the Vertebrate Paleontology Lab at the YPM. She has a BFA from the University of Wisconsin, Milwaukee, and an MFA from the Pratt Institute, Brooklyn, New York. She joined the SVP in 1995 and has worked on *Poposaurus*.

Following in the footsteps of Carrie Barbour, American VP pioneer, Carrie **Herbel** is now chief preparator of fossil vertebrates at the University of Nebraska State Museum, Lincoln. She gained her BS in geology at that university, going on to postgrad work at Bryn Mawr and the University of Pennsylvania. She worked on exhibits and in lab preparation for three years before she gained an MS in 1994 in geology with a vertebrate paleontology and sedimentology focus. From 1995 to 2007, she was museum, collections, and laboratory manager and instructor at the South Dakota School of Mines and Technology, Rapid City, and was curator at the Prehistoric Museum, Utah State University Eastern, Price, from 2013 to 2015. Creating molds and casts of existing fossils is a regular part of her work in the lab. She recently created molds of fossil legs from five fossil horses as backup for a display.

Since 2003 Amelia **Madill** has been a preparator for Research Casting International, a museum services company based in Trenton, Ontario, Canada, that restores, conserves, and prepares fossils for museum exhibition. Among the specimens that she has led preparation on is the most complete ankylosaur *Zuul crurivastator* (named after a monster from the film *Ghostbusters*), on exhibit at Toronto's ROM, and a *Tyrannosaurus rex* at the NMNH's new *Deep Time* exhibit.

Chantal **Montreuil** has been a preparator for the past 10 years at McGill University's Redpath Museum in Montreal. Her most recent (and most exciting) work includes preparing the skull of *Sarcosuchus* (known as the "Super Croc") found in Niger, Africa.

Mary Jean Hildebrandt **Odano** (1924–2017) was a preparator at LACM. She began working there as a volunteer and was quickly hired as a preparator. She also worked at the related George C. Page Museum, La Brea Tar Pits. Well into her career at the museum, Odano and her husband, Tosh, started their own business, Valley Anatomical Preparations, making replications of fossil dinosaurs, as well as rare and endangered animals, until retirement in 2012.

Adele **Panofsky** (1923–unknown) was an amateur paleontologist and self-taught preparator known for helping to collect and prepare a fully mounted cast of the desmostylian *Paleoparadoxia repenningi* found during groundbreaking for the Stanford Linear Accelerator at Stanford University, founded by her husband (Panofsky, 1973). The cast of the skeleton used to be displayed at the Stanford Linear Accelerator Visitors Center and has been moved to the San Mateo County History Museum, together with information about its discovery and preservation.

Michelle **Pinsdorf** is a vertebrate fossil preparator at the NMNH who has assisted in the preparation and assembly of specimens for the museum's paleontology exhibits. She has a BS in geology from the State University of New York, where she minored in archaeology. Her fossil preparation experience started when she was an intern at the Mammoth Site of Hot Springs. She then received an MS in vertebrate paleontology from the South Dakota School of Mines and Technology, describing a skeleton of *Tyrannosaurus rex* for her thesis study. During and after graduate school, she worked as preparator for the South Dakota School of Mines and Technology Museum of Geology, focusing on Oligocene mammals from Badlands National Park and Cretaceous marine reptiles.

Akiko **Shinya** is chief preparator of fossil vertebrates at the FMNH. With a BS in geology from the University of Toronto, she went to work at the ROM and then in Robert Reisz's lab before moving to Illinois in 2001. She spends a few weeks to months each year collecting fossils and has worked in the United States, Antarctica, Argentina, Canada, China, and Romania. Notably, she discovered Late Cretaceous allosaurid *Gualicho shinyae* from Patagonia.

Shino **Sugimoto** is a fossil preparator in the Department of Natural History at the ROM. She received a degree in integrated studies in geology, archaeology, and anthropology from Weber State University, Ogden, Utah, where she learned preparation techniques through internships and volunteer work at the Utah Museum of Natural History and the Utah Geological Survey. In 2008, she immigrated to Canada from Japan and had contract project-based jobs for Research Casting International, the University of Toronto Mississauga, and the ROM until she became a full-time vertebrate paleontology technician with the ROM in 2016.

Constance "Connie" **Van Beek** is a preparator at the Integrative Research Center (FMNH). In recent years she has worked to reveal a rare large articulated Carboniferous gyracanth fish from Delta, Iowa (Snyder et al., 2016).

Other preparators include Kelsie **Adams** (Burke Museum, Seattle, Washington), Becky **Barnes** (North Dakota Geological Survey, Bismarck, North Dakota), Jennifer **Cavin** (John Day Fossil Beds National Monument, Oregon), Lisa **Herzog** (North Carolina Museum of Natural Sciences, Raleigh), Carrie **Howard** (La Brea Tar Pits, Los Angeles, California), Cathy **Lash** (Petrified Forest National Park, Arizona), Christina **Lutz** (YPM), Myria **Perez** (Perot Museum of Nature and Science, Dallas, Texas), Robin **Sissons** (University of Alberta, Edmonton), Natalie **Toth** (Denver Museum of Science and Nature, Colorado), and Vicky **Yarborough** (Duke University, Durham, North Carolina).

Recognition for Preparators

Of the 24 grantees of the Marvin and Beth Hix Preparators' Grant, only six (online appendix S3) (4%) are women. The PREPLIST is an email list open to anyone interested in the exchange of information about preparation of vertebrate fossils. There are annual preparators' conferences associated with the SVP, the SVPCA, and the Association for Materials and Methods in Paleontology, a professional organization for those interested in fossil preparation and collections management.

CT and Other Imaging Scientists

Several imaging techniques exist for characterizing fossils in three-dimensions that have been widely used to aid paleontological research over the last 20 years (Cunningham et al., 2014). The most widely used is x-ray CT, which first appeared in the 1980s (Racicot, 2017). High-energy and high-resolution variants include micro-CT scanners and x-ray synchrotron imaging, with typical resolution in the 1–50 micron range, which is ideally suited to smaller objects. The benefits of CT are many: it is nondestructive and will not damage specimens, and it provides anatomical detail not previously available. For example, morphological details of the brain have been revealed through digital endocasts (e.g., Racicot, 2017), which have provided new insights into neuroanatomical transformations and brought to light unexpected musculature of placoderm eye capsules (Burrow et al., 2005).

Other paleontological applications of three-dimensional data include conservation and education (Tembe and Siddiqui, 2014). The Digital Morphology library (DigiMorph) was created to make visualizations derived from high-resolution CT scans freely available to the public. These visualizations include three-dimensional rotations, reslicings, cutaways, and surface models, mostly of skulls and skeletons. DigiMorph went live in 2002 and now serves high-resolution imagery for more than 1,000 specimens, representing a collaborative effort between the University of Texas X-ray CT Facility and nearly 300 researchers from the world's premiere natural history museums and universities. Other online repositories for CT data include the ESRF (European Synchrotron Radiation Facility) Heritage Database for Palaeontology, Evolutionary Biology and Archaeology and Dryad. Functional analyses are also possible using tomographic methods. An engineering approach, finite element analysis, has been used in paleontology to understand feeding and locomotion in fossil taxa (e.g., Rayfield, 2007). Although many vertebrate paleontologists now employ CT and laser scanning in their research (e.g., Sánchez, 2012), we highlight a few women who were early proponents and developers of these techniques.

Asia

China. Hongyu **Yi** at the IVPP uses high-resolution CT imaging and computer modeling to study the evolution of sensory organs among fossil and living reptiles. She has an undergraduate degree from Beijing University and a PhD from Columbia University, New York, where she studied under Mark Norell at the AMNH.

Europe

Sweden and Switzerland. Swiss-born Sophie **Sanchez** is associate senior lecturer at Uppsala University, Sweden. She studied bone histology beginning in 2002 and received a PhD from the MNHN. She developed a three-dimensional bone histology project studying the fish-tetrapod transition in collaboration with Paul Tafforeau at the European Synchrotron Radiation

Facility, Geneva, and continues that work today. Another recent project revealed the wing-bone geometry that underpinned active flight in *Archaeopteryx* (Voeten et al., 2018).

United Kingdom. Keturah "Ket" **Smithson** is a research assistant in the Department of Zoology at the University of Cambridge. She is working alongside paleontologists from Clack's team on the TWeed Project, specifically focusing on micro-CT scanning of fossil tetrapods.

United States

Amy M. Balanoff is in the Department of Anatomical Sciences at Johns Hopkins University. She received a PhD from Columbia University in 2011. She was one of the first researchers to apply high-resolution CT to the study of fossil embryos and has continued her CT research to study the evolution of the brain in extinct and extant dinosaurs (see Chapter 5). Her collaboration with Mark Norell, Aneila Hogan, and Gabriel Bever for the 2018 special issue of *Brain, Behavior and Evolution* (Balanoff et al., 2018) exemplified the application of digital technology in obtaining virtual endocasts, which permit inference and analysis of brain morphology and evolution in the enigmatic Oviraptorosauria.

Jesse **Maisano** is a research scientist at the University of Texas CT facility in Austin, where she has been facility manager since 2008. She received her BA in geology at Kent State University in 1994 and her PhD in vertebrate paleontology at Yale University in 2000. She then moved to the University of Texas as a postdoc on the DigiMorph project, which she caretakes at present in addition to pursuing research on fossil birds.

Suzanne G. Strait-Holman, in the Department of Biology at Marshall University, Huntington, West Virginia, employs three-dimensional imaging and visualization in her research on the feeding adaptations and diet of early primates. She convened the session "3-D Imaging: New Techniques and Applications" at the 2006 SVP annual meeting. Strait-Holman won the SVP Romer Prize in 1990.

South America

Gabriela **Sobral** is a Brazilian paleontologist working as a postdoctoral researcher at the Museu de Zoologia, Universidade de São Paulo, Brazil, on braincase anatomy, development, and evolution in tetrapods, especially crocodilians, using CT imaging. She received a BS from the Universidade Federal do Rio de Janeiro, an MS from the Universidade de São Paulo, Ribeirão

Preto, and a PhD from Humboldt-Universität in Berlin, Germany, in biology (paleontology).

Curators, Scientific Assistants, and Collections Managers

Curators have, at least since the 18th century, amassed, cared for, and researched specimen collections. A collections management professional, distinct from a scientific assistant to a museum or research curator, has emerged in recent years. Depending on the museum, collection managers sometimes share responsibilities with curators. Collection manager responsibilities include updating the database, improving the storage conditions of fossil specimens, supervising students and volunteers working with the department, facilitating incoming and outgoing loans, and ordering supplies for the collection (the UK Museums Association has handbooks covering such topics on its website). The Society for the Preservation of Natural History Collections, founded in 1985, is an international society whose mission is to improve the preservation, conservation, and management of natural history collections to ensure their continuing value to society (Williams, 1995; Williams and McLaren, 2005). In the mid-1970s, collections began a period of computerization and museums developed searchable databases of their fossil collections.

Natural history museums grew out of the development of private collections (Wyse Jackson and Spencer Jones, 2007). Women during the late 18th century such as Etheldred Benett donated fossil material to various museums in England and as far afield as Philadelphia (see Chapter 2). It was not until the end of the 19th century that women began to take up posts in museums, albeit initially often unpaid (e.g., Dorothea Bate at the BMNH; see Chapter 3). At the beginning of the 20th century, women were employed (although still often unpaid) as assistants to male research and curatorial staff (e.g., Tilly Edinger at the Senckenberg Museum in Germany, Maria Pavlova at MSU in Russia, and Antje Schreuder at the University of Amsterdam Zoological Museum, Chapter 3). By the mid-20th century, women were employed as research scientists and curators of collections (e.g., Hildegarde Howard at LACM and Mary Dawson at the Carnegie Museum, Chapter 4).

We have discussed individual women VP curators in Chapters 3, 4, and 5 (and see Table 7.4; Fig. S7.1). Here, we highlight a few women collection managers and scientific assistants who worked with vertebrate paleontology collections.

Africa

Elize **Butler** is collections manager of the geology and paleontology collections at the National Museum, Bloemfontein, South Africa, where she has been since 1993. She obtained a BS from the University of the Free State. She recently completed an MSc thesis (cum laude) titled "The Post-cranial Skeleton of the Early Triassic Non-mammalian Cynodont *Galesaurus planiceps*: Implications for Biology and Lifestyle."

Australia and New Zealand

Kristen **Spring**, Joanne Wilkinson, and Rochelle **Lawrence** work in different capacities at the Queensland Museum. Spring has worked for two decades and is now collections manager for the geosciences collection, maintaining type and stored material; Wilkinson is in charge of preparation but also manages volunteers in the lab; and Lawrence is technical research assistant to VP curator Scott Hocknull.

Lynne **Bean** is curator in charge of the Australian National University, Canberra, earth sciences fossil collection and has been there since 2014. She completed her BS in geology at Sydney University. A science teacher until 2013, she recently gained her PhD at the Australian National University for her work on Jurassic bony fish of the classic Talbragar site in New South Wales.

Europe

United Kingdom. Lorraine **Cornish** is head of conservation at the NHM. She has over 35 years of expertise in all areas of collections management, preservation, and exhibits. She earned an MS from the University of East London in 1992 and a BA from the British Open University in 2006.

Brazilian-born Martha **Richter** is now deputy director of geology, having been collections manager and then principal curator in charge of fossil vertebrate and anthropology collections at the NHM since 2006. Each section of the Palaeontology Department has had several women curators and, in the last decades, collection managers. For instance, "Sally" V.T. **Young** joined the fossil fish section straight from school in the 1960s and was in charge of fossil fish collections until retirement. Zerina Johanson (see Chapter 5) maintained the position of fossil fish curator until promoted to research scientist in the late 2000s. Alison **Longbottom** and now Emma **Bernard** organize the fossil fish collection. Angela Milner (see Chapter 5) was in charge of the fossil reptiles collection until her retirement. More recently, Sandra D.

Chapman has been in charge of fossil reptiles and dinosaurs, with Nadine **Gabriel** assisting in the management of fossil mammals. Lorna **Steel**, crocodyliform expert, is a former curator in fossil reptiles and is now task force data collector in earth sciences.

Other important museum VP collections in Britain have been managed by women, mostly trained vertebrate paleontologists and geologists—for example, Susan Turner (museum assistant in geology, 1971–1980, Hancock Museum, Newcastle); geology curators Sheila Mahala **Andrews** (curator from the 1970s to 1993) and Roberta "Bobbie" **Paton** (retired in 2000) (Royal Scottish Museum/National Museums of Scotland, Edinburgh); Liz-Ann **Bawden** (honorary curator, Lyme Regis Philpot Museum, in 1999); and Cindy **Howells** (joined the National Museum of Wales as a volunteer in 1985; became permanent geology collections manager in 1999). In Germany in recent years there has been a similar pattern—for example, Verena **Benz** (Hessisches Landesmuseum Darmstadt), Annette **Richter-Broschinski** (Lower Saxony State Museum, Hannover), and Connie **Kurz** (Darmstadt and then Ottoneum Kassel).

Latin America

Belinda Espinosa **Chávez** is the curator of the Paleontology Museum of the Benemérita Escuela Normal de Coahuila, Mexico. Her colleague Claudia **Serrano** works at the Institute of Geology, Universidad Nacional Autónoma de México, on dinosaurs.

United States and Canada

Amy **Henrici** has been collections manager of the Section of Vertebrate Paleontology at the Carnegie Museum since 2003; before that, from 1984 to 2002, she was a preparator. Henrici received her MS in geology from the University of Pittsburgh in 1989. Her research focuses on Permian tetrapods from North America and central Europe and the diversity and evolution of fossil frogs and toads.

Pat **Holroyd** (196?–) has been a museum scientist at the UCB Museum of Paleontology since 1995. After graduating from Kansas University, she received a PhD from Duke University in biological anthropology and anatomy. Her research involves fossil reptiles and mammals and climate change among continental biota.

Charlotte P. **Holton** (1934–2007), from San Francisco, was a scientific assistant in the Department of Vertebrate Paleontology at the AMNH for

40 years (*New York Times*, 2007), where she worked with Bobb Schaeffer until her retirement in 2002. According to her colleagues, she was a classmate of Malcolm McKenna's at UCB (1960) when she joined the SVP, but no one would hire her because she was a woman, except for Marie Hopkins-Healy (see Chapter 4), whom she worked for as a research assistant in the mid-1960s, publishing occasionally (Holton, 1969; McKenna and Holton, 1967). Malcolm McKenna helped secure a position for her at the AMNH, and she is acknowledged in several dinosaur papers. She also worked on the BFV and served as BFV assistant editor (Wilson, 1990).

Russian-born Olga **Potapova** (Mammoth Site of Hot Springs, South Dakota) is curator and collections manager of the World Heritage active paleontological excavation Mammoth Site. She studies the morphology and adaptations of large Pleistocene mammals, including horse, mammoth, steppe bison, and woolly rhinoceros.

Vanessa R. **Ruhe** (LACM) is collections manager at LACM with over a decade of field, laboratory, and collections experience. Previously, she worked as a paleontological field technician, salvaging fossils throughout central and southern California. She is currently cochair of the SVP Preparators Committee.

Ivy **Rutzky** managed the AMNH fossil fish collection and also acted as John Maisey's technical assistant in the 1980s. She then transitioned to become a registrar for VP collection (see the discussion of her earlier in this chapter).

Sally **Shelton** is currently associate director of the South Dakota School of Mines Museum of Geology, with research expertise and experience in collections management. She graduated from Texas A&M University in 1979 and received an MS in paleontology from Texas Tech University in 1984. Among her publications (e.g., Shelton, 1994, 1997) is one specifically focused on preparation techniques (Shelton and Chaney, 1994).

Mary Ann **Turner** (1947–) served as curatorial assistant and later collections manager at the YPM from 1975 to 2010. She received a BS (1969) from Indiana University and an MS (1972) on Pliocene vertebrates of Nebraska and Kansas, and she worked toward a PhD on mastodonts and gomphotheres from the University of Nebraska (e.g., Turner, 1975).

Melissa **Winans** served as collections manager at the University of Texas at Austin Vertebrate Paleontology Laboratory, undertaking much of the primary computerization of the collection in the late 1980s. She is now senior computer systems administrator for the Texas Memorial Museum at the University of Texas.

Other curators, collections managers, and scientific assistants for US and Canadian VP collections include Susan K. **Bell** (AMNH, VP since 1975, assistant to McKenna) and coauthor of McKenna and Bell (1997), *Classification of Mammals above the Species Level*; Sarah **Boessenecker** (collections manager, Mace Brown Museum of Natural History at the College of Charleston, South Carolina); Crystal **Cortez** (John D. Cooper Center, Santa Ana, California); Jessica **Cundiff** (MCZ); Janet Whitmore **Gillette** (MS from the South Dakota School of Mines and Technology, now at the Museum of Northern Arizona); Lynette **Gillette** (curator, Ruth Hall Paleontology Museum, Ghost Ranch, Abiquiu, New Mexico; see Psihoyos and Knoebber, 1994; Colbert, 1996; Eliot, 2000); Kathy **Hollis** (NMNH); Juliet **Hook** (assistant collections manager, LACM); Carolyn **Levitt-Bussian** (Natural History Museum of Utah, Salt Lake City who worked on bone histology of dinosaurs for her MS; Victoria **Markstrom** (Canadian Fossil Discovery Centre, Morden, Manitoba); Amanda **McGee** (Cleveland Museum); Michèle **Morgan** is curator of osteology and paleoanthropology at Peabody Museum, Cambridge, US. She received SVP's Romer Prize in 1991; Hayley **Singleton** (Amherst College); Mary **Thompson** (retired from the Idaho Museum of Natural History); Barbara **Tihen** (née Waters) (UCMP scientist), who, in the 1960s to 1980s, did curatorial work and fieldwork and helped develop the casting exchange program at the UCMP; and Katherine E. "Kathy" **Wolfram** (AMNH VP), who worked as a research assistant and collection manager in fossil fish, publishing on *Notidanus* (Maisey and Wolfram, 1984).

Outreach Educators

Beginning in the mid-1970s, more attention was given to public outreach, with lectures and workshops at schools and museums designed to improve the public's understanding of science and paleontology. This stemmed, in part, from the battle over the teaching of creationism and intelligent design being waged in certain US public schools, which had an obvious connection to paleontology (e.g., Esensten, 2003). This battle spread elsewhere, and entities such as the SVP and the PS, notably Steve J. Gould, responded with position statements on evolution (see National Center for Science Education website; Overton, 1982; Walker, 1984). Consequently, there has been a trend worldwide to use paleontology as a vehicle to improve science education.

Africa

Tsiory **Andrianavalona** is a vertebrate paleontologist working on the taxonomy, paleoecology, and paleoenvironment of Madagascar's fossil sharks. She received her PhD from the University of Antananarivo in Madagascar. She is currently leader of ExploreHome, the first Science Center project in Madagascar, which is focused on inspiring an interest in science among young Malagasy, especially STEM students. She received the 2019 Raymond M. Alf Award for Excellence in Paleontological Research and Education.

Growing up in Nairobi, Louise **Leakey** (1972–) is a third-generation paleontologist and the next of the Leakey family to contribute to Kenyan VP. She has taken on a role in education as director of public education and outreach for the Turkana Basin Institute, Stony Brook University, New York. She is also an assistant research professor at Stony Brook. Along with her mother, Maeve, Louise is a National Geographic Explorer-in-Residence. One of Louise's most memorable early experiences was in a team led by her father, Richard, which found "Turkana Boy," a nearly complete, 1.5-million-year-old skeleton of the hominid *Homo erectus*. She then left Africa for a few years to obtain her PhD in biology at UCL (Thornton, 2012). Louisa is now encouraging younger Kenyans to be involved in their prehistory (see the Turkana Basin Institute's Facebook page).

United States and Canada

Riley **Black** is a freelance science writer, author of *Skeleton Keys, My Beloved Brontosaurus,* and other books about fossils. She has also assisted various academic institutions with paleontological fieldwork across the American west. She is based in Salt Lake City, Utah.

Diane L. **Gabriel** (1946–1994) was a "passionate spokesperson for public understanding of science in general and more specifically vertebrate paleontology" (Coombs et al., 1995:65). A major focus of her work was to co-lead, with fellow paleontologists Mac West and Rolf Johnson, the Milwaukee Public Museum's Dig-a-Dinosaur program, which not only secured significant fossil material from the Hell Creek Formation for the museum's research and exhibits program but also set a standard for the involvement of volunteers engaged in assisting museum staff with paleontological fieldwork. In fact, her many years of work and contributions to the program spawned many similar programs at other museums. In 1990 she moved to Bozeman, Montana, and undertook a PhD under the mentorship of Jack Horner, in addition to continuing her Dig-a-Dinosaur program at the Museum of the Rockies. While at the Museum of the Rockies, Gabriel helped Patrick Leiggi and other VPs with her insights on how to best protect our natural heritage for future generations (Coombs et al., 1995).

Ashley **Hall** is a paleontologist and outreach educator at the Cleveland Museum of Natural History in Ohio and is writing children's books on fossils. Her undergraduate education was in anthropology, but her passion as a child was paleontology. She spent five years as an intern for the US Bureau of Land Management at the Raymond M. Alf Museum in Claremont, California, identifying and cataloging fossil bones and also working at LACM, where she developed tours and educational programming and taught school classes on field trips. She is part of the Bearded Lady project (see Chapter 7).

Kathryn **Hoppe** is a paleontologist and geobiology teacher at Green River Community College near Seattle, Washington. She received a BA in earth and planetary sciences from Washington University, Saint Louis, and a PhD from Princeton University. Her past research involved the use of stable isotope geochemistry of teeth and bones to reconstruct the paleobiology of fossil mammals and paleoenvironmental conditions. Currently she is a professional science writer and the author of several children's books about fossil animals under the pen name of Charlotte Lewis.

Bolortsetseg "Bolor" Minjin is a Mongolian paleontologist well known for her work on fossil repatriation and dinosaur-based outreach, in addition to her research on dinosaurs and early mammals (see Chapter 5). Since moving to the United States, she has focused her work on improving the state of Mongolian paleontology (Plate 39). She collaborated with paleontologist Jack Horner at the Museum of the Rockies, who helped her establish the Institute for the Study of Mongolian Dinosaurs in 2007.

Today, Mongolian laws prohibit the export of vertebrate fossils and require excavators to carry permits, but those laws have been hard to enforce without funding or manpower. As she noted in an interview, "Once a fossil left the country . . . knowledge left with it" (Boodhoo, 2017b). In the last four years, Minjin has helped repatriate more than 30 illegally exported Mongolian dinosaurs that have turned up in the United States alone. In 2013 the AMNH donated to the Institute for the Study of Mongolian Dinosaurs the Moveable Museum, an old Winnebago RV, filled with interactive exhibits including replicas of Mongolian dinosaurs, bringing the museum to grade-school children in small Mongolian towns. In 2016 the Institute for the Study of Mongolian Dinosaurs applied for nonprofit status and launched a fund-raising campaign to keep the Moveable Museum operating. Now Ulaanbaatar has a new Central Museum of Mongolian Dinosaurs, which exhibits many of the repatriated Mongolian dinosaurs. In recognition of her work in fossil repatriation and science outreach, she received the WINGS WorldQuest Women of Discovery Award for the Earth 2009, and she is a National Geographic Explorer and TEDx speaker.

Judith "Judy" G. **Scotchmoor** (194?–) (retired from the UCMP) was director of education and public programs at the UCMP until 2012, with a primary focus on the use of paleontology and technology as vehicles for improving science education in the classroom. Scotchmoor is perhaps best known for her central role in developing Understanding Evolution, one of the premier science education websites in the world, which was initially designed to provide tools for K–12 teachers and for a book she coedited that uses dinosaurs to illustrate the discoveries and methods of science, *Dinosaurs: The Science behind the Stories*, published by the SVP, PS, and AGI (Scotchmoor et al., 2002). She served as the cochair of the SVP Education Committee from 1999 to 2003 and was a recipient of the Gregory Award in 2004. In recognition of her exceptional public service, in 2016 she received the PS's John and Mary Lou Pojeta Award.

Lisa **White** is an American geologist and director of education and outreach at the UCMP. Her initial interest in geology and paleontology was inspired by trips to the California Academy of Science in San Francisco. She received a BA in geology at San Francisco State University and returned there as a faculty member after earning a PhD in earth sciences from the University of California, Santa Cruz, in 1989. She is also a micropaleontologist, and from 1990 until joining the UCMP in 2012, she was professor of geology and later associate dean at San Francisco State University. White has long been a proponent of improving diversity, and she coordinated a number of outreach efforts to encourage the participation of minority students in the geosciences. She received the Bromery Award from the GSA in 2008 in recognition of her work with minorities in the field. She creates what she calls "critical incidents"—experiences that first enlighten young people, especially high school students, about earth science. If you wait too long, she notes, you may miss out on pulling students into the geosciences. By the time students take their first college geology class, many have already decided to do something else with their lives. "If [high school] students can have a positive experience or series of experiences with geoscience—a great class or an outdoor learning opportunity, for example—then that enlightens them," making them think, for the first time perhaps, about becoming an earth scientist themselves, she says (Boodhoo, 2017a). She particularly pointed out that in an interview with us, "Women in VP play such important roles in influencing the next generation of scientists. Promoting our stories, showing women as multidimensional scientists and trailblazing professionals who make unique contributions to VP, are essential storylines that should be communicated. The narratives are likely to draw strong interest from girls and from women in the early stages of their VP careers."

Kate **Wong** is senior editor and writer at *Scientific American*, covering paleontology, archaeology, and the life sciences. She received a BS in physical anthropology and zoology from the University of Michigan. She has led student roundtables at the SVP annual meetings on science communication.

Most museums today have outreach programs, including summer classes, lectures, workshops, and field experiences, that highlight paleontological research. Among VP outreach and science educators are Meaghan **Emery-Wetherell** (PBS science writer, Ellensburg, Washington), Sharon **Holte** (Mammoth Site, Hot Springs, South Dakota), Tara **Lepore** (Raymond Alf Museum, Claremont, California), Alison **Moyer** (Edelman Fossil Park, Mantua Township, New Jersey), Jacqueline M. **Richard** (Delgado Community College, New Orleans, Louisiana), Deborah **Rook** (Geology Museum, University of Wisconsin, Madison), Tanya **Samman** (Canadian Science Publishing), Franziska "Franzi" **Sattler** (Humboldt Museum für Naturkunde, Berlin), Laura **Soul** (NMNH), Ta-Shana **Taylor** (Outdoor Afro, New Mexico), Kelli **Trujillo** (Laramie County Community College, Wyoming; also the SVP 2004 Colbert Poster Prize winner), and Rissa **Westerfield** (Clariden School, Southlake, Texas).

Summary

As noted in Chapter 1 (Figure 1.3), the earliest women VPs in the 18th to mid-19th centuries began as artists and collectors, often illustrating fossil vertebrates for their husbands and other relatives, including, in the United Kingdom, Mary Anning, Mary Buckland, and Mary Mantell. By the late 19th century, women paleoartists began to emerge elsewhere in Europe and the United States. Among the earliest American female scientific illustrators were Orra Hitchcock, Graceanna Lewis, Cecilia Beaux, and Hattie Huntsman, all known for their detailed drawings of fossil vertebrates. In the United Kingdom during this time, Alice Woodward began her work, and she is now known for her skillful, colorful reconstructions of fish, dinosaurs, and mammals for various books and magazines. Later in the 20th century, women began to create their own books and illustrate those of others (e.g., Margaret Colbert, Maggie Lambert, and Jenny Halstead). The "Dinosaur Renaissance" was well under way by the close of the 20th century, influenced by new discoveries and new understandings of their paleobiology. Dinomania ensued, and dinosaur artists such as Ely Kish and Sylvia Czerkas began depicting dinosaurs as active predators with colorful skin and other adornments. By the 21st century, digital art appeared and feathered dinosaurs and mammals were being reconstructed by various artists (e.g., Raven Amos, Karen Carr, Emily Willoughby, Carol Abraczinskas, and Luci Betti-Nash).

Women VP curators began employment during the mid- to late 19th century initially as unpaid museum assistants to largely male curatorial staffs, except for the unusual Carrie Barbour later occupying research positions that still exist today. For all time periods except the earliest (1700s to 1850s) and across all employment categories, women led as researchers in university faculty and as museum curators.

Few women preparators are known from the 19th century, probably the result of the physical effort needed to excavate large fossils using blunt, heavy instruments. SVP charter member Carrie Barbour is among the earliest vertebrate fossil preparators in the United States, as well as being a collector (see Chapter 4). In the 20th century, chemical preparation techniques came into use, and in subsequent decades of the century, noninvasive methods such as scanning electron microscopes and three-dimensional reconstructions of CT scans became common tools in preparation. Women preparators exhibit their highest numbers beginning in the 1970s and continuing to the present.

CHAPTER 7

Challenges and Opportunities

> There is, perhaps, no other branch of science more "international" than vertebrate paleontology. . . . I myself and my colleagues from the IVPP have benefited a lot from the SVP in sharing information, collaborating with colleagues from many countries and in learning from them.
>
> Meeman Chang (1998:3)

In our historical survey in preceding chapters, we have seen how women VPs were rebels, scholars, and explorers from the earliest times, with opportunities and contributions to the field increasing through time so that now we find women vertebrate paleontologists working in many countries. However, there are still major challenges for women entering our discipline, as with any scientific training. Here we provide a framework to address gender parity issues in STEM, as well as data on gender diversity in various areas of geosciences and vertebrate paleontology (i.e., membership, leadership, and recognition in professional societies and scholarship).

Women in STEM, Geosciences, and Paleontology

Gender equality is crucial for producing the best research, and it is everyone's responsibility. Increased diversity of all types (including scientific discipline, work experience, gender, race, ethnicity, and nationality) leads to better science through the benefit of different perspectives and life experiences, which raise unique questions and create new approaches to problem solving (Normile, 2001; Medin and Lee, 2012; Nielsen et al., 2017; Coe et al., 2019; Hooks, 2019).

Women have experienced gender-specific barriers that have impeded their research careers (Greider et al., 2019). Explanations for the underrepresentation of women in science (STEM fields) range from implicit gender bias to historical, social, and institutional barriers (Switek, 2010; Dutt et al., 2016). According to UNESCO data for 2015, less than 30% of the world's researchers are women, representing only 19% in South and West Asia, 23% in East Asia and the Pacific, 30% in sub-Saharan Africa, 32% in North America and western Europe, and 45% in Latin America (UNESCO, 2018).

Undergraduate and Graduate Education

Although women have made significant progress in undergraduate and graduate education in STEM fields, they remain underrepresented overall (Xie et al., 2015). While nearly as many women as men now hold undergraduate degrees in the United States and have done so since the late 1990s, they make up only about 30% of STEM degree holders (Noonan, 2017; NSF, National Center for Science and Engineering, 2017) (Figures 7.1 and 7.2). Outside the United States, however, gender parity in science differs. For the years 2012–2015, based on UNESCO data, the percentage of women among STEM graduates ranged from 12.4% in Macao (China) to 40.7% in Algeria (Stoet and Geary, 2018). The lack of racial and ethnic diversity is especially glaring, as for instance in both the United States and the United Kingdom (Royal Society, 2014; Ball, 2015; NSF, National Center for Science and Engineering, 2017). In 2014–2015 in the United States, women of color earned a small percentage of bachelor's degrees in STEM fields: black women, 2.9%; Latinas, 3.6%; and Asian women, 4.8% (National Center for Education Statistics, 2016). In Australasia, the percentage is also low (nonexistent in VP), but attempts to increase representation of indigenous women in science are under way (CSIRO, undated). Situated within this context of profound underrepresentation, it is unsurprising that the experiences of women vertebrate paleontologists are marred by racial and gender inequity. Renowned paleobiologist Anusuya Chinsamy-Turan recounted in her interview with us, "Growing up in South Africa during the apartheid days, I felt oppressed as a black person (much more so than being a woman). As a black person, I had various restrictions imposed on me (e.g., I needed a study permit to register at Wits University; I could not live close to the university so traveled two hours per day; I was not allowed on most of the buses; I was not able to stay overnight in the Free State, which is the heart of the famous Karoo basin—so [this] directly affected fieldwork). The challenges of being a mother and balancing my career [were] really nothing compared to the oppressiveness of apartheid."

The challenges that social barriers can present are multifaceted. Marisol Montellano-Ballesteros, paleontologist at the Universidad Nacional Autónoma de México, related, "One of the main challenges is being a woman in the field; in my country it was not common to see women in the countryside collecting, working, so the men treated me like a teacher—for them that was [an] easier way to deal with my presence."

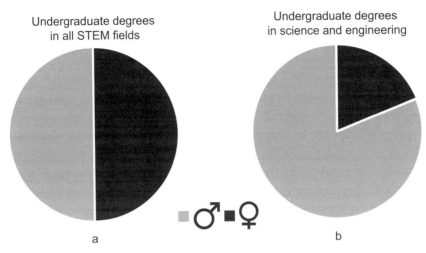

Figure 7.1. *a.* Total degrees in STEM, men versus women. *b.* Total degrees in science and engineering, men versus women.
Source: Data from National Science Board, 2018.

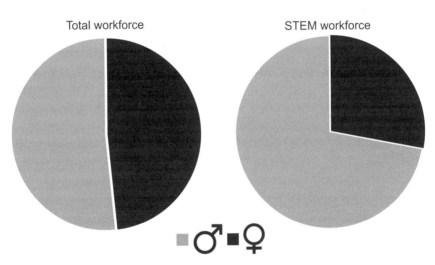

Figure 7.2. Total workforce versus STEM workforce, men versus women.
Source: Data from National Science Board, 2018.

As Anjali Goswami (NHM) noted in her interview with us, "I've been in this field for 20 years now; you definitely feel like as a woman and as a minority, it takes a lot more to get noticed . . . for people to even be aware of who you are. And I definitely feel that . . . at conferences, even as now a fairly senior researcher in my field, I feel invisible at times to people. I definitely feel that you really have to fight to get noticed and for people to pay attention to who you are. And I think one thing that women have to do that . . . men don't really have to do . . . I almost feel like I have to present a CV of myself when I'm meeting new people because until they realize that I am a well-established scientist with a big lab and a healthy record of research, they will be very dismissive of me. Hopefully it will get better. But . . . I don't think it has necessarily changed over the last few decades. Some of the difficulties I've had . . . are very generic. I think women not being included in collaborative networks that are really dominated by groups with men, things like not being cited as much, getting more criticism than your male colleagues. So those things are difficult . . . but I think you just have to push through and I think the solution is for . . . people like me . . . to force our way to the top so that we can make sure that it's easier and better and fairer for all the people that are coming behind us because there is definitely a difference when women are on panels and so that's the solution to get through this period of change."

Paleontology is an interdisciplinary field, which makes obtaining accurate data on women in the various subdisciplines (i.e., micro-, invertebrate, and vertebrate paleontology and paleobotany) challenging, with most of the data and discussion about women in paleontology coming from the geosciences. Whereas nearly 50% of geoscience undergrads in the United States are women, less than 25% of the geoscience workforce is composed of women, based on 2013 data from the NSF and the AGI (Wilson, 2016; Stigall, 2017). Although a higher number of women geoscientists (33%) has been reported for 2018, this category also includes environmental scientists (Figure 7.3; Wilson, 2019). The geosciences lag behind all other STEM fields in terms of both gender and underrepresented minorities (NSF, National Center for Science and Engineering, 2017; Dutt, 2019). In 2016, only 6% of geoscience doctorates awarded to US students and permanent residents went to students from underrepresented minorities, a group who made up 31% of the US population in that year (Kaiser Family Foundation, 2017). Significant gains have been made in terms of gender balance in graduate degrees in geosciences, with women receiving nearly 45% of doctoral degrees. However, the proportion of underrepresented minorities among PhD recipients in the geosciences has not improved in more than four decades despite efforts by the community to try to improve diversity (Bernard and Cooperdock, 2018). The data demonstrate, furthermore, that the geoscience workforce does not match the training pool. Although significant numbers of women earn geoscience degrees in the United States, there is not equivalent representation

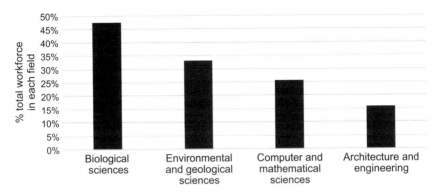

Figure 7.3. STEM workforce by field based on 2018 data. *Source*: Bureau of Labor Statistics, 2019.

of women as faculty. Following graduate school, women in the United States hold only 20% of faculty positions at PhD-granting research universities, less than 15% of full professorial positions, with only 36% as assistant professors (Glass, 2015; Wilson, 2017). In earth sciences in the United Kingdom, nearly one-half of PhD students are women, yet women make up only 10% of professors (Royal Society, 2014, based on 2011–2012 data); in this same document, it was reported that over 95% of UK geoscientists are white. Faculty women in the United States specializing in paleontology represent 34% of women in geoscience subdisciplines, with 17% identified as vertebrate paleontologists based on AGI 2002 data (Holmes and O'Connell, 2003). Data on museum curators is scarce, but a survey by postdoctoral fellow Meghan Balk at the NMNH in 2018 revealed that of 156 US curators in paleontology (i.e., micro-, invertebrate, and vertebrate paleontology and paleobotany), 18% were held by women (Table 7.1; online figure S7.1). This gap between training pool and workforce suggests that the problem of the underrepresentation of women in science will not be resolved solely by more generations of women moving through the pipeline but that, instead, women's advancement within science may be impeded by other factors, such as gender biases that favor male students (Moss-Racusin et al., 2012; see also later discussion).

Graduate programs in VP in the United States and elsewhere listed on the SVP website are provided in our supplementary online appendix S4. Paleontology programs are listed on the Paleontological Society website.

TABLE 7.1

Women curators in paleontology

State	Museum	Department	Number of curators	Number of female curators
Alabama	Alabama Museum of Natural History	Invertebrate paleontology	2	0
Alabama	Anniston Museum of Natural History	Paleontology	1	0
Alaska	University of Alaska Fairbanks	NA	1	0
Arizona	Arizona Museum of Natural History	Paleontology	1	1
California	California Academy of Sciences	Invertebrate zoology and geology	5	1
California	Natural History Museum of Los Angeles County	Dinosaur Institute	2	0
California	Natural History Museum of Los Angeles County	Invertebrate paleontology	1	0
California	Natural History Museum of Los Angeles County	Rancho La Brea	2	1
California	Natural History Museum of Los Angeles County	VP	2	0
California	San Diego Natural History Museum	Paleontology	1	0
Colorado	University of California Museum of Paleontology	Paleontology	14	3
Colorado	Denver Museum of Nature and Science	Paleontology	4	0
Connecticut	Bruce Museum	Science	1	0
Connecticut	Yale Peabody Museum	Invertebrate paleontology	3	0
Connecticut	Yale Peabody Museum	Paleobotany	1	0
Connecticut	Yale Peabody Museum	VP	4	1
Florida	Florida Museum of Natural History	Paleontology	6	0
Hawaii	Bishop Museum	Vertebrate zoology	0	0
Idaho	Idaho Museum of Natural History	Earth sciences	1	1
Illinois	Burpee Museum	Paleontology	1	0
Illinois	Field Museum	Fossils and meteorites	3	0
Indiana	Orma J. Smith Museum of Natural History	Paleontology	3	1
Kansas	Sternberg Museum	Paleontology	4	1
Kansas	University of Kansas Natural History Museum	Invertebrate paleontology	2	0
Kansas	University of Kansas Natural History Museum	Paleobotany	1	1
Kansas	University of Kansas Natural History Museum	VP	2	0
Louisiana	Louisiana State University Museum of Natural Science	Invertebrate paleontology	0	0
Louisiana	Louisiana State University Museum of Natural Science	VP	1	1
Maine	Maine State Museum	Geology	1	0

State	Museum	Department		
Maryland	Calvert Marine Museum	Paleontology	2	0
Massachusetts	Harvard Museum of Comparative Zoology	Invertebrate paleontology	1	0
Massachusetts	Harvard Museum of Comparative Zoology	VP	1	1
Michigan	University of Michigan Museum of Paleontology	Paleontology	5	0
Minnesota	Science Museum of Minnesota	Paleontology	2	1
Mississippi	Mississippi Museum of Natural Science	Paleontology	1	0
Montana	Museum of the Rockies	Invertebrate paleontology	2	1
Nebraska	University of Nebraska	VP	1	0
Nebraska	University of Nebraska	NA	3	0
Nevada	Las Vegas Natural History Museum	Natural history	6	2
Nevada	Nevada State Museum	Paleontology	1	0
New Mexico	New Mexico Museum of Natural History	Paleontology	3	0
New York	American Museum of Natural History	Paleontology	9	1
New York	New York State Museum	Invertebrate paleontology	2	1
New York	New York State Museum	Paleobotany	0	0
New York	New York State Museum	VP	1	0
North Carolina	Aurora Fossil Museum	Director	1	1
North Carolina	North Carolina Museum of Natural Sciences	Paleontology	3	2
Ohio	Cleveland Museum of Natural History	Invertebrate paleontology	1	0
Ohio	Cleveland Museum of Natural History	Paleobotany	1	1
Ohio	Cleveland Museum of Natural History	VP	1	0
Ohio	Ohio History Center	Natural history	2	1
Oklahoma	Sam Noble Museum	Invertebrate paleontology	1	0
Oklahoma	Sam Noble Museum	Paleobotany	2	0
Oklahoma	Sam Noble Museum	VP	1	0
Oregon	University of Oregon	Paleontology	1	1
Pennsylvania	Carnegie Museum	Paleontology	1	0
Rhode Island	Museum of Natural History and Planetarium	Paleontology	1	0
South Carolina	Mace Brown Museum of Natural History	Geology	1	0

(continued)

TABLE 7.1 (continued)

State	Museum	Department	Number of curators	Number of female curators
Tennessee	University of Tennessee, Knoxville, McClung Museum of Natural History	Paleoethnobotany	1	0
Texas	Houston Museum of Natural Science	Paleontology	2	0
Texas	Perot Museum of Nature and Science	Paleontology	2	0
Texas	Whiteside Museum	Paleontology	2	0
Texas	Witte Museum	Paleontology	1	0
Utah	Natural History Museum of Utah	Paleontology	1	0
Virginia	Virginia Museum of Natural History	Paleontology	1	0
Washington	Burke Museum	Micropaleontology and invertebrate paleontology	1	1
Washington	Burke Museum	Paleobotany	1	1
Washington	Burke Museum	VP	**2**	**0**
Washington, DC	Smithsonian	Paleontology	13	1
West Virginia	Mini-Museum of Geology and Natural Science	Paleontology	1	0
Wisconsin	University of Wisconsin–Stevens Point	Paleontology	1	1
Wyoming	University of Wyoming Department of Geology and Geophysics	VP	**1**	**0**

Source: Slightly modified from an unpublished list originally compiled by Meghan Balk (former Smithsonian Institution postdoctoral fellow, 2018) in May 2018.
Note: Number of vertebrate paleontologists are in bold.

Professional Societies: Membership, Awards, and Leadership

Membership. Women are underrepresented in major geoscience professional societies, including paleontological societies (Stigall, 2013b; Popp et al., 2019). Membership records of the SVP were published in the *SVP News Bulletin* beginning in 1941 and continued to appear until 2010. The first Membership Committee was not formed until 1994, cochaired by Catherine Badgley and Emily Buchholtz. The responsibilities of the Membership Committee include reviewing and updating SVP policies related to joining the society, renewing membership, and offering opportunities or honors to particular members (i.e., honorary membership, student travel grants, and Institutional Membership, a program for institutions of developing nations). More recently, under the auspices of this committee, a report examined SVP membership data (2003–2012) to determine the status of women in the field and employment trends, although analyses were only conducted on North American SVP members (Bednarczyk, 2013).

As highlighted earlier, the first organizational meeting for the SVP was held in 1940 at the MCZ at Harvard University. In addition to 31 men, three women were present (see Chapter 4). Not until two decades later, in 1964, did women make up even 10% of SVP membership. By 1975 women represented only 15% of the membership, rising slightly to 18% in 1980. In 1990–2003 slightly more than 25% of the SVP membership was women. By 2012 women SVP members had increased to 36%, with the greatest growth among student members. This same representation of women in the SVP continued in 2017 (Figure 7.4; online table S7.1). A 2017 survey of the membership ($n = 480$) revealed a nearly equal representation of male and female student members, at 25% and 21%, respectively. Although since 2012 women have made up nearly 40% of student membership in the SVP, less than 25% of professional members (e.g., curator or tenure-track faculty) are women (Stigall, 2013a; Plotnick et al., 2014; Whatley and Beck, 2016). The number of women members of the SVP is comparable to that of the PS, where women formed just 23% of full membership in 2013 and 42% of student membership. Diversity data (e.g., gender and orientation, race and ethnicity) has not been tracked in previous years for SVP but is planned in the future. For PS diversity data was tracked beginning in 2017.

Awards. An important aspect of membership in professional societies is recognition of the significant contributions of its members by selection of recipients for society awards (e.g., Holmes, 2011). Again, women in vertebrate

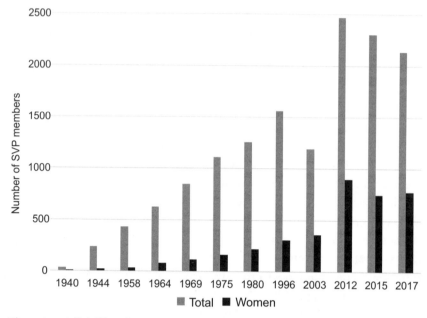

Figure 7.4. Official list of women members of the SVP, 1940–2017.
Sources: Data from SVP Membership Committee reports and the SVP Business Office.
See online table S7.1.

paleontology are underrepresented in terms of awards, especially senior awards. Women have received just 16% of SVP senior awards (Romer-Simpson, Gregory, Skinner, and honorary membership) (*SVP News Bulletins*). It took nearly 30 years for the first women, Nelda Wright and Rachel Nichols, *SVP News Bulletin* and BFV, to gain honorary member status. In terms of student awards, the proportion of SVP women awardees is much higher, with women receiving 39% of student awards since their inception in 1978. Since 2007 this number has held steady at approximately 34% (see online appendix S2).

In comparison, in the PS, only 5.6% of women have received senior awards through 2017 (the PS Medal and the Strimple Award; Paleontological Society website), but women students have received 53% of student research grants since 2007. Among 124 PS fellows, only 4% are women vertebrate paleontologists.

Representation of women vertebrate paleontologists in larger scientific societies is particularly low. Since the 1870s the number of vertebrate paleontologists elected as AAAS fellows is only 3%, with women VPs representing less than 0.5% of the total (data from the AAAS). The representation of women in the National Academy of Sciences (NAS), the nation's leading society of distinguished scientists, was 16% in 2019 (Berenbaum, 2020). Strictly speaking, no women vertebrate paleontologists have been elected to membership in the National Academy of Sciences, although paleoanthropologists Mary and Maeve Leakey (see Chapter 5), with VP contributions, have been elected foreign members. Women vertebrate paleontologists have received other societal awards, including those from the GSL and Linnean Society of London (see Chapters 4 and 5).

Leadership. Women were initially not represented in leadership positions in SVP (i.e., Executive Committees, 1940–1960). As we noted earlier, the first woman to be elected SVP president was Tilly Edinger in 1964. The second woman president did not come until 10 years later, Mary Dawson in 1974. Another 30 years had to pass before another woman, Annalisa Berta, was elected SVP president, although several women in the intervening years were asked to stand for nomination as president but declined owing to other commitments. Women, however, were active in the society and did much of the hard work behind the scenes—*SVP News Bulletin* editorship, BFV compilation, and some foreign (Edinger) and much regional and institutional *SVP News Bulletin* reporting. Encouragingly, since 2006, women have been elected president in 10 of the last 12 years. From 1991 to 1995, women made up less than 25% of Executive Committee (president, vice president, secretary, treasurer, and members at large) membership. Beginning five years later (1996–2000), they made up 30% of the Executive Committee. The proportion of women is 30% at present, with women achieving their highest representation to date, nearly 48%, from 2006 to 2010 (online appendix S3).

Scholarship: Publications, Meeting Presentations, and Grant Success

Publications. One of the most important measures of scientific achievement for career advancement and job retention is authorship of peer-reviewed journal publications. The number and quality of publications are linked with perceived competence in a discipline (Sonnert and Holton, 1996). Previous studies of researchers in related disciplines, evolutionary biology and ecology, indicate a clear difference in the number of publications produced by men and women, with men publishing almost 40% more papers than women (Symonds et al., 2006). These results are consistent with those for paleontology. Since paleontology is such a diverse discipline, vertebrate paleontologists publish their research in a wide range of journals, including those pertaining to both earth and biological sciences. Based on a compilation from the JSTOR database of scholarly research from 1990 to 2011, women authored only 16.6% of paleontology papers (West et al., 2013; Grogan, 2019). Among paleontological journals is the flagship publication of the SVP, the *Journal of Vertebrate Paleontology* (*JVP*), begun by Jiri Zidek (University of Oklahoma) in 1980, with the first issue in 1981. Despite making up 18% of the SVP membership from 1981 to 1985, women with first-authored papers in *JVP* during this same time interval made up only 7% of authors (Figure 7.5; on-line figure S7.2; online table S7.2). This latter number is comparable to the composition of woman-authored papers published in the Argentine paleontological journal *Ameghiniana* during the same period (Miguel et al., 2013).

In some areas of biology, it is now common practice to have the senior author appear last on papers that come out of a principal investigator's lab. From 2010 to 2017, women first- and last-authored papers ranged between 26.4% and 32.4% of all *JVP* articles (Figure 7.5). The numbers for the *Journal of Paleontology* (initiated in 1927) in 2017 are lower, with women as first authors in only 20% of papers published. Similarly, the percentage of women-authored papers in *Palaeontology* is less than 20% during this same time interval (Warnock et al., 2020). These numbers are comparable to the 27% reported by Jory Lerback and Brooks Hanson (2017) for women first-authored papers published in earth science journals. Much has been written about gender differences in research performance, in particular the problem of accounting for the relationship between quantity and quality of publications. One of the most common measures used is the *h* index, which

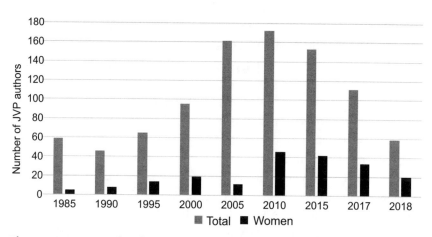

Figure 7.5. Women as first/last authors for papers in *JVP*, 1981–2018.
Sources: Data from *JVP* statistics, courtesy of Lars Werdelin, former *JVP* editor. See online table S7.2.

is the number of papers, *h*, by a scientist where each paper has received *h* or more citations (ideally excluding self-citations). The *h* index is highly correlated with quantity of research output, and it has been shown to be strongly biased against women researchers, which suggests that more equitable measures are needed (Symonds et al., 2006; Cameron et al., 2013). In a study of gender equality in research publications, Michael Bendels et al. (2018) reported underrepresentation of woman-authored papers. Women first- and last-authored papers from 2008 to 2016 for life sciences journals (e.g., *Current Biology, PLoS Biology*) are 35.3%, for multidisciplinary journals (e.g., *Nature, Science, Proceedings of the National Academy of Sciences*) are 30.6%, and for earth and environmental sciences journals (e.g., *Nature Geoscience, Geology*, with no purely paleontology journals included) are 24%. Differences in woman-authored articles in science journals across countries ranged from 17% in Japan to 49.5% in Portugal (Bendels et al., 2018). Another study found that female authorship in ecology and zoology journals has risen marginally since 2002 (from 27% to 31%) (Salerno et al., 2019). Authorship of scientific publications also reflects collaborations among scientists. Eduardo Araújo et al. (2017) discovered substantial gender differences in scientific collaborations among Brazilian scientists. Men were found to collaborate more often with other men than with women across different fields and regardless of the number of collaborators. These results provide quantitative support for future analyses, including specific causes for the patterns.

Related to journal authorship is the peer-review process and solicitation to review manuscripts for publication, important in determining journal priorities. Representation on editorial boards conveys a degree of prestige and provides opportunities for professional growth and networking. Both reviewers and editors influence scholarly publications in many ways, reflecting the differing experiences and values of men and women (Fox et al., 2019). There is clear evidence that women are underrepresented as both reviewers and editors (Fox et al., 2019). Lerback and Hanson (2017) reported that women made up a disproportionately small percentage of reviewers (20%) for American Geophysical Union journals (only one of which is related to paleontology but not specifically VP) from 2012 to 2015. For this same time interval, women reviewers for published papers in *JVP* averaged only 17.5% (online table S7.2). Another study (Fox and Paine, 2019) found that papers authored by women have lower acceptance rates and are less well cited than papers authored by men in ecology and evolution journals. The journal *Evolution* was in the data set; however, no paleontology journals were among the six journals included. A study of the gender diversity of editors among this same data set of ecology and evolution journals revealed lower proportions of women editors. For example, for *Evolution* from 2012 to 2015, both women editors and women reviewers ranged between 30% and 35% (Fox et al., 2019).

Meeting Presentations (Posters and Talks). The presentation of research at conference meetings is another important scholarly activity. Again, since paleontology is a diverse field, workers in different subdisciplines often attend different conferences. The first SVP meeting, in 1941 in Boston, was held jointly with the PS and the GSA. Tilly Edinger was the only woman who made an oral presentation. The number of women making presentations (both talks and posters) during the next decade ranged from 2% to 9% (1950). From 1951 to 1970, this number ranged from 0% to 8% (with an exception of 18% in 1963) and from 1971 to 1980 averaged less than 15%. From 1981 to 1990, women making presentations averaged about 15%. This number had increased to 21% by 2001 and by 2005 to 26%. From 2010 to 2015, approximately 29% of presentations were given by women, still less than one-third of all presentations (Figure 7.6; online table S7.3). In 1987 posters were introduced at annual SVP meetings as a presentation option. The percentage of women as first authors of posters has increased from 16% in 1987, to 23.7% in 2007, and to 32.8% in 2018. Ideally, the proportion of presentations by women should be compared to the proportional membership of women in the society for each year, but unfortunately these numbers are not available for each time period.

As is generally true for the SVP, comparison of all presentations (poster and talk) among various scientific societies, including the American Society of Mammalogists and the American Association of Physical Anthropologists, reveals that the number of women as first authors of posters was higher than their percentage for all presentations (Isbell et al., 2012).

Comparison with the North American Paleontological Convention, held every four to five years beginning in 1969, reveals that women made 2% of presentations as senior authors at the 1st North American Paleontological Convention but by 2014 this same statistic had increased to 41% (Plotnick et al., 2014). Also, a higher proportion of women (31%) first-authored presentations at the Palaeontological Association (founded in the United Kingdom in 1957) annual meeting in 2011. Comparisons of various paleontological organizations with the SVP revealed a slightly lower number of women first-authored presentations for the latter. For example, for the British SVPCA, women first-authored presentations 14% of the time in 1982 (for the SVP in

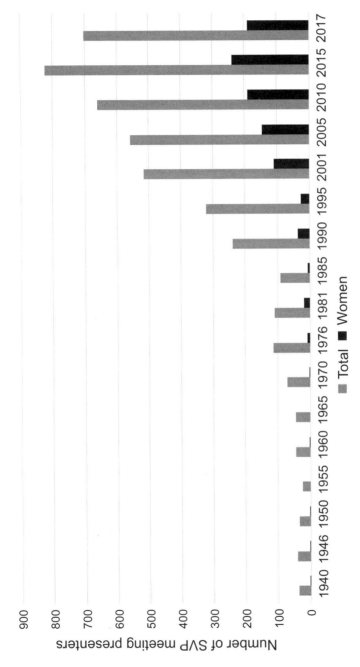

Figure 7.6. Presentations (oral and poster) by women at SVP annual meetings, 1941–2017. *Sources:* Data from SVP annual meeting programs and abstracts. See online table S7.3.

1983, this number was 12%), nearly doubling in 2010 to 25%, a proportion maintained at present (by comparison, the SVP's number was slightly higher) (online table S7.4; the SVPCA is primarily an "informal" symposium and published abstracts were not produced until more recently). Comparison of this metric for the EAVP reveals similar numbers to the SVP's: for 2005, 26.6%; for 2010, 28.5%; and for 2015, 25% (online table S7.4). Women contributed to 36% (27% first authored) of talks and 25% of posters at the first Australasian CAVEPS meeting in 1987 (Turner and Thulborn, 1987).

Grant Success. Grant success is another metric of research performance. Studies have shown that the success rates of research grant applications (e.g., NSF, National Geographic, National Environmental Research Council, Australian Research Council) by male and female applicants are significantly different, with women being less successful (Shen, 2013; Van der Lee and Ellemers, 2015). Gender bias has been shown to affect grant success rates when reviewers evaluated both the science and the investigator. The bias disappeared when science alone was the focus (*Nature*, 2018a). For the Sedimentary Geology and Paleontology Program, an NSF program that funds many vertebrate paleontologists, grant funding rates for women in 1999–2000 were 14%, increasing slightly in 2009–2010 to 19%. By 2014–2015 funding rates for women in the program had risen to 33%, and up to 38% in 2018 (NSF, undated; Table 7.2)—a doubling of the funding rate in 18 years.

Fortunately, there are specialized paleontology and VP funds that provide research support, including those offered by the Dinosaur Foundation, the SVP (the Dawson Grant, the Patterson Memorial Grant, the Estes Memorial Grant, and the Wood Award), and the National Geographic Society. There are also a few fellowships, such as the L'Oréal Fellowships for Women in Science founded 1998 by L'Oréal and UNESCO to advance women's research, which are offered in many countries; in the United States, since 2003, five postdoctoral fellowship grants of $60,000 each have been awarded annually. *Nature* announced two annual awards to recognize inspirational early-career women researchers who in the past 10 years have made exceptional contri-

TABLE 7.2
NSF grant success (sedimentary geology and paleontology)

Awards	Total	Women	%
1999–2000	50	7	14%
2009–2010	21	4	19%
2014–2015	9	3	33%
2017–2018	26	10	38%

butions to science, leadership, tutoring, and mentorship and to recognize individuals or organizations that have led to increased participation by girls and young women in science (*Nature*, 2018c).

Issues Women Face and a Framework to Address Gender Parity

The data presented thus far demonstrates that women are disproportionately underrepresented in the geosciences, including paleontology. Whereas student numbers have significantly increased, postgraduate and tenure-track loss is significant. We offer here a framework that can be used to test hypotheses for why this loss is occurring. Doing so ensures that we focus our attention on relevant issues and generate strategies to increase the numbers of women in vertebrate paleontology.

Implicit Bias

Implicit bias, also termed unconscious bias, refers to allowing stereotypes to unconsciously bias our opinions and decisions. Such influences can be gender based and can negatively affect how we evaluate women for jobs, for admission to graduate school, and for fellowships, grants, and professional awards, even without an awareness of them (Holmes, 2015b; Skibba, 2016). We all have implicit bias and gender stereotypes; it is not simply a men-versus-women issue. The results of a survey done in 2010 by the AAAS found that of the 1,300 respondents, more than 50% of women said that they had encountered gender bias during their careers, compared with just 2% of men (Shen, 2013). For example, stereotypes of women as nurturing and more understanding than men result in them being approached more than male colleagues about personal issues—in short, being expected to be a "mother" rather than a mentor (DeSantis, 2017).

A widely reported nationwide study of gender bias among science faculty in the United States, whose authors were tasked with evaluating the competence of an applicant, found that even with identical résumés, a female student was judged less competent and less worthy of being hired than a male student and was also offered a smaller starting salary and less career mentoring (Moss-Racusin et al., 2012). A study of implicit bias (Nosek et al., 2009) found that 70% of men across 34 countries view science as more male-oriented than female. Common representations of science to children portray men as scientists and women in other roles (Holmes, 2015b). New children's books are helping to counter this (see "Children's Books" in Bibliographic Sources and Further Reading). Gender biases have also been found to affect university student evaluations, and data suggest that male students are more likely to give an excellent score to male teachers than they are to female teachers (Boring, 2015). Malin Kylander (2019) has also addressed the issues related to bias and the need for gender parity in research fields where men outnumber women.

Implicit bias affects letters of recommendation that are written for graduate school applicants and job applicants. Studies of such letters written for faculty applicants in various scientific disciplines have shown that those for men had terms such as "assertive, confident, ambitious, independent, scientific leader, [and] brilliant scientist," whereas letters for women contained terms such as "sympathetic, sensitive, warm, caring, tactful, [and] very knowledgeable" (Holmes, 2015b; Rosen, 2017).

Implicit bias level can be assessed by using a set of online quizzes offered by Project Implicit. Having an awareness of implicit bias is the first step to overcoming it (Chachra, 2017). The next step is to explore ways to reduce its influence on our decisions (Raymond, 2013; Stigall, 2017).

An example of an attempt to overcome unconscious bias is the Bearded Lady Project, which uses film and photography to celebrate the work of women paleontologists and highlight the challenges they face (Bearded Lady Project, undated). The project team includes Lexi Jamieson Marsh, director; paleobotanist Ellen Currano, scientific consultant; and photographer Kelsey Vance (Currano, 2015). They have portrayed women paleontologists in the field and lab wearing fake beards as a way to educate the public on the inequities that exist in the sciences, especially paleontology (Plate 40). The book *The Bearded Lady Project: Challenging the Face of Science*, which includes portraits of bearded women, behind-the-scenes photographs, personal essays, and historical context, edited by Marsh and Currano, was published in May 2020.

Sexual and Gender-Based Harassment

Challenges that women geoscientists faced in the field in the past were various (e.g., Burek and Kölbl-Ebert, 2007; Turner, 2007). In the past few years, the #MeToo and #MeTooSTEM movements have focused awareness on sexual harassment, intimidation, and bullying in geophysical and environmental sciences in both lab and fieldwork situations (Black, 2018). Fieldwork, in particular, has been an aspect of paleontology fraught with challenge, and even danger, for women. In a survey of 666 field scientists across disciplines, Kathryn Clancy and colleagues (2014) reported that 64% had faced some form of sexual harassment in the field (defined as inappropriate behavior or sexual remarks) and more than 20% had been sexually assaulted (defined as sexual contact that was unwanted or nonconsenting or where it was unsafe not to consent). A follow-up study by Robin Nelson and colleagues (2017) found that field expedition organizers fail to provide clear expectations of professional behavior or policies regarding sexual harassment.

Because the discovery and collection of fossil vertebrates is an integral part of vertebrate paleontology, women's traditional exclusion from fieldwork has represented a limiting factor for women trying to develop VP careers. As Emily Buchholz, at Wellesley College, commented in our interview with her, "I'm of an age when it was very difficult to get in the field and I remember asking Nick [Hotton] if I could go out in the field with him and he . . . said, well, we can't make that happen because there are no accommodations for women and I don't take women on my field trips. And I've been eternally sad that I didn't challenge him in some way because he was such a nice guy that I think he could have been convinced. . . . But I think he had no way to make it happen without some trouble. . . . I wish I would have been a little more assertive. It certainly has changed and, I must say, I have some youth envy on that because the young women who are brought up [now] . . . just assume that of course they can be members of field group[s]. I didn't have that access."

More recently, however, this has been changing. Many women have been strongly encouraged to pursue fieldwork and, in some cases, have led field research teams. This is a pivotal evolution. As Catherine Badgley, from the University of Michigan, noted in her interview with us, "[What] I would recommend, . . . because it's been a huge part of my academic life, is fieldwork. Fieldwork is constantly inspiring. And partly it's not just where fossils come from, but it's also a chance to step out . . . of a very busy, often urban, . . . high-stress life into a life that's really out in nature."

Her assertion was reaffirmed by Anjali Goswami of the NHM, who reported in her interview with us, "One thing that I think is really important is fieldwork. . . . We are all vertebrate paleontologists. Our data comes from the field ultimately, and you can do a lot, as I do a lot, with museum collections, but going out and finding new fossils around the world is very, very important, and there's huge parts of the world that have very poor records, especially for the time periods that I'm interested in, the Cretaceous and Paleogene, . . . and that, I think, is one of those things that sometimes women can feel is a very daunting prospect to, to participate in or especially to start and lead fieldwork around the world. And there are lots of barriers to doing that. Certainly other parts of the world have more sexism or racism than some of the Western liberal democracies. . . . But it is really, really worth doing that one because . . . we are the face of science and the more we do that, the more it becomes accepted and the more we are able to inspire young women in those countries to rise up and become leaders in science in those fields. And I think that's been really great. I think it's been really great to have supportive men who have . . . gone out in the field with us or taken me out in the field with them in some cases to really push that and make sure that field crews are more diverse because it does have a really big impact on people to see diversity in science and as a very international field."

Julia Clarke, from the University of Texas, Austin, observed the ambivalent position women in the field may now find themselves in her interview with us: "I'm wholly committed to the idea that we need to still discover new fossils to drive new questions, and being a field paleontologist, as a woman, what are those challenges? . . . Different communities and cultures are more supportive of women scientists or respect our education. As someone who was a highly educated woman, I was kind of like a weird . . . outlier, right? [Even when] I was leading my own field expedition. Sometimes . . . someone turns to a member, a male member, of my team, assuming that they have authority in [my] place as the actual field leader who raised all the money and does all the planning. . . . But I've also been fortunate in many cases with my collaborators . . . that we've worked out those differences. . . . I don't see it as a barrier. I see it as something you have to commit to because you're so passionate about doing this . . . that you will move beyond things that might arise."

While these improvements in the inclusion and safety of women in the field are notable and heartening, the problem of equitable treatment is far from solved. Women in academia reported higher rates of sexual harassment (58%) compared with both private and government sectors, a number topped only by women in the military (69%)—both male-dominated fields (Ilies et al., 2003). Women and men affected by harassment struggle to get jobs, secure tenure, win research funding, get appropriate authorship on papers, and receive scientific recognition (e.g., *Nature Ecology and Evolution*, 2017; National Academies of Sciences, Engineering, and Medicine, 2018). The good news is that codes of conduct for acceptable behavior and the provision of reporting structures to facilitate the resolution of complaints are on the rise in scientific contexts (*Nature*, 2018b; Córdova, 2020). For example, to mitigate sexual harassment, the American Geophysical Union in 2017 adopted a new ethics policy that defines bullying and harassment as scientific misconduct

(Bell and Koenig, 2017), and similar policies exist for the PS and the SVP. The AAAS has joined 77 academic and professional societies in a new group to address sexual harassment in STEM. The consortium will provide research, resources, and guidance to address sexual harassment in the member societies, as well as more broadly in the fields they represent (Ham, 2019).

Gender Pay Gap

According to the 2017 *New Scientist* salary survey, women working in science and engineering in the United Kingdom earn one-fifth of that of their male colleagues. In this same survey, among US scientists, a smaller pay gap of 11% was reported between female and male respondents, whereas in mainland Europe a wider gap, 19%, was found (Fleming, 2018). A study by the NSF across geosciences, atmospheric, and ocean sciences reported in 2016 that the median salary of men was less than 1% higher than the median salary for women, but the pay difference between the sexes in the biological and biomedical sciences was nearly 10% (NSF, 2016).

Dual-Career Couples

Dual-career couples are increasing, with 54% of male scientists and a whopping 83% of female scientists in the United States reported as married to an academic partner (Schiebinger et al., 2008). Such couples are sometimes said to face a "two-body problem," given the difficulty for both partners of finding a job. As we have shown, until recent decades, most women followed their partners who gained jobs, although they often continued their VP work. There is increasing recognition that dual-career couples can be great strengths for departments, and some institutions support dual-career hires (Holmes, 2015a). Although many women today have academic partners, we note that many fewer have partners also in vertebrate paleontology (e.g., Margery and Walter Coombs, Gloria Arratia and Hans Peter Schultze, Pat Vickers-Rich and Tom Rich, Susan Turner and Tony Thulborn, Cathy Forster and Jim Clark, Angela and Andrew Milner, Bolortsetseg Minjin and Jonathan Geisler, Samantha Hopkins and Ed Davis, Maureen O'Leary and Michael Novacek, Martina Kölbl-Ebert and Martin Ebert, Daniela Schwartz and Oliver Wings, Denise Sigogneau-Russell and Don Russell, and Nancy Stevens and Patrick O'Connor).

With regard to the "two-body problem," Emily Rayfield (University of Bristol, United Kingdom) noted in our interview with her, "My husband is not an academic but he's in science as well. And when I finished my PhD . . .

we spent quite a lot of time living in different parts of the country and only seeing each other for some parts of the week, which is not an uncommon problem in academia. And so that was pretty challenging actually. But eventually, we managed to get through it, but you know, people have to make compromises sometimes as well."

Pipeline Problem

Women, especially women of color, in science often experience significant career drop-offs and, in some cases, leave scientific fields entirely at higher rates than men, also known as "pipeline leak." The largest loss of women in academia in general, including the geosciences, occurs early, at the transition from the PhD level to the assistant professor level (Holmes et al., 2008; Newton, 2012). Studies have shown that these "leaks" result from external factors such as implicit bias, pressure from family or society to leave, lack of mentorship and encouragement to proceed in the career, lack of structural support for childcare, and immobility due to partner's position (Holmes 2015a).

Another manifestation of the underrepresentation of women in STEM fields in the highest ranks of academia has been referred to as the "clogged pipeline." Increases in recruitment and promotion rates of women have occurred slowly. For example, according to US faculty census data, the relative proportion of women faculty in STEM disciplines at the associate professor rank increased from 22% in 2008 to 28% in 2014, with a corresponding increase from 8% to 13% for full professor (Van Miegroet et al., 2019).

Family Issues

For any woman scientist, the decision whether and when to have children is a critical one, and it has been cited as responsible for the large decline in women who apply, enter, and advance in tenure-track positions. For this reason, many institutions have adopted tenure-clock-stopping policies over recent decades (de Wet et al., 2002; Williams and Ceci, 2012). These policies allow eligible faculty to stop the tenure clock for one year in response to circumstances (generally reasons related to having children) that would significantly alter their productivity. Little data is available to evaluate whether these policies help women, and one such study found that gender-neutral tenure-clock-stopping policies substantially reduced women's tenure rates while substantially increasing men's tenure rates (Antecol et al., 2016).

A recent study found that being a parent in STEM is not just a mother's issue but in fact has broad effects on both parents. Results based on career trajectories of an NSF database over eight years (2003–2010) revealed that after STEM professionals become parents, 43% of women and 23% of men switch fields, transition to part time, or leave the workforce entirely (Cech and Blair-Loy, 2019). These differences make it clear that, even though the attrition rate for fathers is higher than expected, mothers still face particular career challenges (e.g. Turner, 2008). A challenge for both men and women is now the cost of childcare—a major drain on income (Shen, 2016; Calisi, Working Group of Mothers in Science, 2018).

A study conducted on Australian female researchers examined whether caring for children affects women's research output and collaboration (Sewell and Barnett, 2019). It found that caring for children did not greatly affect a woman's publication counts over her career. Women experienced a small increase in publications after the first child, followed by a decrease after the second child. This result has been corroborated by some studies (Cole and Zuckerman, 1987), although other studies (e.g., Bentley, 2012) found a negative correlation. Further research is warranted to examine factors contributing to these disparities, but our look at women in VP will add to the data.

Women vertebrate paleontologists have dealt with the challenge of family issues in various ways. Jaelyn Eberle, at the University of Colorado, Boulder, explained in our interview with her, "I take my kids in the field with me. . . . I bring my graduate students out and often have one or two of my children. . . . For graduate students coming through, I think there is this idea that you're either a scientist or a mom. . . . I think that's a myth. I think that it's quite doable to be both a good scientist and a mom, and I think it's been important to me to bring my kids [into the field]. They come and my graduate students see that and go, 'Oh, well, I can do that.'"

British paleontologist Emily Rayfield is also enthusiastic about the positive impacts of life as a mother and a scientist in our interview with her: "What's been interesting and unexpected about [balancing motherhood and work] is that I've actually become more organized and more productive after having the children than I was before because I've had to focus and organize my life."

Kristi Curry Rogers, at Macalester College in Minnesota, who, like Eberle, chooses to take her daughter into the field with her, wryly commented in our interview with her, "I have a 15-year-old daughter now, and when I was pregnant with her I remember thinking, I am going to just continue doing everything I normally do. I'm not [going to] let having a child stop me and my husband [who] is also in the field. We work together, we collaborate a lot on projects, and that meant that I was going to have to take my kid to the field with me. Starting when she was three months old, we brought her out with us and it was much harder than the first three years. The last page of every field notebook says . . . 'never do this again.' . . . We did it again the next summer."

In addition to caring for children, women often care for aging parents or ill relatives (viz. Elaine Anderson, Chapter 4), which also places them at a competitive disadvantage in the workplace. As yet, there is little data on this factor apart from anecdotes (Turner, 2008; Appendix 1).

Mentoring

Success in one's profession is the result of a combination of many factors, and help from mentors is fundamental to the process (O'Connell, 2015). A mentor is "someone who takes a special interest in helping another person develop into a successful professional" (National Research Council, 1997). Choosing the right mentor is important for everyone, and perhaps even more so for any woman entering the sciences, including VP. As perhaps best expressed by Catherine Early, a postdoc at the University of Florida, in our interview with her, "I think [a mentor] is someone who actually has the time to think about and invest in people in the future." Many societies (e.g., the GSL, the PS) now foster mentorship opportunities, and the Facebook page in Paleontological Education began a mentor drive in 2019. Both women and men have mentored women vertebrate paleontologists. As noted by vertebrate paleontologist Jaelyn Eberle, "You need an excellent mentor of any gender. Get to know people, see how they react to situations and find somebody whom you have respect for and get along with" (Powell, 2018:423). In fact, male VPs have specifically expressed their appreciation of having women mentees. Phil Currie (Royal Tyrrell Museum, Drumheller, Canada) noted in our interview with him, "Having women in the lab (at times, half of my grad students have been female, and last summer the undergraduate field school class was entirely female) has always kept a balance, stability, and tranquility that is not always evident in an all-male lab." Similarly, the indefatigable American paleontologist Don Prothero stated, "I've noticed that my female students tended to be harder working and more serious and dedicated than my male students."

Indeed, women in vertebrate paleontology greatly enjoy recounting great mentoring experiences and are eager to lavish credit on those who have given them a hand up (see Appendixes 1–3). Larisa DeSantis, at Vanderbilt University, enthusiastically shared in our interview with her, "I've had a lot of really important mentors at all the different institutions I've been. My PhD adviser [Bruce MacFadden] was really helpful in providing me with good professional development and . . . making me feel as if I could . . . do it all. I could have a family and I could be a paleontologist. As an undergraduate, Mark Goodwin helped get me a job in the prep lab. He got me to my first SVP meeting. He got me out in the field and he really helped facilitate a lot of these things. He also connected me with a lot of the graduate students at the university and they . . . then connected me with other folks, so [he] was extremely helpful early on."

Kristi Curry Rogers, at Macalester College, expressed gratitude for many mentors who supported her at various points in her career in her interview with us, "When I first started out in college, I went to . . . Montana State University and one of my mentors [was] Jack Horner. . . . I think he had 19 graduate students, but I was the only undergrad and that gave me some sort of special opportunities to do fieldwork or to work in the museum on research questions. I started off as a fossil preparator and Jack was always very open about letting me try new things. . . . [His] trusting me to ask my own questions and to find out whether I could answer them or not with the methods I was trying to use was really inspirational. I also had a lot of really wonderful female mentors during that time of my life . . . because Jack had so many wonderful women graduate students and . . . they helped me . . . navigate the early parts of my career, including things like . . . what do you do when you're in the field, what are field norms, and how is it to be a female member of an all-male field crew. And those women were very important people like Mary Schweitzer, Karen Chin, . . . Carrie Ancell. . . . And then I went on to graduate school at Stony Brook University and I picked that program in part because my adviser was . . . Cathy Forster. And I thought it was really a wonderful . . . idea [to have] a female mentor after being in [a] . . . male-dominated program. And it was actually some of those female grad students that put me in touch with Cathy and helped me make that connection and I had a great time at Stony Brook being her student."

Ana Valenzuela-Toro, at the University of California, Santa Cruz, in her interview with us felt similarly that it takes multiple mentors to help build a successful woman VP: "I will say that many people have been very instrumental in my incipient career as a paleontologist. One of them is Carolina Gutstein, a paleontologist from Brazil . . . [working] in Chile. Actually she was the first paleontologist . . . I ever met. Another person who has a really important role for me is Dr. Nicholas Pyenson. He has been a mentor . . . and he has provided a lot of opportunities. . . . I am sure I [would] not be here right now if he was not involved with me."

Romala Govender, curator of Cenozoic paleontology at the Iziko South African Museum, commented in her interview with us, "I met the person who has taught me about what mentorship was during my second postdoctoral fellowship. This was when I got to know Anusuya Chinsamy-Turan. She showed me that one inspires others by allowing them to explore new areas of research even if you are not familiar with it. You inspire by encouraging people to do things that they probably did not think they were capable of. She also inspires through her career as a paleohistologist. She has become a leader in her field by not being afraid to take on a challenge. She also encourages the students and people around her to take on new challenges that not only make you realize that you are capable but also that [allow] you to learn new skills."

A legendary mentor herself, Kay Behrensmeyer, of the Smithsonian, noted in her interview with us, "I went to Harvard and Brian Patterson took me on as a graduate student. He himself was an iconoclast, a kind of a self-made person, and he was very supportive of unusual graduate students like me. There weren't . . . many women at the time and I wanted to do something different and so he was a very, very big influence on me as well as just being a tremendous scholar."

Catherine Badgley's experiences interestingly demonstrate the way supported mentees go on to become committed mentors. She recalled in her interview with us, "As an undergraduate, my adviser was Stephen Gould and it was actually his amazing lectures in my very first science course that made me realize that it's not just about geology, it's all about a whole history of ideas. Also, Kay Behrensmeyer, although she was not formally on the faculty at Yale when I was a graduate student, she was part of our research team . . . working in Pakistan. I had done fieldwork with her earlier in Kenya, and here was a very . . . bright, interesting, successful paleontologist who was also striking out in new directions with her work on taphonomy and paleoecology. And [in] that . . . sense [she] showed me that there is nothing to fear about being a woman in this field. And I've tried to encourage my students that way too. I want them to . . . know I'm happy if they are part of my research program. And I'm also happy if they take off and want to go in another direction. Mainly I just want them to realize that whatever they want to do, I want to be there for them."

This kind of lineage forms a crucial network that does not just build up women scientists but also builds an entire heritage of women in science. A useful way to track the lineages of academic mentorship is through the online database the Academic Family Tree (Evolutionary Biology and Geoscience).

Given the pivotal role that women VPs assign to mentors in establishing and growing successful careers, it makes sense that when asked about the most important advice to impart to a woman beginning work in vertebrate paleontology, the topic of mentorship was high on the list.

Advice for Girls and Young Women Interested in VP
(See Also Appendixes 1–3)

1. Find Good Mentors and Collaborators

"Seek out good mentors, which is a harder piece of advice to give because you don't always know who the good mentors are. . . . Just meet as many people as you can because in that process you will probably run into some people who are good mentors and maybe some who aren't. But if you meet as many as you can, you start to become a known entity. . . . Volunteer to do things in your society. . . . If you want to do it, you should be able to do it, and people are always happy to have the help. . . . I think just putting yourself out there and . . . getting involved as much as you can is probably the best advice."
—Catherine Early, postdoc, University of Florida, Gainesville

"Ask for advice, seek out your elders in the field. The people who have been doing it longer than you because they've been there, they've done that. They know . . . what works and what doesn't. . . . But also figure out what works best for you and follow your gut. A lot of times you've got to trust yourself. That's one of the things that I've had to learn the hard way . . . but it's all worked out in the long run." —ReBecca Hunt-Foster, Dinosaur National Monument, Utah and Colorado

"One key insight that I've benefited greatly from in my own career, and feel is so essential for my students and postdocs: professionally, build a network of reliable collaborators, and find trusted friends and mentors that you can speak openly with about ideas and concerns—these allies will provide both essential support and honest opinions (even those that may not agree with yours, as sincere advice is invaluable)." —John Flynn, AMNH

2. Possess Self-Confidence, Focus, and Flexibility

"Be prepared to be flexible, learn as many skills as you can, take advice wherever you can find it. Some of the most important skills to have are keen observation and good writing." —Carole Burrow, Queensland, Australia

"When choosing one's career, I believe that it is very important to follow one's passion. Once you've chosen your field of research, it is important to work hard and do the best that you can and don't feel that you have to fit the stereotype but rather be compelled to challenge the paradigms." —Anusuya Chinsamy-Turan, University of Cape Town, Cape Town, South Africa

"Be confident in yourself and your abilities. You have to enjoy your work and have fun, even on days when it seems like everything is going wrong. Even if you don't find the answers you set out to find, there is always an interesting discovery. I have realized that what sometimes seems like a difficult and negative experience has a positive outcome as you end up learning a new skill or realize you have a strength you never had before. One of the important lessons that I have learned is that one's VP career is not set in stone and that opportunities come along that [see] the course of your career change." —Romala Govender, Iziko South African Museum, Cape Town, South Africa

"Be persistent in the face of obstacles when you have a worthwhile goal, and maintain strong will in the face of adversity—some of the best ideas or projects are resisted early on because they challenge assumptions or the status quo (or simply because people or systems may be biased or unnecessarily obstructionist), and it's often valuable to figure out routes around such roadblocks. Many times we judge our own work too harshly and underestimate its significance—so seek the highest-profile publication outlets for your work, whenever appropriate and possible, even if it means trying a few different journals and takes a bit longer to publish. But also develop a 'thick skin' to rejection, given the lower odds of success in such submissions, so that you remain willing to strive for the highest-impact outcomes of your research and outreach efforts." —John Flynn, AMNH

"Even though you are your worst critic, believe in yourself, despite what you're thinking, believe in yourself and give those talks and present your work and publish your work because it will be well received. And this is a very supportive, encouraging environment." —Blaire Van Valkenburgh, UCLA

3. Know the Science

"I would say it's important to really familiarize yourself with what the field actually is. A good way to do that is to attend Society of Vertebrate Paleontology meetings to really learn what the science is. . . . You might find a good mentor in another area of paleontology or another area of geology that really is inspiring." —Jaelyn Eberle, University of Colorado, Boulder

"In this field there are many techniques, both computational and things like scans, and they are great, but . . . it's still really important to know your organism and to know your data because these techniques, they can give you results but they can't give you the questions. You come up with the questions and . . . it's important to have a good stack of pencils and backs of envelopes [to] work things out first. To know what to expect to find first, to understand the animals and the bones, everything. And then you can use those techniques to their best development. . . . Because you can get some very precise results but you have to somehow know that that's going in the right direction, otherwise there is also room for error." —Christine Janis, Brown University, Providence, Rhode Island

"I encourage my students and postdocs to develop a breadth of expertise across disciplines and methodologies, and integrating fieldwork, collections, and innovative laboratory research whenever feasible. But also to have areas of special expertise that clearly distinguish their work and contributions, ideally in frontier disciplines/methods. In fieldwork, there are amazing opportunities all around the world, continuing to make paleontology such an exciting field; when in the field, respect diverse backgrounds, cultures, and norms, but exercise appropriate leadership and demand respect." —John Flynn, AMNH

"[You] need to really work very hard and also think broadly about the subjects that [you] study because paleo and especially the kind of stuff that I do is not simply biology and anatomy, and looking at fossils, it encompasses so many different subjects, areas, physics, math, chemistry, geology as well too. . . . I often say to the students, get a broad background, as much breadth as you can, read in the subjects that you study, . . . read around, don't just read about the particular group of animals that you're interested in." —Emily Rayfield, University of Bristol, Bristol, United Kingdom

4. Learn from Failure and Cultivate Resilience

"Don't be afraid to make mistakes. Don't be afraid to fail, because you're going to succeed more often than you're going to fail. And do you learn from mistakes. I think the whole course of my career has been just seeing opportunities and taking them on." —Zerina Johanson, NHM

"I think being resilient [is] one of the most important things, being persistent. . . . The hardest thing is a getting a job as a vertebrate paleontologist. There aren't that many of them. . . . If as a graduate student you're at a place where you have an opportunity to teach human anatomy, . . . that opens up an avenue for postdocs that you can have or ultimately even positions . . . that broaden your opportunity. . . . But also one of the hard[est] things is funding. It's . . . difficult to get funding and so you have to be willing to try and try and try again and you have to also avoid distractions." —Blaire Van Valkenburgh, UCLA

"Never give up. It is hard sometimes, . . . but you have to continue working. . . . I think that it's the most important thing for me." —Ana Valenzuela-Toro, PhD candidate, University of California, Santa Cruz

5. Insist on Equality and Respect

"Work with people who promote you and actively work with you to help you achieve your career goals. Do not feel pressured to pursue career goals that are not your own; your career is for you, not for the aggrandizement of your supervisor(s). Avoid people who seek to work with you solely because it will benefit them. Expect and demand recognition for the work that you do. Maintain high standards for yourself, in your work as well as in your professional behavior and ethics. Expect and demand recognition for the work that you do. If you witness or experience harassment, report it to those who will listen and take action. The only way to kill that behavior is to eliminate it at its source." —Christopher Bell, University of Texas, Austin

"I often tell young students, especially females, to never let anyone tell them how far they can go." —Carolina Loch S. Silva, University of Otago, Dunedin, New Zealand

"Never be afraid to go and talk to people about your ambitions. Many people think that because dinosaurs were big animals that are difficult to excavate, then it really is not an appropriate place for females. The reality is that our field camps tend to be equally made up of both males and females, that females enjoy the camping and work as much as the males, and find just as many specimens." —Phil Currie, Royal Tyrrell Museum, Drumheller, Alberta, Canada

"Stand by your convictions, as this will show people that you will not be intimidated." —Romala Govender, Iziko South African Museum, Cape Town, South Africa

6. Maintain a Good Work-Life Balance

"As you achieve your goals, however large or small they might be, take the time to celebrate. Celebrate with others. Strive to achieve a good 'work-life' balance. It is always a challenge, but that balance can happen only if you make the effort.

"Be happy in your work; life is too short to waste your time being unhappy. Avoid negative people and poisonous personalities; they will drag you down. Choose supportive partners." —Christopher Bell, University of Texas, Austin

"I have . . . seen a few young paleontologists waste marvelous chances by being overambitious, unethical, unpleasant, or simply too bent on overwork. I have been in it for the long haul, have published carefully researched and substantial papers, and have cared about life balance and human relationships. In the end, that is what counts for me." —Margery Coombs, University of Massachusetts, Amherst

Imposter Syndrome

The feeling of self-doubt, that regardless of your accomplishment, you are about to be unmasked as a fraud, is all too common among academics, especially women in science, and has been termed "imposter syndrome" (Woolston, 2016). The term dates back to the late 1970s, when it was identified by two psychologists among "high-achieving" women, and it has been reported that up to 70% of women have suffered from it at one time or another. Ways to combat it include finding a supportive, understanding mentor; looking at examples of your own successful work; and accepting that some tasks will not be done perfectly (Kaplan, 2009).

For a list of positive changes that have occurred in recent years, as well as work that remains to be done to achieve gender parity in paleontology, see Boxes 7.1 and 7.2.

Conclusions

Women VPs have contributed significantly to the history of vertebrate paleontology. Beginning in the 18th century, the science of vertebrate paleontology developed with strong geologic roots and was primarily an observational science. Women in the late 18th and early 19th centuries were, mostly in Britain, contributing to the field as collectors, preparators, and artists of fish and reptiles (Chapters 2 and 6). As women began to gain degrees and new avenues opened up in the late 19th and early 20th centuries, they started to tackle research on dinosaurs and other large extinct animals, such as mammoths and mastodons (Chapters 3 and 4). During this time, fossil hunting by women spread to other countries in Europe, North America, South America, Asia, Africa, Australia, and Russia/USSR. Women vertebrate paleontologists were describing and interpreting fossils in terms of their spatial and temporal context

Box 7.1 **Positive Changes That Have Occurred in Recent Years in Paleontology**

1. Increase in women members of professional societies including the Society of Integrative and Comparative Biology, the SVP, the PS, the SVPCA, and the EAVP. Many of these also offer emerita status to those who have been members for 30 years or more (e.g., the SVP). In the SVP since 2012, although female student membership has increased to 40%, less than 25% of professional members (tenure-track faculty and curators) are women.

2. Increase in number of women receiving societal awards at all levels. For the SVP, although nearly 40% of student awards since their inception in 1978 have gone to women, women have received just 16% of SVP senior awards.

3. Increase in number of women serving leadership roles on society executive committees. Women leaders in the SVP over the last 20 years have increased considerably, averaging 30% or more.

4. Increase in number of women giving presentations at meetings (e.g., SVP, North American Paleontological Convention, PS, SVPCA, EAVP, CAVEPS). In the SVP, the number of women giving meeting presentations has increased, especially in the last 10 years, averaging nearly one-third of all presentations.

5. Increased awareness of challenges to equal opportunities for women (i.e., struggle to get jobs, authorship on papers, receive scientific recognition, secure tenure, and obtain research funding) and steps to remedy them. For the SVP, student and postdoc roundtables held at the annual meeting and early-career conferences (e.g., International Meeting of Early-Stage Researchers in Palaeontology, Spain) provide opportunities for students to learn about grants, grant writing, and networking and to receive career advice.

6. Efforts to foster work-life balance through the establishment of family-friendly policies (i.e., providing childcare assistance, extending tenure clocks for new parents, and offering flexible work arrangements). For the SVP, childcare will be available at the annual meeting beginning in 2020.

7. Increase in networking opportunities at meetings. Since the early 2000s (originally organized by Robin Whatley, Allison Beck, and Karen Sears), the SVP has supported Women in Vertebrate Paleontology Luncheons (renamed Diversity Luncheons) at the annual meeting to discuss ways to

increase equity and diversity in the SVP. All meeting attendees are welcome to attend this event.

8. Increase in organizations to encourage women to enter science careers and to provide support. Social media groups have been established that foster collegiality (e.g., the Facebook groups Women in Vertebrate Paleontology, since 2009, and Women in Geosciences, since 2012; the Twitter group SVP [@SVP_vertpaleo]; and the Revolvy site Women in Paleontology).

The Association for Women Geoscientists, is a nonprofit member organization of the AGI founded in 1977 with chapters in the United States and internationally. They provide career advice and sponsor field trips, lectures, outreach, and social functions (Coates, 1986; Schneider et al., 2018). The Earth Science Women's Network is an organization dedicated to improving the retention of women by promoting career development, building communities, providing opportunities for mentoring and support, and facilitating collaborations (Hastings and Wiedinmeyer, 2015; Barnes et al., 2018). There are other such groups in other countries (e.g., the African Association of Women in Geoscience).

In the United States, several NSF programs support women in science. In 2001 the NSF launched the ADVANCE program to develop systemic approaches to increase the participation and advancement of women in academic STEM careers. While prior programs were designed to help individual women (e.g., the NSF program Professional Opportunities for Women in Research and Education, 1997–2000), the bulk of funding for the ADVANCE program (which continues at present) has gone to institutions to address the institutional barriers that, if removed, could improve hiring, retention, and climate for women faculty (Laursen et al., 2015; de Aro et al., 2019). An NSF-sponsored program, Promoting Geoscience, Research, Education and Success, developed in 2015, provides resources (e.g., professional development workshops, online discussions) and women mentors to support undergraduate women entering the geosciences (Fischer et al., 2018). Evaluation of this program reveals higher rates of persistence in geoscience-related disciplines (95% vs. 73%) among participants (Hernandez et al., 2019). The US-based National Association of Black Geoscientists provides inspiration, practical advice, funding, and mentorship to black students.

Recognition of gender inequities in the United Kingdom led to establishment of the Athena project and Scientific Women's Academic Network (SWAN), known as the Athena SWAN program, in the early 2000s, which continues today with the inclusion of professional, support, and technical staff and expansion to academic disciplines outside of STEM (Rosser, 2019). Another web-based resource, Sci Sisters, was initiated in 2017 as a Goggle map through which senior women scientists in STEM in Scotland could find each other (Arnold, 2018).

Based on the successes of SWAN in 2016, the Australian government established two programs to take on gender equity issues: Science in Australia Gender Equity and Male Champions of Change. The latter, the only program targeting men, has been designed to challenge male leaders to accelerate the progress of gender equality (Latimer et al., 2019).

Literary and online celebrations of women scientists, including vertebrate paleontologists—for example, in books (Burek and Higgs, 2007; Johnson, 2018), social media, and blogs such as *TrowelBlazers* and *Lady Science*—have increased greatly in recent years. There are modern heroes such as Fossil Hunter Lottie and Paleontologist Barbie (Pat Vickers-Rich claims it was based on her: S. Turner et al., 2010a; Carlson, 2013), the latter of which has her own Facebook page. All encourage young girls to consider paleontology as a career.

Box 7.2 **Work That Remains to Be Done to Achieve Gender**
           Parity in Paleontology

1. Collect and publish more current and historical data within our departments, journals, societies, conferences, and funding agencies on gender representation and our success rates. One problem with this is that, as a result of the rise of electronic communication, written records are often no longer kept. Individuals and societies, including the SVP, need to consider the availability of such records in the future.

2. Continue to incorporate diversity (gender, ethnicity, abilities, ideas, disciplines), inclusion, and respectful environments in our departments and societies. Since 2017 the SVP Membership Committee has conducted annual surveys collecting member information—gender, citizenship, institutional affiliation, primary role, and interest areas. This effort should continue, and the data should be archived and analyses performed to develop and implement recommendations for the future, with the goal of fostering a diverse and inclusive professional environment.

3. Promote and ensure effective mentorship opportunities (i.e., surround yourself with a network of friends and colleagues who support you) and community outreach (i.e., establish professional relationships and engage in outreach events) to help foster an interest in science and in paleontology among girls and young women. Notable in SVP history have been women who were early nominators of student membership (see the *SVP News Bulletin*).

4. Encourage and invite women—especially women from minority groups—as keynote, plenary, and symposium speakers and moderators. One way that the underrepresentation of women at conferences has been addressed is with the creation of databases in many STEM disciplines. For example, a large database called Request a Woman Scientist contains a list of more than 800 scientists from 133 countries (Langin, 2019; McCullagh et al., 2019). There is a Women in Paleontology Directory with a list (growing) of more than 276 women, of which more than 40% are VPs (Women in Paleontology Directory, undated).

5. Recommend more women, especially women of color and LGBTQ women, as reviewers for manuscripts. For *JVP*, from 2012 to 2015, women reviewers for published papers averaged less than 20%.

6. Provide training for individual scientists as well as field and research teams and professional societies on ethical leadership—that is, develop rules and enforce codes of conduct to prevent sexual and gender-based harassment in the field and workplace. The SVP has recently revised their meeting code of conduct statement to define unacceptable behavior, as well as procedures for reporting violations and disciplinary actions (SVP, undated).

7. Support better recognition of women and underrepresented minorities for awards.

8. Facilitate more equitable and transparent hiring and promotion practices (including start-up packages, salaries, and internal grant funding), improving the representation of women at all levels. Develop a rubric for assessing diversity statements early in the evaluation process. Develop a mentoring plan for early-career faculty (Bhalla, 2019).

9. Provide professional development for work success (i.e., grant writing, pay negotiation and equity).

10. Share the spotlight—credit the work of colleagues and amplify their voices.

11. Reinstate (for all SVP members) and encourage published obituaries for vertebrate paleontologists (Falk, 2000). Help your historians!

and ecological relationships. By the late 20th and early 21st centuries (Chapter 5), paleontology had entered the modern era, becoming a multidisciplinary and integrative science, with women VPs studying a wide range of vertebrate taxa using a variety of techniques, investigating them from both evolutionary and ecological perspectives. Although in recent years the number of women VPs has increased worldwide, there are still places where women researchers are low in number (e.g., East Asia and the Pacific, many African countries).

As is true for other STEM disciplines, women have been underrepresented in vertebrate paleontology. Although progress toward gender equality has been made, change has been slow. The lack of women in paleontology and science in general is due to past and present systemic, organizational, institutional, cultural, and societal barriers to equity and inclusion. These barriers must be identified and removed through increased awareness of the challenges, combined with evidence-based, data-driven approaches leading to measurable targets and outcomes.

In the more than 200 years since women began actively pursuing the science of vertebrate paleontology, many changes have occurred—opportunities have expanded, challenges have been confronted, and experiences have improved going forward. Until late in the 20th century, women were almost certain to give up their science when marriage or children were in the offing. This is no longer necessarily the case, and the diversity we note today allows women to conduct vertebrate paleontology in different ways. As the 21st century unfolds, we look forward to sustained and improved encouragement and support of women and underrepresented groups in vertebrate paleontology, thus broadening the reach and quality of our science.

Excerpts from Women VPs' Responses to Oral Interviews

1. What drew you to/interests you/excites you about VP?

I knew I loved the natural sciences, but I like geology. I like biology. I thought I wanted to learn about astronomy. I wanted to do it all. . . . As an undergraduate I did . . . some geology and biology and then . . . there was a class in vertebrate paleontology which I took and I realized that vertebrate paleontology . . . allows you to go in many directions and so . . . that was the kind of flexibility I wanted.

—BLAIRE VAN VALKENBURGH, University of California, Los Angeles

I grew up on a grain farm in southeastern Saskatchewan, and as a child I spent a lot of time on the land, particularly with my grandmother. We would walk the land, we collected rocks. . . . I was always looking for fossils.

—JAELYN EBERLE, University of Colorado, Boulder

I first got interested in vertebrate paleontology as a freshman at the University of Chicago. I actually went there to be a political science major and happened upon a paleobiology class, and the rest was history. I got the bug. I started preparing fossils and I ended up transferring to Berkeley, which was an equally great place to study paleontology, and pursued my interest there for a little while.

—LARISA R.H. DESANTIS, Vanderbilt University, Nashville, Tennessee

Well, the way I became interested in paleontology is a bit different than I think your typical paleontologist. You often hear that people love or [have been] interested in fossil organisms since they were young. Typically, dinosaurs. . . . For me it was different. My mom went back to do a PhD in geology when I was . . . 8 to 11, so I became really interested in geology from my mom. And then I went to Occidental College. I was . . . the only declared freshman geology major. And at the time I didn't really know what my interests would be, so I picked volcanology because that sounded cool, but my adviser was Dr. Don Prothero and . . . he's the reason I'm a paleontologist. I had no interest in it until we crossed paths, and he's an absolutely amazing teacher, . . . the best teacher I've ever experienced. I was . . . completely . . . fascinated with evolution and paleontology. So right away

I told Don . . . I wanted to be a paleontologist. Next semester I took paleontology and I couldn't even get through the class before I [knew I was] serious about this, I want to do it. And he was very supportive. . . . I've always been interested in my Chinese culture, and there [were] amazing fossils coming out of China at the time and still [are] today. I asked him to help me combine these two passions, and he contacted all the paleontologists who work in China in the LA area and I ended up working with Xiaoming Wang [at the University of Southern California] for my undergrad research.

—JINGMAI O'CONNOR, Institute of Vertebrate Paleontology and Paleoanthropology at the Chinese Academy of Sciences, Beijing

I have always been interested in history and I've always been interested in biology. My father was a botanist and I remember picking . . . coral fossils out of my Ordovician limestone driveway when I was a child and being incredibly intrigued. I started college as a music major and then I became a history major and finally I became a biology major and I think in a lot of ways paleontology integrates many of my interests. . . .

I actually think that I had a fairly atypical entry into the field because I came in through biology and at that time I think many people came in through geology. I had a major . . . in evolution at George Washington University. But I was lucky enough to have courses taught by Nick Hotton at the Smithsonian [Institution, Washington, DC]. He served as a mentor even though he wasn't formally my mentor for my PhD, but he was my touchstone and I really enjoyed him very much.

—EMILY BUCHHOLTZ, Wellesley College, Wellesley, Massachusetts

When I was in graduate school, I started out saying I want it to be interdisciplinary. . . . I came from a geological background, [but I wanted] to learn about paleontology and biology, anthropology, ecology. And they . . . said, well, you really have to specialize . . . , but I went ahead and became interdisciplinary, . . . interested in how things became fossilized. So that's the field of taphonomy [science of burial]. And then I was fortunate to be able to do a dissertation on that in relation to the story of human evolution in East Africa. And so I really have my ancestors long ago to thank for my early career boost because the anthropologists were very interested in taphonomy. They were trying to separate what humans did to bones from what everything else was, bones. So I pursued that all the rest of my career. I realized that I could solve puzzles outdoors. I love being outdoors and learning a whole new language by reading the rocks. And then I had a great experiences and field camps. I was the first woman teaching assistant at the Indiana University field camp and that helped to build up my confidence.

—ANNA "KAY" BEHRENSMEYER, Smithsonian Institution, Washington, DC

I was never one of those people who wanted to be a paleontologist when I grew up. I always loved the outdoors, . . . camping, . . . hiking, . . . animals. So when I went to . . . the University of Michigan, I thought I would go into biology. . . . One evening I was hanging out with some people and we were talking about how much we like camping and somebody mentioned that there was a class that was eight weeks of camping . . . in Wyoming and it was a geology class. So I signed up for it and not knowing anything about geology. And one of the instructors was a paleontologist.

—ANJALI GOSWAMI, Natural History Museum, London

I became interested in paleontology in general when I was an undergraduate and I took both biology and geology courses at my first year in college and I was enthralled with my geology courses and less impressed with my biology courses. What eventually got me interested to go into paleontology as a field was the fact that studying paleontology would help me accomplish a kind of grand vision about how the history of life fit into the history of the earth.

—CATHERINE BADGLEY, University of Michigan, Ann Arbor

2. What has been your most exciting field experience/discovery/contribution in VP?

I'm most excited [about] discovery of the . . . macroevolutionary ratchet that seems to affect large, highly carnivorous mammals. And . . . selection for something at the individual level that ultimately is negative at the clade level. . . . [This is] a really exciting discovery. And . . . also the . . . realization that the study of something as simple as tooth fracture . . . gives you insights into predator-prey levels in ancient ecosystems and the whole dynamics of an ecosystem. That was really a surprise to me.

—BLAIRE VAN VALKENBURGH, University of California, Los Angeles

Most of my research is centered [on] looking at mammalian responses to climate change in the past, and instead of looking at . . . modern systems, we can use that fossil record and go back thousands or even millions of years to see how animals have moved, adapted, or what they've done in response to changing climates. And . . . this can actually allow us to extract lessons learned about climate change. Some of the things that we've been able to determine so far is that aridification has really played a large role . . . in the potential extinction of some megafauna in Australia. . . . . We've seen the direct effects of aridification on . . . these animals, . . . potentially making them more vulnerable.

—LARISA R.H. DESANTIS, Vanderbilt University, Nashville, Tennessee

The most remarkable experience was my excavation of fossil whales in the Atacama Desert in Chile—my country. It was particularly difficult because the discovery of these whales was in the context of construction of a highway. We had a few months before leaving the site for the construction to be done. We had to put all our efforts to get all the data that we could.

—ANA VALENZUELA-TORO, PhD student, University of California, Santa Cruz

The research that I do looks at the form and function of the skeleton, and mainly what we're interested in is looking at the function of the skull and the evolution of the skull. I use computational methods and x-ray scanning to reconstruct 3-D images of the skull to get an idea of the external and the internal anatomy of the skull. [I] also use computational biomechanical methods to look at how the skull functions, how it's stressed and strained, while the animal was undertaking behaviors like feeding, for example.

—EMILY RAYFIELD, University of Bristol, United Kingdom

My current research is on the origin and evolution of birds, but writ broadly, I'm interested in biological novelty. I'm interested in how, in evolution, simultaneously everything is old, but there are some innovations that are . . . out of [the] norm, . . . they stand out against the background tick of innovation that we see. How did those arise and how do they influence future biodiversity? . . . I discovered the oldest known fossilized remains of the bird vocal organ in the late Cretaceous . . . (from the age of dinosaurs), but I realized we didn't have any fundamental data to make sense of that structure. We didn't know how it arose and in the ontogeny of living birds; we didn't know how it related to acoustic function and communication and directly in terms of the structure. So it was a whole new playground for ideas, for collaborations in terms of looking at this novelty and what it teaches us about biological novelty more generally.

—JULIA CLARKE, University of Texas, Austin

My research is really about quantifying how shape evolves or how, for me, [it] evolves through deep time using high-resolution imaging tools to quantify shape. . . . I'm using our tools like 3-D geometric morphometrics, which is a powerful approach . . . to look at how different organisms evolve and captures diversity in a way that gives you a lot more information about their ecology and life history. . . . That's what my group at the museum works on mainly. . . . We try and do this across all vertebrates, at this point we're even doing some work in invertebrates . . . trying to look at the full diversity of life on the planet and how it's evolved through deep time. . . . One of my most significant contributions is what we do with all of the data—we test genetic and developmental models of how evolution has happened, why we see certain forms evolving over and over. Why we don't see other forms evolving. . . . If you read the genetics and developmental literature, there are lots of hypotheses, but they don't have the time angle to actually test them robustly. So [that's] what my research has been [focused on] throughout my career; . . . it's trying to test these developmental ideas about how organisms evolve with actual data on how they have evolved. . . . It's something that I think will take up the rest of my career.

—ANJALI GOSWAMI, Natural History Museum, London

What I've been focusing on the last 10 years is [the] repatriation of Mongolian fossils. There [has been] illegal collecting . . . of fossils in Mongolia. . . . We have a

[one-]hundred-year history of discovering vertebrate fossils. We had discoveries ... starting from the 1920s when [the] American Museum of Natural History ... went to Mongolia. We have very rare, beautiful specimens being sold on
the open market instead of being available for other people to study. This is
bringing negative [attention] to the country. I'm fighting and trying to reduce
that problem. And public awareness is really important in in my country. So
that's why my ... current project is focusing on build[ing] the museum so that
fossils can be cared for properly.

—BOLORTSETSEG MINJIN, Institute for the Study of Mongolian Dinosaurs,
Ulaanbaatar / c/o Museum of Rockies, Bozeman

3. *Which mentor(s)/role model(s) was important in your taking up VP? Why?*

The first opportunity he [mentor Nick Pyenson] provided ... was participat-
ing in the excavation of whales [in the] Atacama Desert and thanks to that, I
[got] ... involved with other paleontologists ... from different countries ... and
then, thanks [to] him also, I [did] an internship [at] the Smithsonian [Institution,
Washington, DC]. The Smithsonian is like ... a Christmas tree because they have
a lot of wonderful fossils and they have a wonderful collection and for marine
mammals the ... collection is super, super important.

—ANA VALENZUELA-TORO, PhD candidate, University of California, Santa Cruz

When I was at university as an undergraduate, the person who was lecturing
in vertebrate paleontology was Tom Kemp [Prof. Oxford University]. And [he]
really got me interested in paleo. And then, moving on to do my PhD, I think it
was an interesting time in vertebrate paleontology in the UK because we had
some really strong female vertebrate paleontology role models. People like (the
late) Jenny Clack, Angela Milner, and Susan Evans, they were ... people to look
up to, [since they] had forged a career as women in vertebrate paleontology.

—EMILY RAYFIELD, University of Bristol, United Kingdom

And then I had lots of other mentors along the way. Glenn Isaac was one of
them; ... they're ... almost all men. ... But you know, Richard Leakey was also
very important and Mary Leakey who I connected with in East Africa. And just
the fact that she was the towering figure in the field made it possible for others to
follow along too.

—ANNA "KAY" BEHRENSMEYER, Smithsonian Institution, Washington, DC

I was co-taught by ... Dr. Mary Schweitzer and Dr. Dan Ksepka (curator of
science, Bruce Museum, [Greenwich,] Connecticut). Dr. Ksepka ... invited me to
start doing research in his lab and he has ... remained a very important mentor. ... I think ... it's important to recognize the distinction between a mentor
and a role model. ... For example, Dr. Schweitzer might be a role model, ...
seeing myself in another, a woman in paleontology. [But] if you only see role

models or just members of a certain profession that don't look anything like you, you potentially can't identify yourself as someone who could be part of that profession. . . . But you also have the mentoring aspect . . . and Dr. Ksepka at the time didn't have many students in his lab. . . . He had two other undergraduates who . . . he was working with and he was able to take a really active role in . . . our development as scientists. And it has really continued . . . , even as I've been in graduate school, we are still working on projects together.

—CATHERINE EARLY, postdoc, University of Florida, Gainesville

4. Have there been any difficult challenges (i.e., motherhood, etc.)/negative influences in your career so far? What did you learn from them?

I've done a lot of . . . varied fieldwork. . . . I enjoy working with a variety of researchers . . . and interacting with other students and volunteers. I never really had too many problems [doing] fieldwork until I started having children and it was a little more difficult to take a baby out on the outcrop with you. [Now] I have . . . quite a bit of experience taking my kids out in the field with us, whether it's been with my husband digging up invertebrate fossils . . . or . . . excavating dinosaurs.

—REBECCA HUNT-FOSTER, Dinosaur National Monument, Utah and Colorado

Women are often called upon to serve on committees and do various outreach activities, which we want to do because we want to serve as role models, but it can be a bit excessive. For young women especially, I think they have to keep their eyes on the prize and go further, focus on their research and do some of those opportunities, but try to keep them . . . at a minimum, always focusing on the goal, because . . . getting a PhD is much harder than you think it is.

—BLAIRE VAN VALKENBURGH, University of California, Los Angeles

Well, certainly not being able to get out into the field was the big challenge. I think the other challenge was trying to get back into academia as a whole after my children were born and that was a very difficult. . . . I would like to give . . . a callout to Jim Hopson because he was at the University of Chicago and I was . . . in Milwaukee and I didn't know what to do. So I went to visit [him]. I almost can't speak of it without tearing up because he didn't do anything in particular other than to let me know that it was OK. He was encouraging and supportive. I did happen to have an NSF [National Science Foundation] grant, . . . while I was an adjunct at the University of Wisconsin–Milwaukee, but [it] turned out to be the key to everything because it gave me credibility. I've been at Wellesley twice, but in the first time it was as a replacement in the Geology Department because at the time I wasn't adjunct. I had been teaching intro[duction] geology at the University of Wisconsin–Milwaukee and so I had . . . adjunct credit . . . and that basically got me in the door and . . . I think he told me that I shouldn't give up. And it meant the world to me.

—EMILY BUCHHOLTZ, Wellesley College, Wellesley, Massachusetts

I think some of the greatest challenges that I've had in science . . . I've overcome by having a network of peer supporters, . . . I'd really call them great colleagues, that have offered advice [and] support, that can talk about their experiences dealing with a particular professional situation, and people who've been around longer who dealt [with] a lot of the same things that I've dealt with. . . . My peer colleagues that are supporters [of] those more challenging times I've had have often been women faculty that are not in my area. They have dealt with some of the challenges [in] an [an] academic . . . or a scientific context, but maybe in different disciplines. And we've supported each other through . . . challenging times.

    —JULIA CLARKE, University of Texas, Austin

I've been very lucky. I've had great mentors. I've had the opportunity to be involved with lots of amazing projects and do fieldwork all over the world. But I would say it's definitely been difficult at times too, . . . in ways that are very specific to, I think, being a woman, being a minority. And I didn't necessarily appreciate those things when I was younger. I would say when I was in graduate school, I was almost unaware of them . . . and maybe very lucky actually. I certainly . . . think as a graduate student I was aware of the fact that women who had children would not get tenure at times or would have a hard time getting through tenure. I did notice those things, but I wasn't directly affected by them. I, of course, moved to Europe right after my PhD. . . . so it's a very different situation. It's . . . more supportive in terms of things like maternity leave and childcare. And so, I think I've had an easier time with that. . . . I have two young children and I've been able to . . . manage a work-life balance that works for me anyway . . . [and] hopefully for my husband too.

    —ANJALI GOSWAMI, Natural History Museum, London

5. What lessons learned in the course of your career can you offer as advice to those considering entering the field?

  Based on my experiences, I would tell anybody interested in joining the field of vertebrate paleontology to follow what they believe.

    —REBECCA HUNT-FOSTER, Dinosaur National Monument, Utah and Colorado

I think that if I were [going to] give advice to someone, . . . you really have to know who you are and you have to be able to stand up for yourself and then it's not always the easiest thing to do and especially in a field environment. I think that there can be real challenges . . . at times. I always encourage people to find mentors that they really trust that they can really talk to honestly and openly and to be confident in what they know and also to realize that they don't know everything and to be open to people. I think being as confident as possible and making sure that you know who you are and you're willing and able to stand up for yourself. It's probably one of the most important things that I've learned over the course of my career.

—KRISTI CURRY ROGERS, Macalester College, Saint Paul, Minnesota

If you can connect with people who have made it through, who have become faculty members, are working at museums who you see and you are inspired by, those are the people to go to and ask questions and they're more than happy to sort of share their knowledge and help you on this . . . big, amazing adventure.

—LARISA R.G. DESANTIS, Vanderbilt University, Nashville, Tennessee

If I'm giving advice to somebody, . . . you really need to get experience and decide this is something that's for you. I always tell people to go to the local museum, volunteer in the prep lab. If you're getting good with the team, then they can take you out into the field. There [are] so many opportunities for you to actually get out there, do research, go see some collections, go in the field, . . . and then actually decide for yourself that this is for you. And then if [you] are certain that [you] want to be a paleontologist, . . . there are infinite research questions out there. I know that I've been very, very lucky . . . [to have] people who have offered me guidance in my path that . . . it's important to me to give back. . . . I hope in the future I will be a good mentor to . . . a whole new generation of scientists.

—JINGMAI O'CONNOR, Institute of Vertebrate Paleontology and Paleoanthropology at the Chinese Academy of Sciences, Beijing

In terms of advice I would give, ... I think there [are] a number of different things as a student [that] you need to consider. Your interest in the subject will take you so far in this day and age, [but] it's so difficult and it's competitive to get into graduate school.

—EMILY RAYFIELD, University of Bristol, United Kingdom

I guess my only advice is to follow your heart and do what works because everybody's trajectory through children and partners and spouses and stuff, it's so completely different that, how could you possibly give advice other than . . . saying hang in there.

—EMILY BUCHHOLTZ, Wellesley College, Wellesley, Massachusetts

I try to be very honest with students. . . . When they say they want to pursue a career, I want them to think about why they want to do it because of course it is really hard to do a PhD and it's really hard to pursue a career in academia. There [are] not a lot of jobs . . . and it's a very competitive field. . . . It can be really difficult, I think, for a lot of people because you're constantly open to criticism. In fact, that is the scientific process. We have to be critical. We have to constantly be trying to look for the errors in what we're doing, and that can be difficult, as I have certainly found that difficult at times and I know everybody else does as well. It's a job where you need to be very robust and you have to be open to those sorts of criticisms, but it also is incredibly rewarding. And I tell this to students, . . . in my view, it's worth it because you get to do whatever you want to think about. You wake up in the morning and you think this is interesting and then you can go to your office and try and figure it out. It's just the kind of life and career that most people don't get to have. And so it's a really rewarding one, but it does take a lot of dedication and a lot of time and . . . a thick skin at times. I tell them that they have to be prepared for all that. And then they also need to make sure they go out and meet people and draw ideas from lots of different fields that requires a certain amount of networking and being social in the field. And also that it's really important for people to think about what they want to achieve, what they want to do next in their career and whether they stay in academia or whether they go, . . . all the different things that you can do with a PhD, and to make sure that the choices that we're making . . . will help them get the kind of careers that they want.

—ANJALI GOSWAMI, Natural History Museum, London

If I talk to undergraduates or even high school students who haven't quite figured out what they want to do, I think my advice would be to explore all the different aspects of paleontology, not just academic research, but also teaching, public outreach, illustration, museum work. Because all of those different aspects of paleontology are necessary in order to keep the field moving forward. And frankly, as we all know, . . . there's always a dearth of jobs, particularly academic jobs, compared to the number of highly qualified people. And yet there are other ways, many other ways, of being highly engaged in the field that are also very fulfilling. And I think it's important for people to understand, at least before they go to graduate school, whether a research career is for them or whether one of these other, more supporting roles would be [a better fit] . . . I think what I would recommend that people explore. I know a lot of students who start out just working in the prep lab because they think it's interesting. And then they . . . realize that the whole field is interesting and some of them have gone on to be, you know, very successful researchers and others are just very happy as preparators.

—CATHERINE BADGLEY, University of Michigan, Ann Arbor

6. Is there anything that we haven't discussed that you would like to add about your experiences as a vertebrate paleontologist?

There are going to be times when you feel very down about your PhD, either that's never going to get done or we all set such high expectations for what we want to discover and we think we're going to discover. And then when . . . you didn't solve all the world's problems or you didn't find that great thing that you hoped you would find, students get deflated and then they don't think they want

to publish or they don't even want to talk about it. . . . You have to go beyond that and just give your talks. Because what always happens is [that] people come up to you and say, wow, that was really interesting.
—BLAIRE VAN VALKENBURGH, University of California, Los Angeles

One thing I haven't mentioned . . . and most people don't know about me is I actually used to have epilepsy and had 30 to 40 seizures every day. . . . I think it's important to mention that because there are lots of people out there who have a disability who may not believe that they can go out in the field and become a paleontologist and embark in this area. But in fact, you can. There's a lot of things that you can do. And I . . . want to be a role model for those [people].
—LARISA R.G. DESANTIS, Vanderbilt University, Nashville, Tennessee

I have discovered that there is a stereotype of . . . paleontologists in general and even [more so] for women in paleontology. For example, it's not very common for the general public to see women in paleontology So sometimes I have to explain myself and say yes and I'm a paleontologist and I'm real. I'm doing paleontology.
—ANA VALENZUELA-TORO, PhD candidate, University of California, Santa Cruz

I think that we are becoming more and more aware of the issues that women are facing in STEM disciplines, and vertebrate paleontology unfortunately is no exception to that. And I . . . think a woman like myself, in the stage of my career [I am in], it's so important for us to be mentors to young women in vertebrate paleontology and in STEM disciplines as a whole. And it doesn't even have to be . . . a formal role. You can even just . . . say to somebody . . . , that was a great talk, or I really enjoyed the poster, but just to be there to do that mentoring, and I think women in paleontology have to be there for each other. They have to support each other. And I think that [is] one message that I would like to get across to people. But I have to say [at] this meeting, I've seen so many young women so confident and speaking so well not only at the presentations [as they] are standing in front of their posters but also in meetings. I think we, women, have a bright future in paleontology, but I do think it's up to people like myself and yourself [A. Berta] to be mentors.
—ZERINA JOHANSON, Natural History Museum, London

I . . . just want to say that the field is so diverse and more diverse now than it ever has been. That it's possible for not just women but anyone to sample different parts of it. For instance, I have primarily a biology background and I don't feel categorized or excluded at all. Early in my career when I was trying to do fieldwork, I thought of myself as this is what this is. In the last 15 years or so I've been trying to integrate developmental biology with the fossil record, and now it seems like it's the most reasonable thing in the world for me with my particular background, so you can see that I've been in different niches either involved or at least changed. I've been at different kinds of parts of the field at different times, and I think looking back you can see that not only can different people be in different subparts of the discipline, but you yourself could be in different parts at different times in your career.

—EMILY BUCHHOLTZ, Wellesley College, Wellesley, Massachusetts

There are so many opportunities now to go in interdisciplinary directions and so it's not so hard to fight for those, but it's also, I think it is hard to carve out a really individual . . . niche and you just have to think you can do it and look for creative ways to, to engage with the fossils. Then also with all the modern analogues and the wonderful, quantitative methods and persistence, you know, . . . you get an idea, it's a dream, perhaps, and then you just go for it and you get the people around you to help with that.

—ANNA "KAY" BEHRENSMEYER, Smithsonian Institution, Washington, DC

One thing that I think is really important is fieldwork. . . . We are all vertebrate paleontologists. Our data comes from the field ultimately, and you can do a lot, as I do a lot, with museum collections, but going out and finding new fossils around the world is very, very important and there's huge parts of the world that have very poor records, especially for the time periods that I'm interested in, the Cretaceous and Paleogene, . . . and that, I think, is one of those things that sometimes women can feel is a very daunting prospect to, to participate in or especially to start and lead fieldwork around the world. And there are lots of barriers to doing that. Certainly other parts of the world have more sexism or racism than some of the Western liberal democracies. . . . But it is really, really worth doing that one because . . . we are the face of science and the more we do that, the more it becomes accepted and the more we are able to inspire young women in those countries to rise up and become leaders in science in those fields. And I think that's been really great. I think it's been really great to have supportive men who have . . . gone out in the field with us or taken me out in the field with them in some cases to really push that and make sure that field crews are more diverse because it does have a really big impact on people to see diversity in science and as a very international field. Paleontology has really captured the public attention. I think it's really important that we portray that around the world. I think it can have more of an impact in effecting change in a lot of other sciences.

—ANJALI GOSWAMI, Natural History Museum, London

I would just say that [my] whole career in vertebrate paleontology has been very fulfilling for me . . . , and it's really still the same reasons that I got into it in the first place. Partly it's the grand vision of how humans relate to the rest of life. And I think one of the most interesting things that I think about more than I used to is now how insights from the fossil record contribute to our understanding of environmental crises today and how we can use some of the insights from the past too, for conservation, for changes in society that we should make in order not to make a big mess of the rest of nature and even preserve our own well-being. So I'm very grateful to have that deep time perspective and I think that that's a wonderful thing, a wonderful big idea to try to pass on to others.

—CATHERINE BADGLEY, University of Michigan, Ann Arbor

Excerpts from Women VPs' Responses to Written Interviews

1. What drew you to/interests you/excites you about VP?

The books of Roy Chapman Andrews were a main influence. Discovery in the field and in collections is always a thrill.

—JUDITH SCHIEBOUT, Louisiana State University, Baton Rouge

I really loved this *Diplodocus* (a large, herbivorous dinosaur) that my parents bought for me when I was a kindergartener. All of the other girls were playing with cute dolls, but I was the only one playing with dinosaurs.

—TAMAKI SATO, Tokyo Gakugei University, Japan

I did grow up on a dairy farm and spent a lot of time outdoors with animals. My older sister wanted to be a biologist, and so did I. In college (Oberlin) I considered premed but decided against it; I liked anatomy but was disillusioned by the rat race of premeds. A course in evolution exposed me to vertebrate history, and I was hooked from there. The other thing that attracted me to VP was my interest in geography and travel. I quickly realized that paleontologists often went in the field and visited museums all around the world. Thus anatomy, the outdoors, and travel were big lures for me.

—MARGERY COOMBS, University of Massachusetts, Amherst

The transition from land to sea during the evolution of aquatic mammals has caught my attention since high school. This was a big motivation for my work in VP.

—CAROLINA LOCH S. SILVA, University of Otago, Dunedin, New Zealand

When I first heard about fossils, I was completely enthralled by these remnants of past life. I recall being thrilled to learn that besides the actual anatomy of the fossil bones that are preserved, the histological structure is also preserved, and that this holds clues to the biology of these extinct vertebrates. I was so taken up by the latter that I ended up specializing in fossil bone micro-structure to try and understand the life history of extinct vertebrates.

—ANUSUYA CHINSAMY-TURAN, University of Cape Town,
Cape Town, South Africa

So many things, going to field, finding and collecting fossil remains, and studying them. Trying to figure out to whom the remains belong, who they lived with, finding out about their uniqueness, how they fit in the whole evolutionary picture of the group, just to think that the remains belong to an extinct taxon, that is amazing!

—MARISOL MONTELLANO-BALLESTEROS, Universidad Nacional Autónoma de México, Mexico City

Fieldwork. It is pretty exciting knowing that you are the first person to see something that has been dead for, e.g., 300 million years.

—ANNE WARREN, retired, La Trobe University, Melbourne, Australia

The "oldness" of it all. Deep time is an interesting concept to try and wrap your head around. Initially I was interested in ancient history, then archaeology. But, once I was introduced to fossils, well, there was no turning back.

—KATHY LEHTOLA (Michigan State University), retired, Land Use and Transportation Department, Oregon

I have always liked the question why. It has got me into hot water occasionally. This is an important question when working in VP. It helps to move your thinking forward. A love for jigsaw puzzles as well brought me to VP. The process of putting the pieces of a VP project together and not always having all the pieces. The most exciting thing about VP is that you never know what direction a project will take once you begin; especially when working in collections as you never know what you will encounter on any given day.

VP projects can be challenging and once you accept the challenge it can be an exciting and satisfying experience. You also learn about yourself and your ability to learn new things and build your confidence as a scientist.

—ROMALA GOVENDER, Iziko South African Museum, Cape Town, South Africa

2. What has been your most exciting field experience/discovery/contribution in VP?

Each time I go to the field is exciting, because there is always the uncertainty of what is going to be found, and in each of them there is a "good or extraordinary" fossil collected.

—MARISOL MONTELLANO-BALLESTEROS, Universidad Nacional Autónoma de México, Mexico City

[My] most interesting and profitable field contribution was the excavation of Morava Ranch Quarry, near Agate in northwest Nebraska. Along with my husband and a student field crew, I spent most of the summer of 1975 excavating and mapping this quarry. The matrix was like cement, so all our finds were hard wrought. One thing I hadn't really calculated in the equation was how long it would take to prepare all the fossils we collected. [The] University of Massachusetts and Amherst College have no staffs to do this kind of work, so I had to rely on myself and on student labor. Then of course I had to do quite a bit of legwork to identify the material (mostly isolated postcrania) that we had collected. The main paper derived from this work (Coombs and Coombs, 1997, *Palaios*) investigated the taphonomy and paleoecology of the locale. I'm proud of it, partly because I was responsible for the whole thing from start to finish.

—MARGERY COOMBS, University of Massachusetts, Amherst

My most exciting experience was . . . discovering my first fossil in the field. . . . We were standing discussing what we were going to do and which directions we were going to walk in. I was going to turn around and walk toward a hill behind me. As I turned and took my first step I looked down and was about to step onto this small complete *Diictodon* [therapsid] skull. That is how I found my first fossil and I have been lucky ever since in being able to find fossils.

—ROMALA GOVENDER, Iziko South African Museum, Cape Town, South Africa

Finding the equivalent of Walcott's fossil fish across the North American inland sea from his location. He published his finds in 1892; I found mine in 1970.

—KATHY LEHTOLA (Michigan State University), retired, Land Use and Transportation Department, Oregon

3. Which mentor(s)/role model(s) was important in your taking up VP? Why?

I only came into paleontology in my third year of university when I took a paleontology module that was taught by Mike Raath and Judy Maguire. They were both passionate about their research, and I was drawn into the field. I was initially very keen on archaeology, but then Mike Raath introduced me to the field of bone histology and I was hooked! I had read all about the exciting fieldwork that Zofia Kielan-Jaworowska [Chapter 4] had done in Mongolia, and I felt quite inspired by her. I am delighted to have once met her, and to have realized that "in person" she was such an incredibly dynamic, vivacious woman. Through Peter Dodson as my wonderful postdoc supervisor, I had the opportunity of working in the USA for two years and visiting several amazing collections and museums there and in Canada.

—ANUSUYA CHINSAMY-TURAN, University of Cape Town,
Cape Town, South Africa

When I started graduate school (at Columbia University) I didn't know much about what it took to be a vertebrate paleontologist. I think my best mentors were the fellow graduate students who were my contemporaries or just a bit ahead. These were a wonderful group: Bob Hunt, Gene Gaffney, Tom and Pat Rich, Bob Emry, and Walter Coombs. All of them, by virtue of their greater experience, seemed intimidating to me at first, but they were also helpful and good role models. Tom and Pat took me on my first summer field expedition (to Como Bluff,

Wyoming) in 1969, Bob Hunt later helped me get started with fieldwork in Nebraska, and Walter [Coombs's husband] (of course) was the best mentor of them all. As my work progressed, the Frick staff [personnel] at the American Museum of Natural History (particularly Morris and Marie Skinner, Ted and Marian Galusha) were helpful for my finding a footing in research; they were also good examples of married couples working together cooperatively in paleontology. My adviser, Malcolm McKenna, was a pioneer in being supportive of women in paleontology. He was also an endless mine of original ideas.
    —MARGERY COOMBS, University of Massachusetts, Amherst

(1) Richard C. Harington (Canadian Museum of Nature). He was my mentor in high school. Introduced me to vertebrate paleontology (Ice Age mammals). When in university he invited me to join his team to Strathcona Fiord, Beaver Pond site. This was my first experience in the High Arctic and I was hooked.
    (2) Mary Dawson. Vert[ebrate] paleo[ntology] pioneer of vert paleo research in the North American High Arctic. She joined my team in 2007, and we found *Puijila* [fossil pinnipedimorph].
    These two people are great explorers, and have been great inspiration to me. Their mentorship has been invaluable as I developed my own Arctic research program.
    —NATALIA RYBCZYNSKI, Canadian Museum of Nature, Ottawa, Ontario

My PhD supervisor Prof. Ewan Fordyce was very influential in my career and a great mentor in VP and in science in general. He is one of the most dedicated and thorough scientists I know and I have learnt so much from him. I am also very proud to have been the first female student to complete a PhD under his supervision and help[ing] pave the way [for] others who followed.
    —CAROLINA LOCH S. SILVA, University of Otago, Dunedin, New Zealand

My main mentor was the excitement of science and just getting out in the field and searching for things that I was the first to find. Ruben Arthur Stirton and Alden Miller at the University of California were true mentors, as was Walter Bock and Malcolm McKenna at the American Museum and Columbia. They were important in that they gave me material to work on, occasional conversations, but gave me the feeling that I was able to move forward with my research as a colleague rather than someone that had to have their hand held. They also gave me monetary support with scholarships and when an undergraduate, education was free! For a poor farm girl from a rural background this opened the doors to paradise for me.
    —PAT VICKERS-RICH, Monash University, Melbourne, Australia

My inspirers and mentors have all been male, . . . not surprising since [paleontology is] a male-dominated field. These people include the late George Lammers (mammalian paleontologist, previously at the Manitoba Museum), the late Karl

Hirsch (previously affiliated with the University of Colorado at Boulder), and Philip Currie (dinosaur paleontologist at the University of Alberta and former supervisor).
    —DARLA ZELENITSKY, University of Calgary, Canada

4. Have there been any difficult challenges (i.e., motherhood, etc.)/negative influences in your career so far? What did you learn from them?

In South Africa, in particular, being a woman of color breaking into VP has been challenging. One constantly finds [oneself] under pressure to prove that you are capable of doing the work and that you know what you are talking about. The most important lesson I have taken from this is perseverance and to have confidence in your own abilities. To remember that you are the person doing the work and making the discoveries. Some days that is a difficult thing to remember. Learn to trust your instincts and decision making.
    —ROMALA GOVENDER, Iziko South African Museum, Cape Town, South Africa

Two things came close to ending my academic studies and paleontology research: lack of confidence in myself because of earlier failures, and intimidation and bullying by a male paleontologist in a position of power. The support and suggestions of my two mentors, as well as encouragement from everyone in the Earth Science Department to enter the PhD program, were critical in helping me deal with the problems.
    —FRANKIE D. JACKSON, Montana State University, Bozeman

From an early age, if confronted by anything negative, I just walked away from it and got on with my life of curiosity. I have just kept in mind that the main thrust of my life is to understand the world around me, to engage others in that as well. . . . If this means working with a salary, or just working with no salary and making enough to survive in one way or other so that one can keep up the standards of the university . . . —so that students are still coming out with the skills they need—I just have done it.
    —PAT VICKERS-RICH, Monash University, Melbourne, Australia

I was born with a genetic defect called Klippel-Trenaunay syndrome, which made me increasingly lame during adulthood. It was not diagnosed correctly until I was middle aged. Knowing what it was did not help a lot, because it is rare and genetic. I got some of it out of the way by having a BTK [below-the-knee] amputation of my left leg in 2014. Sometimes you walk, sometimes you crawl, and sometimes you buzz around on wheels.
    —JUDITH SCHIEBOUT, Louisiana State University, Baton Rouge

My children were born in 1970 (when I was a graduate student) and 1980 (shortly after I received tenure). Naturally, both my husband, Walter, and I worried about how we would manage both a family and careers. Walter was both extremely supportive and helpful; our families picked up the slack when possible.

Nonetheless, it is very difficult to multitask a family and a career. The challenges don't stop when the kids start school either; there are school projects, ferrying to extracurricular activities, crises with friends, you name it. Sacrifices were necessary, and I'm sure I would have completed more projects and been promoted faster without these distractions. But the distractions were also enriching. I am a better, more empathetic, and happier person than I would have been without our kids (now age 47 and 37).

Some other challenges were work related. In many ways I was a pioneer in my job, because I was the only tenure-track woman in my department when I was hired. In my experience, conscientious women often become critical to the way departments operate, but time-consuming committee work is not much rewarded. It is only frosting on the cake of research. One might say that women should avoid committees when possible to concentrate on research. Personally, I have always cared about a balance among teaching, research, and service, so I think the reward system should recognize the three more equally.

—MARGERY COOMBS, University of Massachusetts, Amherst

The challenges I have faced so far were mostly in the sense of breaking glass ceilings and stereotypes. I was only the second person in three generations of my family to complete a university degree. Coming from a working-class family in Brazil and having only studied in public schools, there was a lot of hard work and dedication put through my studies to get me where I am. I am the only person in my family to have a doctorate degree, and of course doing it on the other side of the globe and in a foreign language was challenging, nevertheless very rewarding. Being Latin American, young, and female sometimes comes attached to perceived negative stereotypes but those have never held me back.

—CAROLINA LOCH S. SILVA, University of Otago, Dunedin, New Zealand

Motherhood was a challenge, because when my husband and I decided to have a baby, I was certain that I was going to take care of her. My decision was that during the early years, I was going to be a part-time researcher-teacher and half-time mother. I was fortunate to enjoy and see my daughter growing, share with her little and big things. My professional career slowed down for many years, with some frustrations and disappointments, but looking back I believe I did the right thing. She is now a happy and healthy young lady. But at the same time I enjoy my job. I am lucky because I do what I like.

—MARISOL MONTELLANO-BALLESTEROS, Universidad Nacional Autónoma de México, Mexico City

5. *What lessons learned in the course of your career can you offer as advice to those considering entering the field?*

My first piece of advice for students is that the VP mansion has many connected rooms. It is not necessary to get a PhD and become a university professor or museum curator. Those are great careers, but they are not for everyone, and those committed in that direction should be aware of what the heavy-duty expectations on them might be. Students with VP interests can also do great work as teachers/docents, preparators, field geologists/collectors, park rangers, artists, exhibit designers, and writers/science journalists. It is very important for them to assess their talents, skills, and interests and to seek collateral training accordingly. I highly recommend field and museum internships to undergrads. Those who have done internships have often found unexpected opportunities arising from them. For those heading toward a research career, it is important to assess what kinds of niches are available in academe. Few young PhDs are hired strictly as vertebrate paleontologists these days. They need other skills, such as genomics, functional morphology, paleoecology, geologic mapping, whatever. Fortunately, there are many ancillary areas that people can choose. Finding a good match of a graduate school and graduate adviser is a very important step in choosing directions.

In many cases I have observed, there is some component of luck in being successful in paleontology, i.e., being in the right place at the right time. We all have to accept that and not totally blame ourselves if things don't go quite as planned (or hyper-congratulate ourselves on successes).
—MARGERY COOMBS, University of Massachusetts, Amherst

From a fieldwork perspective, I think it can be very helpful to build interdisciplinary field teams. In the case of the Arctic, our interdisciplinary field teams have allowed us to collect data on flora, fauna, landscapes, paleoclimate, etc., providing a critical context for understanding vertebrate ecology and evolution.
—NATALIA RYBCZYNSKI, Canadian Museum of Nature, Ottawa, Ontario

Do not give up! Pursue your dreams, but do not die in the attempt, it does not make any sense, life is full of experiences and it offers so many things. In Spanish there is a phrase, "Not all the roads are for everybody," there is a place for each of us, one has to find it, where one feels OK. In our times, being a woman is challenging and demanding, during all our lives we are asking ourselves which way and what would be the best. It is important to set priorities, a difficult task, but necessary to survive and enjoy life.
—MARISOL MONTELLANO-BALLESTEROS, Universidad
Nacional Autónoma de México, Mexico City

I think another important lesson I have learned is that life requires balance. VP is not the only thing I do. I have my own horses that I ride and compete on at dressage and show jumping. VP is one of the most interesting jobs you can pursue but do not let it consume you!!
—ROMALA GOVENDER, Iziko South African Museum, Cape Town, South Africa

6. *Is there anything that we haven't discussed that you would like to add about your experiences as a vertebrate paleontologist?*

The subject has given me the most marvelous career imaginable—and not one I would ever have imagined as a kid.

—JENNY CLACK, Cambridge University, Cambridge, United Kingdom

I believe it is vitally important for paleontologists to share their research with the wider community, especially to encourage women and diverse (historically disadvantaged) communities to become more involved in VP.

—ANUSUYA CHINSAMY-TURAN, University of Cape Town,
Cape Town, South Africa

We are encouraged by today's society to live in the moment, but as paleontologists we have a uniquely longer perspective that is oftentimes very valuable.

—KATHY LEHTOLA (Michigan State University), retired, Land Use and Transportation Department, Oregon

Literature Mentioned by Interviewees

Andrews, R.C. 1943. *Under a Lucky Star: A Lifetime of Adventure*. Viking Press, New York.

Andrews, R.C. 1953. *All about Dinosaurs*. Ill Thomas W. Voter. Random House, New York.

Coombs, M.C., Coombs, W.P. 1997. Analysis of the geology, fauna and taphonomy of Morava Ranch Quarry, early Miocene of northwest Nebraska. *Palaios* 12: 165–187.

Gallenkamp, C. 2001. *Dragon Hunter Roy Chapman Andrews and the Central Asiatic Expeditions*. Foreword Michael J. Novacek. Viking with the cooperation of the American Museum of Natural History, Penguin Group, New York.

Gould, S.J. 1989. *Wonderful Life: The Burgess Shale and the Nature of History*. Norton, New York.

Yochelson, E.L. 1998. *Charles Doolittle Walcott, Paleontologist*. Kent State University Press, Kent, OH.

Yochelson, E.L. 2001. *Smithsonian Institution Secretary, Charles Doolittle Walcott*. Kent State University Press, Kent, OH.

Excerpts from Male Mentors' Responses to Written Questions

1. Are there any specific things that you learned from the experience of women students in your lab/field?

I've learned a great deal from all my students, and their varied interests have broadened my experience in ways I never would have imagined. This is particularly true of my female students, who have taught me about such diverse topics (outside of my usual focus) as primate and artiodactyl ears, whale origins, spinal flexibility and function, biomechanics of leaping, and anatomy of aardvarks and elephant shrews.

—KEN ROSE, Johns Hopkins University, Baltimore, Maryland

Although recognizing the difference, I don't think I have ever categorized graduate students by gender; perhaps this is the product of having two academically successful daughters. My approach has been to try to identify academically talented applicants, identify their areas of research interest, and then support them to the best of my ability.

—W.A. CLEMENS, University of California, Berkeley

I see no difference in mentoring female and male students in the lab or in the field. If they are eager, intelligent, hardworking and follow through on their commitments, I treat them all the same.

—DON PROTHERO, Occidental College, Los Angeles, California

Honestly, I try to treat all students the same, regardless of gender. I do not get too friendly and try to maintain a professional distance. In terms of work ethic, I hope that I lead by example. I also realize that not one size fits all for grad students, and some require special mentoring, e.g., those experiencing profound family problems during their time with me, or a single mom, recently divorced, raising two children. I am there to listen for them.

—BRUCE MACFADDEN, Florida Museum of Natural History, Gainesville

The lesson I take from Kate A. Andrzejewski [his recent PhD student] is that ours is a dynamic science, always changing and improving, adding new techniques, always exciting.

—LOUIS JACOBS, Southern Methodist University and Shuler Museum of Paleontology, Dallas, Texas

[Louis Jacobs also received the following from a former student:] "Today, I [Yuri Kimura] was interviewed for a kid's magazine, talking about how I made my dream come true (I am still on my way, though!). I mentioned you and Yuki [Tomida] as those who inspired me. I wore my lab coat and demonstrated how to make casts of small mammals. My lab coat is quite messy with lots of paints and yellowish with my sweat (yuck!), but I am still wearing it. Before the interview, I deeply looked back [on] my way and thought how lucky I have been. It was a real fortune to be your student;)."

While doing fieldwork in Pakistan, Alisa [Winkler], Michelle Morgan, and Catherine Badgley dubbed themselves the "Buckettes" to celebrate their screen-washing for small specimens. I also very fondly remember a meeting at Lake Baikal organized by Margarita Erbaeva and the Siberian Academy of Science. Alisa, Larry Flynn, and I were there from the US. . . . The crew set out nets in the evening, and in the morning, we had a huge supply of Lake Baikal trout. Alisa was always ready and happy to pitch in, every step of the way. There was a pile of fish on one side, and fish guts from the cleaning on the other. It turned into a wonderful feast, a wonderful symposium, and a wonderful memory of the fun of our science.

—LOUIS JACOBS, Southern Methodist University and Shuler Museum of Paleontology, Dallas, Texas

I have learned both positive and uplifting things, and also some disappointing and frustrating things about vertebrate paleontology and about some vertebrate paleontologists. On the positive and uplifting side, I have seen the emergence of a serious dialogue about gender issues and inclusivity within the community of vertebrate paleontologists. I have seen women ascend to leadership positions in the Society of Vertebrate Paleontology and help to guide the society toward being a more healthy and respectful scientific organization. I have seen female graduate students move into careers of their choice and succeed in those careers. I have seen female students form working collaborations and a support network that maintained itself long after graduation. I was fortunate to be able to work with excellent female students who helped me to detect, understand, and address my own inherent biases, and helped me to see our discipline through the eyes and experiences of young women. I had, and continue to have, fruitful conversations with them about my strengths and weaknesses as a supervisor; those conversations helped me to grow and, I hope, improve as a supervisor.

But I have also seen the darker side, and watched as my own students struggled under stereotypical challenges. A few examples will suffice to demonstrate that we still have a long way to go in creating a supportive and inclusive community in vertebrate paleontology. I witnessed students forced to make decisions about who would be on their committees, based not upon the scientific qualifications of the candidate committee members, but upon their behavioral patterns, especially with respect to women. I witnessed an established male scientist, for the benefit of his own selfish motives, ask a young female graduate student to commit data fraud. I witnessed committee members who wrote sabotaging letters as female students applied for jobs. I witnessed female scientists push female students hard to achieve high standards, but did so at the expense of becoming emotionally abusive. I witnessed male members of search committees commenting on qualifications of job candidates, but their first words addressed physical attributes of the candidates. I also watched as rising professionals struggled with spousal-hire issues in academic institutions; whether the male or the female is the primary negotiator, this remains a thorny problem, and one with which it seems many universities continue to struggle.

—CHRISTOPHER BELL, University of Texas, Austin

My advice to women I've mentored is much the same that I give anyone (mostly things I've had to learn for myself, or by consulting with or observing other colleagues), but informed by recognizing that women have faced many long-standing barriers to success and should not have to or be willing to accept those. Insist on equality and respect, if in a situation or institution that doesn't provide those obvious, fundamental rights; while striving to be kind, friendly, and collaborative, it also is essential to remain confident in your abilities, assertive in support of your ideas, and unafraid to stand up for and advocate for yourself. Most importantly, family and personal friends are essential priorities, and maintaining those relationships provides foundations for personal satisfaction and professional successes.

—JOHN FLYNN, American Museum of Natural History, New York

2. Can you offer any recommendations for women interested in pursuing a career in VP?

I give all students the same advice: careers in VP are very limited, so they need to be realistic about their plans once they have decided on such a career. I tell them about the terrible odds and overabundance of students, the fact that fewer than 1 in 20 VP PhDs gets a good job doing research at a museum or research university. I tell them that if they are serious about such a career, and cannot imagine doing anything else, then they need to work harder than anyone else, and publish like crazy as an undergrad and as a grad student to have even a remote chance of getting such a job. Despite the fact that most of my students were only undergrads with me, and did their grad work elsewhere, a remarkable number of them have survived and gone on to get jobs. A few are in research VP positions, but most went into teaching, environmental consulting, and oil, where most of the jobs are.

—DON PROTHERO, Occidental College, Los Angeles, California

I'd offer the same recommendations that I would to any student, regardless of gender. Work hard, follow your passions, ask interesting questions, and be prepared for a very tight job market.

—BRUCE MACFADDEN, Florida Museum of Natural History, Gainesville

As you move through your academic training, from undergraduate through graduate degrees, and postdoctoral experiences, pay attention to what is happening around you. Learn from your experiences and the experiences of others. Talk with professionals in your career of choice, especially women, and learn about the challenges you will face and the skills you will need to succeed in that career. Ask yourself and others questions about how academic and other institutions work, how professionals in your career of choice operate and how they conduct themselves. Adopt the positive attitudes and behaviors you witness and experience, and reject the ones that were unjust, unfair, annoying, or drove you nuts.

Avoid strict hierarchical thinking, and especially hierarchical behavior—everyone has something to offer, and it takes many people to form a healthy and productive department, lab, research group, or team. Avoid taking advantage of others, and being taken advantage of, as you pursue your career goals.

Be happy in your work; life is too short to waste your time being unhappy. Avoid negative people and poisonous personalities; they will drag you down.

Take time to make connections early and continuously, and forge strong, healthy professional relationships. Your relationships with your fellow students, and with staff, faculty, research scientists, and others, will provide a support network for you and will help to shape opportunities for you in the future. Take advantage of those opportunities.

I would like to say that you will be judged solely on the basis of your science, but I don't think that is true. Vertebrate paleontology remains, to a large extent, an old-boy network, and this means you will face additional challenges that your male colleagues do not. Go in with eyes wide open, and be creative in finding solutions to those challenges. Talk to women. I'm not one, so I don't really know what is going on. My comments and suggestions are based on what I have seen and experienced across 30 years of engagement with academia, but those are necessarily the observations and experiences of a man.

—CHRISTOPHER BELL, University of Texas, Austin

Be resilient, using constructive critique to improve your work and avoiding spending excessive energy or time when frustrated by unwarranted criticism (but also be willing to challenge a clearly unreasonable or unfair evaluation).

Give credit to others for their contributions, but also insist on adequate recognition of yours (e.g., in authorship); addressing expectations very early in any potential collaboration can help minimize misunderstandings and problems later. Pursue funding assertively, as this provides a necessary foundation for your work (and later, for your students and collaborators if in academia). And be adventurous in seeking support from sources that initially might not seem within your "comfort zone," as this can open new avenues of research if you can clearly explain to and excite a new audience about your work. Be willing to take advantage of interesting, unexpected opportunities (even when not clearly "timely" to do so, as there's rarely a truly "good time" to take on any major new endeavor or commitment), but also be sure to continue to meet essential commitments you've made to yourself and to others. Most importantly, always act with integrity!

    —JOHN FLYNN, American Museum of Natural History, New York

Examples of Taxa Named for and by Women

Details on author, date, age, reference, etc. of most taxa can be found at the Paleobiology Database, Fossilworks, and Fossiilid.info websites.

Taxa Named for Women

Acrodus anningiae Agassiz, 1837. Hybodont shark named for Mary Anning

Aegialodon dawsoni Kermack, Lees, and Mussett, 1965. Cretaceous stem mammal named for Mary Dawson

Aletrimyti gaskillae Szostakiwskyj, Pardo and Anderson, 2015. Permian lepospondyl amphibian named for Pamela Gaskill

Anningasaura Vincent and Benson, 2012. Stem plesiosaurian named for Mary Anning

Anningia megalops Broom, 1927. Permian Karroo reptile named for Mary Anning

Anningiamorpha (order) Broom, 1932. Permian reptiles named for Mary Anning

Archaeornithura meemannae Wang et al., 2015. Cretaceous bird named for Meeman Chang

Arcticanodon dawsonae Rose et al., 2004. Eocene palaeanodont named for Mary Dawson

Avimaia schweitzerae Bailleul, 2019. Cretaceous fossil bird named for Mary Schweitzer

Balaenoptera bertae Boessenecker, 2013. Miocene-Pliocene baleen whale named for Annalisa Berta

Basilodon Kammerer, Angielczyk, and Fröbisch, 2011. "King tooth" dicynodont named for Gillian King

Belenostomus anningiae Agassiz, 1844. Jurassic bony fish named for Mary Anning

Borealestes mussetti Sigogneau-Russell, 2003. Mesozoic stem mammal (should be mussettae) named for Frances Mussett

Bridetherium dorisae Clemens, 2011. Early Jurassic stem mammal named for Doris Kermack

Cheirolepis cummingiae Agassiz, 1835. Devonian bony fish named for Lady Gordon Cumming

Citipati osmolskae Clark, Norell, and Barsbold, 2001. Mesozoic oviraptorosaur dinosaur named for Halszka Osmólska

Coccosteus markae Obrucheva, 1962. Devonian placoderm named for Elga Mark-Kurik

Corveolepis elgae (Tarlo, 1965), Blieck and Karatajūtė-Talimaa, 2001. Devonian heterostracan named for Elga Mark-Kurik

Crocodilus hastingsae Owen, 1848. Crocodile named for Barbara Hastings

Dalmehodus turnerae Long and Hairapetian, 2000. Devonian shark named for Susan Turner

Dipnorhynchus kurikae Campbell and Barwick, 1985. Devonian lungfish named for Elga Mark-Kurik

Dryosaurus elderae Carpenter and Galton, 2018. Ornithopod named for Ruth Elder?

Eugnathus philpotae Agassiz, 1839. Jurassic bony fish named for Elizabeth Philpot

Faberaspis elgae Elliott, Lassiter, and Blieck, 2018. Cyathaspid named for Elga Mark-Kurik

Fallacosteus turnerae Long, 1990. Devonian placoderm named for Susan Turner

Franimys amherstensis A.E. Wood, 1962. Paleocene rodent named for Florence "Fran" Wood

Gacella soriae Alcala and Morales, 2006. Fossil antelope named for Maria Dolores Soria Mayor

Gomphonchus turnerae Burrow and Simpson, 1995. Silurian acanthodian named for Susan Turner

Grenfellacanthus zerinae Long, Burrow, and Ritchie, 2006. Devonian placoderm named for Zerina Johanson

Gualicho shinyae Apesteguia et al., 2016. Cretaceous theropod dinosaur named for Elga Mark-Kurik

Gyroptychius elgae Vorobyeva, 1977. Devonian sarcopterygian named for Elga Mark-Kurik

Hainina vianeyae Pelaez-Campomanes et al., 2000. Tertiary mammal named for Monique Vianey-Liaud

Hariosteus kielanae Tarlo, 1964. Devonian agnathan (psammosteid) named for Zofia Kielan-Jaworowska

Hexaprotodon coryndonae Geze, 1985. Plio-Pleistocene hippopotamus named for Shirley Coryndon

Ichthyoconodon jaworowskom Sigogneau, 1995. Mesozoic mammal named for Zofia Kielan-Jaworowska

Kenyapotamus coryndonae Pickford, 1983. Miocene hippopotamus named for Shirley Coryndon

Konzhukovia Gubin, 1991. Mesozoic melosaurid temnospondyl named for Yelena Dometyevna Konzhukova

Koolasuchus Warren, Rich, and Rich, 1997. Youngest Cretaceous temnospondyl amphibian named for Lesley Kool

Lamplughsaura Kutty et al., 2007. Indian Jurassic prosauropod named for Pamela Robinson

Llankibatrachus truebae Báez and Pugener, 2003. Eocene frog named for Linda Trueb

Lopadichthys colwellae Murray, Zelenitsky, et al., 2018. Tertiary ray-finned fish named for Jane Colwell-Danis

Lucashyus coombsaei Prothero, 2015. Miocene peccary named for Margery Coombs

Malerisaurus robinsonae Chatterjee, 1980. Triassic dicynodont reptile named for Pamela Robinson

Markacanthus Valiukevicius, 1985. Devonian acanthodian named for Elga Mark-Kurik

Meemania Zhu et al., 2006. Devonian sarcopterygian fish named for Meeman Chang

Megadonichthys kurikae Vorobyeva, 2004. Devonian rhipidistian named for Elga Mark-Kurik

Microwhaitsia mendrezi Huttenlocker and Smith, 2017. Permian therapsid named for Christiane Mendrez-Carroll

Myotragus palomboi Bover et al., 2010. Balearic "mouse goat" named for Rita Palombo

Neldasaurus wrighti Chase, 1965. Permian temnospondyl amphibian from Texas named for Nelda Wright

Nichollsaura borealis (Druckenmiller and Russell, 2008). Early Cretaceous plesiosaur, originally *Nichollsemys baieri* Brinkman, 2006. Stem Cretaceous cheloniid turtle named for Betsy Nicholls

Nichollsia, name preoccupied, named for Betsy Nicholls

Niurolepis susanae Hairapetian, Blom, and Miller, 2008. Silurian thelodont named for Susan Turner

Nothosaurus edingerae Schultze, 1970. Triassic nothosaur (marine reptile) named for Tilly Edinger

Novogonatodus kasantsevae Long, 1988. Carboniferous palaeoniscoid fish named for A. A. Kazantseva

Orthechinorhinus davidae Welton, 2016. Oligocene thorn shark named for Lore David

Osmolskina czatkowicensis Borsuk-Bialynicka and Evans, 2003. Early Triassic euparkeriid (stem archosauriform) from Poland named for Halszka Osmólska

Pamelaria dolichotrachela Sen, 2003. Triassic prolacertid reptile named for Pamela Robinson

Pamelina Evans, 2009. Crown diapsid reptile named for Pamela Robinson

Parabos soriae Morales, 1984. Pliocene bovine named for Maria Dolores Soria Mayor

Pavlodaria Vislobokova, 1980. Early Pliocene cervid from Kazakhstan named for Maria Pavlova

Phoebodus turnerae Ginter and Ivanov, 1992. Devonian shark from Poland named for Susan Turner

Psammosteus markae Tarlo, 1961. Devonian heterostracan named for Elga Mark-Kurik

Raynerius splendens Giles, Darras, Clement, Blieck and Friedman, 2015. Devonian ray-finned fish named for Dorothy Rayner

Ruthiromia elcobriensis Eberth and Brinkman, 1983. Permian pelycosaur named for Ruth Romer

Sinovenator changii Xu, 2002. Cretaceous theropod dinosaur named for Meeman Chang

Sphenonectris turnerae Wilson and Caldwell, 1998. Thelodont named for Susan Turner

Terenolepis turnerae Burrow, 1995. Devonian ray-finned fish named for Susan Turner

Trionyx barbarae Owen and Bell, 1849. Eocene turtle named for Barbara Hastings

Ugandax coryndonae Gentry, 2006. Pliocene bovine named for Shirley Coryndon

Valyalepis crista Turner, 1995. Silurian thelodont named for Valentina
Karatajūtė-Talimaa
Vastanagama susani Prasad and Bajpal, 2008. Eocene lizard named for Susan E.
Evans
Wapuskanectes betysnichollsae Druckenmiller and Russell, 2003. Early Cretaceous
plesiosaur named for Betsy Nicholls
Yizhousaurus sunae Zhang, You, Wang, and Chatterjee, 2018. Yunnan Jurassic
dinosaur named for Sun Ai-lin

Taxa Named by Women

Aberrosquama occidens Burrow et al., 2019. Silurian gnathostome (jawed fish)
Acanthodes guizhouensis Wang and Turner, 1985. Carboniferous acanthodian
Acherontiscus caledoniae Clack et al., 2019. Carboniferous amphibian
Actinolepis spinosa n. sp. (Arthrodira) Mark-Kurik, 1985. Devonian placoderm
Aethia rossmoori Howard, 1968. Miocene bird
Agriocharis anza Howard, 1963. Pleistocene bird
Albionbaatar denisae Kielan-Jaworowska and Ensom, 1994. Stem mammal
Alcodes ulnulus Howard, 1968. Miocene bird
Amphilagus magnus Erbajeva, 2013. Fossil rabbit
Amphilagus orientalis Erbajeva, 2013. Fossil rabbit
Amphilagus plicadentis Erbajeva, 2013. Fossil rabbit
Amphilagus tomidai Erbajeva, Angelone, and Alexeeva, 2016. Fossil rabbit
Anabernicula oregonensis Howard, 1964. Quaternary bird
Angonisaurus cruickshanki Cox and Li, 1983. Tanzanian dicynodont
Ankylacanthus Burrow and Turner, 2008. Late Devonian acanthodian
Apalolepis toombsi Turner, 1973. Devonian thelodont
Arcadia myriadens Warren and Black, 1985. Triassic amphibian
Argentodites coloniensis Kielan-Jaworowska et al., 2007. Cretaceous stem mammal
Arianalepis megacosta Hairapetian et al., 2015. Devonian thelodont
Arikarornis macdonaldi Howard, 1966. Miocene bird
Asio priscus Howard, 1964. Pleistocene owl
Australolepis seddoni Turner and Dring, 1981. Late Devonian thelodont
Braclzyramphus pliocenus Howard, 1949. Pliocene bird
Brantadorna downsi Howard, 1963. Pleistocene bird
Bucephala fossilis Howard, 1963. Late Pliocene bird
Bulganbaatar Kielan-Jaworowska, 1974. Stem mammal
Bulganbaatar nemegtbaataroides Kielan-Jaworowska, 1974. Stem mammal
Callorhinus gilmorei Berta and Deméré, 1984. Fossil otariid
Canonia Vieth, 1980. Devonian thelodont (jawless fish)
Carolowillhemina Mark-Kurik and Carls, 2002. Devonian placoderm
Catopsbaatar Kielan-Jaworowska, 1994. Stem mammal
Catopsbaatar catopsaloides Kielan-Jaworowska, 1974. Stem mammal
Celsiodon ahlbergi Clack et al., 2018. Carboniferous amphibian
Cerorhinca minor Howard, 1971. Miocene auklet (bird)
Chendytes milleri Howard, 1955. Pleistocene flightless sea duck (bird)

Chimaerasuchus paradoxus Wu, Sues, and Sun, 1995. Cretaceous crocodyliform
Chulsanbaatar vulgaris Kielan-Jaworowska, 1974. Cretaceous stem mammal
Daphoenus demilo Dawson, 1980. Eocene bear-dog
Desmatolagus chinensis Erbajeva and Sen, 1998. Fossil rabbit
Diabolepis Chang and Xu, 1984. Devonian sarcopterygian
Diomedea milleri Howard, 1936. Miocene albatross (bird)
Dissaeus progressus Golpe-Posse and Crusafont-Pairo, 1968. First creodont mammal
Djadochtatherium catopsaloides Kielan-Jaworowska, 1974. Stem mammal
"Edestus" (Protopirata) karpinskyi Missuna, 1907. Carboniferous shark
Elephas criticus Bate, 1907. Pleistocene elephant
Elephas wusti Pavlova, 1899. Pleistocene elephant
Emsolepis hanspeteri Turner, 2004. Devonian shark
Enaliarctos barnesi Berta, 1991. Stem pinnipedimorph
Enaliarctos emlongi Berta, 1991. Stem pinnipedimorph
Enaliarctos tedfordi Berta, 1991. Stem pinnipedimorph
Eobaatar Kielan-Jaworowska, Dashzeveg, and Trofimov, 1987. Stem mammal
Eobaatar magnus Kielan-Jaworowska, Dashzeveg, and Trofimov, 1987. Stem mammal
Eopelobates leptocolaptus Borsuk-Bialynicka, 1978. Cretaceous frog
Falco oregonus Howard, 1946. Pleistocene falcon
Fulica hesterna Howard, 1963. Fossil rail (bird)
Fulmarus hammeri Howard, 1968. Miocene petrel (bird)
Gavia brodkorbi Howard, 1978. Miocene loon (bird)
Genettia plesictoides Bate, 1903. Cyprus Pleistocene gennet
Geococcyx conklingi Howard, 1931. Quaternary bird
Gromovia daamsi Erbajeva, Khenzykhenova, and Alexeeva, 1993. Fossil rabbit
Gyracanthides hawkinsi Turner, Burrow, and Warren, 2005. Carboniferous
    acanthodian
Hainina belgica Vianey-Liaud, 1979. Paleocene multituberculate
Heptranchias howelli (Reed, 1946) Leriche, 1938. Tertiary cow shark
Hypolagus multiplicatus Erbajeva, 1976. Fossil rabbit
Hypolagus transbaikalicus Erbajeva, 1976. Fossil rabbit
Jesslepis johnsoni Turner, 1995. Devonian thelodont
Kentuckia Rayner, 1951. Carboniferous palaeoniscid fish
Keratobrachyops australis Warren, 1981. Triassic amphibian
Kimaspis tienshanica Mark-Kurik, 1973. Devonian placoderm
Loganellia Turner, 1991 [usurped by Fredholm, 1990]. Silurian thelodont
Loganellia avonia Turner, 2000. Silurian thelodont
Longodus Märss, 2006. Silurian thelodont
Machaeracanthus hunsrueckianum Südkamp and Burrow, 2007. Devonian
    acanthodian
Mancalla cedrosenisis Howard, 1971. Pliocene shorebird
Mancalla milleri Howard, 1970. Pliocene shorebird
Marasuchus Sereno and Arcucci, 1994. Basal dinosauriform archosaur
Morus magnus Howard, 1966. Miocene gannet (bird)
Mcmurdodus whitei Turner and Young, 1987. Devonian shark
Miohierax stocki Howard, 1944. Miocene hawk

Miortyx aldeni Howard, 1966. Miocene bird
Miosula recentior Howard, 1949. Pliocene bird
Moris reyana Howard, 1936. Pleistocene bird
Mycteria wetmorei Howard, 1935. Pleistocene stork
Myotragus balearicus Bate, 1909. Pliocene-Holocene deer
Neophrontops vallecitoensis Howard, 1963. Pleistocene vulture
Neoturinia Hairapetian, Turner and Blom, 2016. Devonian thelodont
Neoturinia gondwana (Turner, 1988) Turner in Hairapetian, Turner and Blom, 2016.
    Devonian thelodont
Neoturinia pagoda (Wang, Dong and Turner, 1985) Turner in Hairapetian, Turner
    and Blom, 2016. Devonian thelodont
Nerepisacanthus denisoni Burrow, 2011. Silurian acanthodian
Nessovbaatar multicostatus Kielan-Jaworowska and Harum, 1997. Cretaceous
    mammal
Nethertonodus Märss and Miller, 2004. Silurian thelodont
Nikolivia milesi Turner, 1982. Devonian thelodont
Nostolepis scotica (Newton, 1892) Burrow and Turner, 2010. Devonian stem shark
Opisthocoelicaudia skarzynskii Borsuk-Bialynicka, 1977. Cretaceous dinosaur
Ossinodus pueri Warren and Turner, 2004. Carboniferous amphibian
Osteodontornis orri Howard, 1957. Miocene bird
Oxyura bessomi Howard, 1963. Pleistocene seaduck
Palaeoscinis turdirostris Howard, 1957. Miocene perching bird
Paralogania? "Turinia" fuscina (Turner, 1986) Turner in Burrow, Turner and Young,
    2010. Silurian thelodont
Paralycoptera Xu and Chang, 2009. Cretaceous fish
Paratosuchus gungani Warren, 1980. Triassic amphibian
Paratosuchus rewanensis Warren, 1980. Triassic amphibian
Pederpes finneyae Clack, 2002. Carboniferous amphibian
Phalacrocorax goletensis Howard, 1965. Pliocene cormorant (bird)
Phalacrocorax kennelli Howard, 1949. Pliocene cormorant (bird)
Phalacrocorax rogersi Howard, 1932. Pliocene cormorant (bird)
Phoenicopterus minutus Howard, 1955. Pleistocene flamingo
Pinnarctidion rayi Berta, 1994. Stem pinnipedimorph
Plagiobatrachius australis Warren, 1985. Triassic amphibian
Plotopterum joaquinensis Howard, 1969. Oligocene bird
Podokesaurus holyokenis Talbot, 1911. Theropod dinosaur
Polyborus prelutoms Howard, 1938. Fossil caracara (bird)
Praemancalla lagunensis Howard, 1966. Miocene auk (bird)
Praemancalla wetmorei (Howard, 1966) Storer, 1976. Miocene auk (bird)
Prolatodon contrarius (Johns, Barnes, and Orchard, 1997). Triassic shark
Protostrix californiensis Howard and Willett, 1934. Eocene owl
Pseudodontichthys whitei Skeels, 1962. Devonian holocephalian
Pseudodontornis stirtoni Howard, 1969. Miocene bird
Puffinus barnesi Howard, 1978. Miocene shearwater (bird)
Puffinus calhounii Howard, 1968. Miocene shearwater (bird)
Puffinus felthami Howard, 1949. Pliocene shearwater (bird)

Puffinus kanakoffi Howard, 1949. Pliocene shearwater (bird)
Puffinus tedfordi Howard, 1971. Pliocene shearwater (bird)
Reginaselache morrisi Burrow and Turner, 2011. Early Carboniferous shark
Rewana quadricuneata Howie, 1972. Triassic amphibian
Rhombus serbicus Andelkovic, 1966. Sarmatian bony fish
Sardinella beogradensis Andelkovic, 1967. Sarmatian bony fish (sardine)
Sloanbaatar Kielan-Jaworowska, 1970. Stem mammal
Spizaetus willetti Howard, 1935. Pliocene hawk-eagle
Stercorarius shufeldti Howard, 1946. Pliocene skua (bird)
Stethacanthus thomasi Turner, 1982. Carboniferous shark
Strix brea Howard, 1933. Eocene waterfowl
Sula pohli Howard, 1958. Miocene booby (bird)
Sylvisorex olduvaiensis Butler and Greenwood, 1979. Fossil shrew
Synaptotylus Echols, 1963. Sarcopterygian fish
Teltmabates antiquus Howard, 1955. Eocene waterfowl
Teratomis incredibilis Howard, 1952. Fossil raptor
Tesseraspis mosaica Karatajūtė-Talimaa, 1983. Devonian heterostracan
Thrinacodus ferox (Turner, 1982) Long, 1990. Carboniferous shark
Turinia antarctica Turner and Young, 1992. Devonian thelodont
Tylocephalonyx skinneri Coombs, 1979. Fossil chalicothere
Urubitinga milleri Howard, 1932. Fossil hawk
Wasonaka yepomerae Howard, 1966. Fossil duck
Xenobrachyops allos Howie, 1972. Triassic amphibian
Youngolepis Chang and Xu, 1981. Devonian fish
Zuegelepis Turner, 1999. Early Silurian thelodont named for the Zuege family women

Contributions to Taxonomy: Higher Taxa

Anisodontini Coombs and Goehlich, 2019
Arandaspididae Ritchie and Gilbert-Tomlinson, 1977
Carolinidae Borsuk-Bialynicka, 1985
Djadochtatheroidea Kielan Jaworowska and Hurum, 2001
Euvertebrata sensu Turner et al., 2010
Evansauria Conrad, 2008
LagerpetonidaeArcucci, 1986—Thecodontia
Loganiidae/-iformes Karatajūtė-Talimaa, 1993
Omalodontida = Omalodontiformes Turner, 1998
Parabuchanosteidae Mark-Kurik in Long et al., 2014
Priscagaminae Borsuk-Bialynicka and Moody, 1984

AAAS	American Association for the Advancement of Science
AGI	American Geosciences Institute, Alexandria, Virginia
AMNH	American Museum of Natural History, New York
ANSP	Academy of Natural Sciences of Philadelphia
BFA	bachelor of fine arts
BFV	Bibliography of Fossil Vertebrates
BMNH	British Museum (Natural History), London
BS and BSC	bachelor of science
CAVEPS	Conference on Australasian Vertebrate Evolution, Palaeontology and Systematics
CONICET	Consejo Nacional de Investigaciones Científicas y Técnicas (National Council of Scientific and Technical Research), Argentina
CT	computed tomography
CV	curriculum vitae
EAVP	European Association of Vertebrate Paleontologists
FMNH	Field Museum of Natural History, Chicago, Illinois
GSA	Geological Society of America
GSL	Geological Society of London
IVPP	Institute of Vertebrate Paleontology and Paleoanthropology of the Chinese Academy of Sciences, Beijing
JVP	*Journal of Vertebrate Paleontology*
LACM	Los Angeles County Natural History Museum
MCZ	Museum of Comparative Zoology, Harvard University, Cambridge, Massachusetts
MNHN	Muséum national d'Historie naturelle, Paris
MS	master of science
MSU	Moscow State University
NHM	Natural History Museum, London
NMNH	National Museum of Natural History, Smithsonian Institution, Washington, DC
NSF	National Science Foundation, Washington, DC

PhD	doctor of philosophy
PIN	Paleontological Institute of the Russian Academy of Sciences, Moscow
PS	Paleontological Society
ROM	Royal Ontario Museum, Toronto
SAVP	Society of American Vertebrate Paleontology
SSR	Soviet Socialist Republic
STEM	science, technology, engineering, and math
SVP	Society of Vertebrate Paleontology
SVPCA	Symposium of Vertebrate Palaeontology and Comparative Anatomy
SWAN	Scientific Women's Academic Network
UCB	University of California, Berkeley
UCL	University College London
UCLA	University of California, Los Angeles
UCMP	University of California Museum of Paleontology, Berkeley, California
UNSW	University of New South Wales, Sydney
USSR	Union of Soviet Socialist Republics
VP	vertebrate paleontology/ist
YPM	Yale Peabody Museum of Natural History, New Haven, Connecticut

Abramson, N.I. 2003. Igor Mikhailovich Gromov, a prominent Russian theriologist. *Russian Journal of Theriology* 2: 2–4.

Adams, F.D. 1938. *The Birth and Development of the Geological Sciences.* 1954 edition. Dover, New York.

Agadjanian, A.K., Vislobokova, I.A., Godina, A.Y., Dmitrieva, E.L., Dubrovo, I.A., Karhu, A.A., Lopatin, A.V., Mashchenko, E.N., Rozanv, A.Y., Sukacheva, I.D., Sychevskaya, E.K., Trofimov, B.A., Shishkin M.A., et al. 1998. Memorial of Nina Semenovna Shevyreva (1931–1996). *Paleontologicheskii Zhurnal* 32(2): 214.

Agassiz, J.L.R. 1833–1844. *Recherches sur les Poissons Fossiles.* Imprimerie de Petitpierre, Neuchâtel, 5 vols.

Agassiz, J.L.R. 1844–1845. *Monographie des poissons fossiles du Vieux Gres Rouge ou systeme Devonien des Iles Britanniques et de Russie.* xxxvi + 1–171.

Ahlberg, P.E., Luksevics, E., Mark-Kurik, E. 2000. A near-tetrapod from the Baltic Middle Devonian. *Paleontology* 43(3): 533–548.

Alden, D. Undated. *Dinosaur Researchers.* Dinosaur Society, New Bedford, MA.

Aldrich, M.L. 2001. Orra White Hitchcock (1796–1863), geological illustrator: another belle of Amherst. *Geological Society of America Annual Meeting, November 5–8, Program and Abstracts*, 103-0. http://gsa.confex.com/gsa/2001AM/finalprogram /abstract_25979.htm.

Alexander, A.M. 1905. Saurian Expedition Scrapbook 1905, A.M. Alexander Papers 1(3), University of California Museum of Paleontology Archives. http://www.ucmp .berkeley.edu/about-ucmp/history-of-ucmp/saurian-expedition-1905/.

Allmon, W.D. 2006. The pre-modern history of the post-modern dinosaur: phases and causes in post-Darwinian dinosaur art. *Earth Sciences History* 25(1): 5–36.

Amherst College Digital Collections. Undated. Orra White Hitchcock drawing of five lines of ornithichnites fossil footprints. Accessed December 18, 2019. https://acdc .amherst.edu/view/asc:20161.

Anderson, E. 1992. Björn Kurtén, an eminent paleotheriologist. *Annales Zoologii Fennici* 28: 123–126.

Andrews, S.M. 1982. *The Discovery of Fossil Fishes in Scotland up to 1845 with Checklists of Agassiz's Figured Specimens.* Royal Scottish Museum Studies, Edinburgh.

Anning, M. 1839. Note on the supposed frontal spine of the genus *Hybodus: Magazine of Natural History* 12(3): 605.

Anon. 1857. The fossil finder of Lyme Regis. *Chambers's Journal of Popular Literature, Science and Arts*, December 12, 7: 382–384.

Anon. [?Dickens, C.]. 1865. Mary Anning, the fossil-finder. *All the Year Round* 13, 60–63.

Anon. 1925. *Illustrated London News*, January 17: 98–99.

Anon. 1941. Membership List. *SVP News Bulletin* 2: 6–13.

Anon. 1942. Geographical List of Members. *SVP News Bulletin* 7: 23.

Anon. 1945. Report from Europe. *SVP News Bulletin* 15: 6–8.

Anon. 1951. *SVP News Bulletin* 31.

Anon. 1952. Membership List. *SVP News Bulletin* 35: 2, 6.

Anon. 1959–1960. *SVP News Bulletins* 55–59.

Anon. 1963. Lucille Bailey Baird 1926–1963. *SVP News Bulletin* 68: 35.

Anon. 1966. University of California at Berkeley. Museum of Paleontology. Department of Paleotology. *SVP News Bulletin* 76: 2–10.

Anon. 1985. Priscilla Turnbull obit. *Chicago Tribune*, December 10, 1985.

Anon. 1989. Obituary Florence D. Wood. *SVP News Bulletin* 145: 78.

Anon. 1992. Obituary Vivian (Vivienne) Leota Castleberry. *Ekalaka Eagle*, August 21, 1992.

Anon. 2014. Indian mammalian fossils. *Stuttgarter Zeitung*, November 21, 2014.

Anon. 2015. Bull, Edith H., obit. *New York Times*, July 5, 2015.

Antecol, H., Bedard, K., Stearns, J. 2016. Equal but inequitable: who benefits from gender-neutral tenure clock stopping policies? IZA Discussion Paper No. 9904. Discussion Series. Institute for the Study of Labor, Bonn, Germany.

Araújo, E.B., Araújo, N.A.M., Moreira, A.A., Herrmann, H.J., Andrade, J.S., Jr. 2017. Gender differences in scientific collaborations: women are more egalitarian than men. *PLoS ONE* 12(5): e0176791. https://doi.org/10.1371/journal.pone.0176791.

Archer, M., Clayton, G. (eds.). 1984. *Vertebrate Zoogeography and Evolution in Australia*. Hesperian Press, Carlisle, Western Australia.

Archibald, J.D. 2014. *Aristotle's Ladder, Darwin's Tree*. Columbia University Press, New York.

Arnold, P. 2018. She persisted. *Nature* 561: 421.

Bailey, M.J. 1994. *American Women in Science: A Biographical Dictionary*. ABC-CLIO, Denver, CO.

Balanoff, A.M., Norell, M.A., Hogan, A.V.C., Bever, G.S. 2018. The endocranial cavity of oviraptorosaur dinosaurs and the increasingly complex, deep history of the avian brain. *Brain, Behavior and Evolution* 91: 125–135. https://doi.org/10.1159/000488890.

Ball, R. 2015. STEM the gap: science belongs to us mob too. *Australian Quarterly* 86(1): 13–19.

Barber, L. 1980. *The Heyday of Natural History*. Jonathon Cape, London.

Barbour, C.A. 1898. Some methods of collecting, preparing, and mounting fossils. *Nebraska State Historical Society: Proceedings and Collections*, 2nd ser., 2: 258–264.

Barbu, V. 1959. Contribuții la cunoașterea genului *Hipparion. Academica Romania, Biblioteca de Geologie și Paleontologie* 5: 1–83.

Barnes, R.T., Marin-Spiotta, E., Morris, A.R. 2018. Building community to advance women in the geosciences through the Earth Science Women's Network. In: Johnson, B.A. (ed.), *Women in Geology: Who Are We, Where Have We Come from, and Where Are We Going?* Memoir 214. Geological Society of America, Boulder, CO, 121–128.

Barrett, P., Naish, D. 2016. *Dinosaurs: How They Lived and Evolved.* Natural History Museum, London.

Barsbold, R. 2016. Paleontology of Mongolia: Studies by Russian scientists (1920–1960). *Studies in the History of Biology* 8(4): 7–32.

Bate, D.M.A. 1901. A short account of a bone cave in the Carboniferous limestone of the Wye valley. *Geological Magazine,* new ser., 4th decade, 8: 101–106.

Bate, D.M.A. 1932. A note on the fauna of the Athlit Caves. *Journal of the Royal Anthropological Institute of Great Britain and Ireland* 62: 277–279.

Bayer, J. 1927. *Der Mensch im Eiszeitalter.* Deuticke, Vienna.

Beard, K.C. 2007. Mammalian paleontology on a global stage: a tribute to Mary R. Dawson. *Bulletin of the Carnegie Museum of Natural History* 39: 1–5.

Bearded Lady Project. Undated. Homepage. Accessed March 12, 2020. https://thebeardedladyproject.com.

Bednarczyk, S. 2013. Women in SVP. Data summary prepared for the SVP Membership Committee, March. PowerPoint presentation. SVP Archives.

Behrensmeyer, A.K., Hill, A.P. (eds.). 1980. *Fossils in the Making: Vertebrate Taphonomy and Paleoecology.* Prehistoric Archaeology and Ecology Series. University of Chicago Press, Chicago.

Bell, R.E., Koenig, L.S. 2017. Harassment in science is real. *Science* 358(6368): 1223.

Bendels, M.H.K., Mueller, R., Brueggmann, D., Groneberg, D.A. 2018. Gender disparities in high-quality research revealed by Nature Index journals. *PLoS ONE* 13(1): e0189136. https://doi.org/10.1371/journal.pone.0189136.

Bentley, P. 2012. Gender differences and factors affecting publication productivity among Australian university academics. *Journal of Sociology* 48(1): 85–103.

Berenbaum, M. 2019. Speaking of gender bias. *Proceedings of the National Academy of Sciences* 116(17): 8086–8088.

Bernard, R.E., Cooperdock, E.H.G. 2018. No progress on diversity in 40 years. *Nature Geoscience* 11: 292–295.

Berta, A. 2017. *The Rise of Marine Mammals.* Johns Hopkins University Press, Baltimore.

Bessudnova, Z. 2006. Eduard Suess' letters to the first Russian woman-geologist Maria Pavlova in the Archive of the Russian Academy of Sciences. *Jahrbuch der Geologischen Bundesanstalt* 146(3–4): 155–161.

Bhalla, N. 2019. Strategies to improve equity in faculty hiring. *Molecular Biology of the Cell* 30: 2744–2749.

Black, R. 2018. The many ways women get left out of paleontology. *Smithsonian Magazine,* June 7, 2018. https://www.smithsonianmag.com/science-nature/many-ways-women-get-left-out-paleontology-180969239.

Blanche, L. 1959. *The Wilder Shores of Love.* Penguin, Harmondsworth, UK.

Blieck, A., Turner, S. (eds.). 2000. Palaeozoic Vertebrate Biochronology and Global Marine/Non-Marine Correlation. Final Report of IGCP 328 (1991–1996). *Courier Forschungsinstitut Senckenberg* 223: 1–575.

Blum, S.D., Maisey, J.G., Rutzky, I.S. 1989. A method for chemical reduction and removal of ferric iron applied to vertebrate fossils. *Journal of Vertebrate Paleontology* 9(1): 119–121.

Bodylevskaya, I.V. 2007. The Paleontological Institute during World War II: Academician A.A. Borissiak and the Moscow Group. *Paleontological Journal* 41(2): 212–221.

Boishenko, A.F. 2001. 70 years of the scientific and technical collaboration between geologists of Russia and Mongolia. *Otech. Geol.* 4: 62–63.

Boissoneault, L. 2017. The forgotten women scientists who fled the Holocaust for the United States. November 9, 2017. https://www.smithsonianmag.com/history /forgotten-women-scientists-who-fled-holocaust-united-states-180967166/.

Boisvert, C., Mark-Kurik, E., Ahlberg, P.E. 2008. The pectoral fin of *Panderichthys* and the origin of digits. *Nature* 456: 636–638.

Boodhoo, T. 2017a. Down to earth with: paleontologist Lisa White. *Earth Magazine*, September 2017. https://www.earthmagazine.org/article/down-earth -paleontologist-lisa-d-white.

Boodhoo, T. 2017b. Saving Mongolia's dinosaurs and inspiring the next generation of paleontologists. *Earth Magazine*, March 2017, https://www.earthmagazine.org /article/saving-mongolias-dinosaurs-and-inspiring-next-generation -paleontologists.

Boring, A. 2015. Gender biases in student evaluations of teachers. Working paper. Observatoire français des conjunctures économiques, Paris, France.

Borisyaka, A.A., Severtsova, A.H. 2017. Emiliya Ivanovna Vorobyeva (1934–2016). *Paleontologicheski Zhurnal* 3, 107–108. https://doi.org/10.7868 /S0031031X17030084.

Botha-Brink. J. 2016. How looking 250 million years into the past could save modern species. *The Conversation*, June 7, 2016.

Bottjer, D.J. (ed.). 2006. *Future Directions in Paleontology*. Report of a workshop held April 8–9, 2006, at the National Museum of Natural History, Washington DC, sponsored by the National Science Foundation. Paleontological Society, Boulder, CO.

Boule, M. 1923. *Fossil Men: Elements of Human Palaeontology*. Oliver and Boyd, Edinburgh.

Bowden, A.J., Tresise, G.R., Simkiss, W. 2010. *Chirotherium*, the Liverpool footprint hunters and their interpretation of the Middle Trias environment. In: Moody, R.T.J., Buffetaut, J.E., Naish, D., Martill, D.M. (eds.), *Dinosaurs and Other Extinct Saurians: A Historical Perspective*. Special Publication 343. Geological Society, London, 209–228.

Brett-Surman, M., Farlow, J., Holtz, T. 2012. *The Complete Dinosaur*, second edition. Indiana University Press, Bloomington.

Brinkman, D.R. 2013. *Rhinoceros Giants: The Paleobiology of Indricotheres*. Indiana University Press, Bloomington.

Brongersma-Sanders, M. 1957. Mass mortality in the sea. In: Hedgpeth, J.W. (ed.), *Treatise on Marine Ecology and Paleoecology*, Volume 1. Memoir 67. Geological Society of America, Boulder, CO, 941–1010..

Broom, R., George, M. 1950. Two new gorgonopsian genera from the Bernard Price collection. *South African Journal of Science* 46: 188–190.

Bruner, J. 2004. Barbara Jaffe Stahl. *Complete Mesoangler* 10(1): 1–4.

Buchholtz, E.A., Seyfarth, E.-A. 1999. The gospel of the fossil brain: Tilly Edinger and the science of paleoneurology. *Brain Research Bulletin* 48: 351–361.

Buchholtz, E.A., Seyfarth, E.-A. 2001. The study of "fossil brains": Tilly Edinger (1897–1967) and the beginnings of paleoneurology. *BioScience* 51(8): 674–682.

Buckland, W. 1823. *Reliquiae Diluvianae; or, Observations on the Organic Remains Contained in the Caves, Fissures and Diluvial Gravel, and on Other Geological Phenomena Attesting the Action of a Universal Deluge.* John Murray, London.

Buckland, W. 1824. Notice on the *Megalosaurus,* or great fossil lizard of Stonesfield. *Transactions of the Geological Society London* 2: 390–396.

Buckland, W. 1829a. Additional remarks on coprolites and fossil sepia. *Proceedings of the Geological Society of London* 11: 142–143.

Buckland, W. 1829b. On the discovery of a new species of pterodactyle; and also of the faeces of the *Ichthyosaurus;* and of a black substance resembling sepia, or Indian ink, in the Lias at Lyme Regis. *Proceedings of the Geological Society of London* 1: 96–98.

Buckland, W. 1835. On the discovery of coprolites, or fossil faeces, in the Lias at Lyme Regis, and in other formations. *Transactions of the Geological Society of London,* 2nd ser., 3: 223–236.

Buckland, W. 1836. *Geology and Mineralogy, Considered with Reference to Natural Theology,* Volume 1. Bridgewater Treatise No. 6. William Pickering, London.

Bureau of Labor Statistics. 2019. Celebrating women in STEM occupations. March 6, 2019. https://blogs.bls.gov/blog/2019/03/06/celebrating-women-in-stem -occupations/.

Burek, C.V. 2001. The first lady geologist, or collector par excellence? *Geology Today* 17(5): 192–194.

Burek, C.V. 2004. Benett, Etheldred Anna Maria (1776–1845). In: Lightman, B. (ed.), *The Dictionary of Nineteenth-Century British Scientists,* Volume 1, *A–C.* Thoemmes Continuum Press, Bristol, 179–181.

Burek, C.V., Higgs, B. 2007. Preface. In: Burek, C.V., Higgs, B. (eds.), *The Role of Women in the History of Geology.* Special Publication 281. Geological Society, London, vii–viii.

Burek, C.V., Kölbl-Ebert, M. 2007. The historical problems of travel for women undertaking geological fieldwork. In: Burek, C.V., Higgs, B. (eds.), *The Role of Women in the History of Geology.* Special Publication 281. Geological Society, London, 115–122.

Burgess, G.H.O. 1900. *The Curious World of Frank Buckland: The Man Who Tried Everything.* John Baker, London.

Burleson, G. 1948. A Pliocene pinniped from the San Diego Formation of southern California. *University of California Publications in Zoology* 47(10): 247–254.

Burrow, C.J. 2019. Exceptional preservation in the Emsian/?Eifelian Cravens Peak Beds limestone deposit, western Queensland, Australia. *Ichthyolith Issues Special Publication* 14: 11–12.

Burrow, C.J., Jones, A.S., Young, G.C. 2005. X-ray microtomography of 410 million-year-old optic capsules from placoderm fishes. *Micron* 36: 551–557.

Butler, P.M., Greenwood, M. 1965. Insectivora. Chiroptera. In: Leakey, L.S.B., et al., *Olduvai Gorge, 1951–61.* Cambridge University Press, Cambridge, 13–15.

Butts, S. 2010. A general view of the Animal Kingdom: conservation of the Anna Maria Redfield wall chart. *Yale Environmental News* 16: 10–11.

Byron, G.G. 1973. *Byron's Letters and Journals*. Volume 2, *"Famous in My Time," 1810–1812*. Edited by L.A. Marchand. John Murray, London.

Cadbury, D. 2000. *The Dinosaur Hunters: A True Story of Scientific Rivalry and the Discovery of the Prehistoric World*. Fourth Estate, London.

Cain, J.A. 1990. George Gaylord Simpson's "History of the Section of Vertebrate Paleontology in the Paleontological Society." *Journal of Vertebrate Paleontology* 10(1): 40–48.

Calisi, R.M., Working Group of Mothers in Science. 2018. Opinion: how to tackle the childcare–conference conundrum. *Proceedings of the National Academy of Sciences* 115(12): 2845–2849.

Cameron, E.Z., Gray, M.E., White, A.M. 2013. Is publication rate an equal opportunity metric? *Trends in Ecology and Evolution* 28(1): 7–8.

Camp, C. 1944. Query? *SVP News Bulletin* 12: 23.

Campbell, K.E. 1980. The contributions of Hildegarde Howard. In: Campbell, K.E., Jr. (ed.), *Papers in Avian Paleontology Honoring Hildegarde Howard*. Contributions in Science 330. Natural History Museum of Los Angeles County, Los Angeles, xi–xv.

Campbell, K.E. 2000a. Hildegarde Howard obituary. *SVP News Bulletin* 178: 131–133.

Campbell, K.E. 2000b. In memoriam: Hildegarde Howard (1901–1998). *Auk* 117(3): 775–779.

Campbell, K.S.W., Jell, J.S. 1998. Dorothy Hill 1907–1997. *Historical Records of Australian Science* 12(2): 205–228.

Carlson, S.J. 2013. President's address: paleontologist Barbie. *Priscum* 20: 4, 11.

Carpenter, K. 1999. *Eggs, Nests and Baby Dinosaurs*. Indiana University Press, Bloomington, Indiana.

Carpenter, K. 2001. Karen Alf obituary. *SVP News Bulletin* 180: 134.

Carpenter, K. 2002. Elaine Anderson obituary. *SVP News Bulletin* 183: 24.

Carpenter, K., Hirsch, K.F., Horner, J.R. 1994. *Dinosaur Eggs and Babies*. Cambridge University Press, Cambridge.

Carpenter, M. 2000. The bone collector. *Pittsburgh Post-Gazette*, March 19, 2000.

Carr, K. 2002. *Dinosaur Hunt: Texas 115 Million Years Ago*. HarperCollins, New York.

Carroll, R.L. 1988. *Vertebrate Paleontology and Evolution*. W.H. Freeman, Oxford.

Carroll, R.L. 2004. 2004 (Romer-Simpson Medal) Robert L. Carroll. Society of Vertebrate Paleontology. http://vertpaleo.org/the-Society/Awards/Past-Award -Winners/2004-(Romer-Simpson-Medal)-Robert-L-Carroll.aspx.

Carroll, R.L., Gaskill, P. 1978. The order Microsauria. *American Philosophical Society* 126: 1–211.

Caster, K.E. 1983. Memorial to Katherine Van Winkle 1895–1982. *Journal of Paleontology* 57(5): 1141–1144.

Cattoi, N.V. 1943. Otras observaciones sobre Typotheriidae. *Physis* 19: 355–362.

Cattoi, N.V. 1951. El "statud" de *Tapirus dupuyi* (C. Amegh.). *Comunicaciones del Museo Argentino de Ciencias Naturalies "Bernardino Rivadavia" e Instituto Nacional Investigaciónes Ciencias Naturales* 2: 103–112.

Cattoi, N.V. 1957. Una especie extingua de *Tapirus* Brisson. (*T. rioplatensis* nov. sp.). *Ameghiniana* 1: 15–21.

Cech, E.A., Blair-Loy, M. 2019. The changing career trajectories of new parents. *Proceedings of the National Academy of Sciences* 116(10): 4182–4187.

Chachra, D. 2017. To reduce gender biases, acknowledge them. *Nature* 548: 373.

Chang, M.-M. 1998. Honorary membership. *SVP News Bulletin* 173: 2–3.

Chang, P.H. 1969. China's scientists in the Cultural Revolution. *Bulletin of the Atomic Scientist* 25(5): 19–40.

Chauhan, P.R., Patnaik, P. 2008. Gudrun Corvinus (1932–2006)—pioneer paleoanthropologist. *Quaternary International* 192(1): 1–5.

Chernov, V. 1989. *Geologists of Moscow University*. Publishing House of Moscow State University, Moscow.

Chin, K., Holmes, T. 2005. *Dino Dung*. Step into Reading. Random House, New York.

Chinsamy, A. 1995. Within the bone. *Natural History* 104(6): 62–63.

Chinsamy, A. 2002. Bone microstructure of early birds. In: Chiappe, L.M., Witmer, L., (eds.), *Mesozoic birds: Above the heads of dinosaurs*. Berkeley: University of California Press, 421–431.

Chinsamy-Turan, A. 2005. *The Microstructure of Dinosaur Bone: Deciphering Biology with Fine-Scale Techniques*. John Wiley and Sons, New York.

Chinsamy-Turan, A. (ed.). 2011. *Forerunners of Mammals: Radiation, Histology, Biology*. Indiana University Press, Bloomington, Indiana.

Churcher, C.S. 2000. Loris Shano Russell 1904–1998. *Proceedings of the Royal Society of Canada* 12(193): 265–270.

Cifelli, R. 2015. Zofia Kielan-Jaworowska (1925–2015). *Nature* 520: 158.

Cifelli, R., Hurum, J., Borsuk-Bialynicka, M., Luo, Z.H., Kaim, A. 2015. Zofia Kielan-Jaworowska (1925–2015). *Palaeontologica Polonica* 60(2): 287–290.

Clack, J.A. 2002. *Gaining Ground: The Origin and Evolution of Tetrapods*. Life of the Past. Indiana University Press, Bloomington.

Clancy, K.B.H., Nelson, R.G., Rutherford, J.N., Hinde, K. 2014. Survey of academic field experiences (SAFE): trainees report harassment and assault. *PLoS ONE* 9(7): e102172. https://doi.org/10.1371/journal.pone.0102172.

Cleevely, R.J. 1983. *World Palaeontological Collections*. British Museum (Natural History), Trustees of the British Museum (Natural History), and Mansell, London.

Coates, M.S. 1986. Women geologists work toward equality. *Geotimes* 31(11): 11–14.

Coe, I.R., Wiley, R., Bekker, L.-G. 2019. Organisational best practices toward gender equality in science and medicine. *Lancet* Special Issue 393: 587–593.

Cohen, C. 1994. *The Fate of the Mammoth: Fossils, Myth, and History*. University of Chicago Press, Chicago.

Colbert, E.H. 1951. *The Dinosaur Book: The Ruling Reptiles and Their Relatives*. McGraw-Hill / American Museum of Natural History, New York.

Colbert, E.H. 1955. *Evolution of the Vertebrates: A History of Back-Boned Animals through Time*, first edition. John Wiley and Sons, New York.

Colbert, E.H. 1965. *The Age of Reptiles*. Weidenfeld and Nicolson, London.

Colbert, E.H. 1968. *Men and Dinosaurs: The Search in Field and Laboratory*. Evans Brothers, London.

Colbert, E.H. 1980. *A Fossil Hunter's Notebook. My Life with Dinosaurs and Other Friends*. E.P. Dutton, New York.

Colbert, E.H. 1983. *Dinosaurs: An Illustrated History.* Dembner Book. Hammond, Maplewood, NJ.

Colbert, E.H. 1984. *The Great Dinosaur Hunters and Their Discoveries.* Dover, New York.

Colbert, E.H. 1985. Rachel E. Nichols obituary. *SVP News Bulletin* 134: 61–62.

Colbert, E.H. 1996. *The Little Dinosaurs of Ghost Ranch.* Columbia University Press, New York.

Colbert, E.H., Morales, M., Minkoff, E.C. 2001. *Evolution of the Vertebrates: A History of Back-Boned Animals through Time,* 5th edition. Wiley-Liss, New York.

Cole, J.R., Zuckerman, H. 1987. Marriage, motherhood, and research performance in Science. *Scientific American* 255: 119–125.

Connelly, M.P. 2015. Women in the labour force. The Canadian Encyclopedia. https://www.thecanadianencyclopedia.ca/en/article/women-in-the-labour-force.

Connelly, W. 2009. The life and work of Helen Haywood (1907–1995). In: Beare, G. (ed.), Imaginative Book Illustration Society, *IBIS Journal 3—Diverse Talents:* 98–143.

Conniff, R. 2016. *House of Lost Worlds: Dinosaurs, Dynasties, and the Story of Life on Earth.* Yale University Press, New Haven, CT.

Conservation Paleobiology Workshop. 2012. *Conservation Paleobiology: Opportunities for the Earth Sciences.* Report of a National Science Foundation–funded workshop held at the Paleontological Research Institution, June 3–5, Ithaca, NY. Paleontological Research Institution, Ithaca, NY.

Conybeare, W.D., De La Beche, H.T. 1821. Notice of the discovery of a new fossil animal, forming a link between the *Ichthyosaurus* and crocodile, together with general remarks on the osteology of the *Ichthyosaurus. Transactions of the Geological Society of London* 1st ser., 5(2): 559–94.

Conybeare, W.D. 1824a. Additional notices on the fossil genera *Ichthyosaurus* and *Plesiosaurus. Transactions of the Geological Society of London* 2nd ser., 1: 103–123.

Conybeare, W.D. 1824b. On the discovery of an almost perfect skeleton of the *Plesiosaurus. Transactions of the Geological Society of London* 2nd ser., 1: 381–389.

Cook, A. 2005. Dr Mary Wade. *TAG Newsletter* 137(12): 45.

Coombs, M., Johnson, R., Leiggi, P., Horner, J.R. 1995. Diane Gabriel obituary. *SVP News Bulletin* 164: 65–66.

Córdova, F. 2020. Leadership to change a culture of sexual harassment. *Science* 367: 1430–1431.

Courtenay-Latimer, M. 1948. *Kannemeyeria wilsoni* Broom; how it came to the East London Museum. In: Du Toit, A.L. (ed.), *Special Publications of the Royal Society of South Africa: The Robert Broom Commemorative Volume.* Royal Society of South Africa, Cape Town, 107–109.

Cox, G., Wiffen, J. 2002. *Dinosaur New Zealand.* HarperCollins, Auckland.

Crane, A. 1877. On certain genera of living fishes and their fossil affinities. *Geological Magazine* 4(5): 209–219.

Creese, M.R.S. 1998. *Ladies in the Laboratory? American and British Women in Science, 1800–1900.* Scarecrow Press, Lanham, MD.

Creese, M.R.S. 2007. Fossil hunters, a cave explorer and a rock analyst: notes on some early women contributors to geology. In: Burek, C.V., Higgs, B. (eds.), *The Role of*

Women in the History of Geology. Special Publication 281. Geological Society, London, 39–49.

Creese, M.R.S., Creese, T.M. 1994. British women who contributed to research in the geological sciences in the nineteenth century. *British Journal for the History of Science* 27(1): 23–54.

Creese, M.R.S., Creese, T.M. 2006. British women who contributed to research in the geological sciences in the nineteenth century. *Proceedings of the Geologists' Association* 117: 53–83.

Creese, M.R.S., Creese, T.M. 2015. Geologists, paleontologists, and other explorers: imperial Russia's women in science. In: Creese, M.R.S., Creese, T.M. (eds.), *Ladies in the Laboratory,* Volume 4, *Imperial Russia's Women in Science, 1800–1900.* Rowman and Littlefield, Lanham, MD, 99–106.

Critchley, E. 2010. *Dinosaur Doctor: The Life and Work of Gideon Mantell.* Amberley, Chalford, UK.

CSIRO (Commonwealth Scientific and Industrial Research Organisation). Undated. Indigenous STEM Education Project. Canberra, ACT. Accessed August 5, 2019. https://www.csiro.au/en/Education/Programs/Indigenous-STEM.

Cucchi, R.J. 2016. La Revista del Centro de Estudiantes del Doctorado en Ciencias Naturales. *Revista del Museo de La Plata* 1, Número Especial: 111–118.

Cunningham, J.A., Rahman, I.A., Lautenschlager, S., Rayfield, E.J., Donoghue, P.C.J. 2014. A virtual world of paleontology. *Trends in Ecology and Evolution* 29(6): 301–368.

Currano, E. 2014. *Dr. Kay Behrensmeyer: An Insatiable Field Scientist.* Retrieved from http://www.ellencurrano.me/dr-kay-behrensmeyer-an-insatiable-field-scientist/.

Currano, E. 2015. Life as a palaeontologist: a thoroughly suitable job for a woman. *Palaeontology Online* 5(3). https://www.palaeontologyonline.com/articles/2015 /life-palaeontologist-thoroughly-suitable-job-woman/.

Curwen, E.C. (ed. and intro.). 1940. *The Journal of Gideon Mantell, Surgeon and Geologist, Covering the Years 1818–1852.* Oxford University Press, London.

Czerkas, S.A., Czerkas, S.M. 1989. *My Life with the Dinosaurs.* Pocket Books, New York.

Czerkas, S.A., Glut, D.F. 1982. *Dinosaurs, Mammoths and Cavemen: The Art of Charles R. Knight.* E.P. Dutton, New York.

Czerkas, S.M. 1988. Windows to the past: the combined views of art and science in images of Alberta dinosaurs. *Alberta Studies in the Arts and Sciences* 1(1): 131–140.

Czerkas, S.M., Czerkas, S.A. 1990. *Dinosaurs: A Complete World History.* B. Mitchell, Limpsfield, Surrey, UK.

Czerkas, S.M., Olson, E.C. (eds.). 1987. *Dinosaurs Past and Present.* Natural History Museum of Los Angeles County, Los Angeles, and University of Washington Press, Seattle, 2 vols.

Dadley, P. 2004. Hastings, Barbara Rawdon [née Barbara Yelverton], marchioness of Hastings and suo jure Baroness Grey de Ruthin (1810–1858), fossil collector and geological author. In: *Oxford Dictionary of National Biography.* Oxford University Press. https://www.oxforddnb.com/view/10.1093/ref:odnb/9780198614128.001.0001 /odnb-9780198614128-e-56505.

Danis, J. 1973. *The Age of Dinosaurs in Canada.* National Museums of Canada, National Museum of Natural Sciences, Ottawa.

Darling, L., Darling, L. 1959. *Before and after Dinosaurs.* William Morrow, New York.

Darling, L., Darling, L. 1960. *Sixty Million Years of Horses.* William Morrow, New York.

Darwin, C. 1859. *On the Origin of Species by Means of Natural Selection.* John Murray, London.

David, L.R. 1942. Use of fossil fish scales in micropaleontology. *Bulletin of the Geological Society of America* 53: 1816–1817.

David, L.R. 1943. Miocene fishes of southern California. *Geological Society of America Special Paper* 43: 1–224.

David, L.R. 1957. Fishes (other than Agnatha). In: Hedgpeth, J. (ed.), *Treatise on Marine Ecology and Paleoecology,* Volume 2. Memoir 67. Geological Society of America, Boulder, CO, 999–1010.

Davidson, J.P. 2000. Fish tales: attributing the first illustration of a fossil shark's tooth to Richard Verstegan (1605) and Nichola Steno (1667). *Proceedings of the Academy of Natural Sciences of Philadelphia* 150: 329–344.

Davidson, J.P. 2008. *A History of Paleontology Illustration.* Indiana University Press, Bloomington.

Davis, L.E. 2009. Mary Anning of Lyme Regis: 19th century pioneer in British palaeontology. *Headwaters: The Faculty Journal of the College of Saint Benedict and Saint John's University* 26: 96–126. https://digitalcommons.csbsju.edu/headwaters/vol26/iss1/14.

Dean, D.R. 1999. *Gideon Mantell and the Discovery of Dinosaurs.* Cambridge University Press, Cambridge, UK.

De Aro, J., Bird, S., Ryan, M. 2019. NSF ADVANCE and gender equity: past, present and future of systemic institutional transformation strategies. *Equality, Diversity and Inclusion* 38(2): 131–139.

Dechaseaux, C. 1962. Cerveaux d'animaux disparus. Essai de Paleoneurologie. *Colloquies du Evolutionaire Sciences* 24: 1–151.

Deforzh, H. 2019. "Essay on the history of fossil ungulates" by academician M. V. Pavlova as phenomenon in the world palaeozoological science (end of the nineteenth and early twentieth centuries). *History of Science and Technology* 9(2) (15): 197–210. https://doi.org/10.32703/2415-7422-2019-9-2(15)-197-210.

De La Beche, H. 1848. Obituary notices [Anning]. *Quarterly Journal of the Geological Society of London* 5(4): xxiv–xxv.

DeSantis, L.R.G. 2017. I'm not your mother. *Science* 358: 690.

Det, M.E. van, Hoek-Ostende, L.W. van den. 2002. Antje Schreuder (1887–1952): een bescheiden pionier. *Cranium* 19: 123–129.

De Wet, C.B., Ashley, G.M., Kegel, D.P. 2002. Biological clocks and tenure timetables: restructuring the academic timeline. *GSA Today,* Supplement: 1–7.

Dickins, J.M. 1982. Joyce Gilbert Tomlinson. *nomen nudum,* no. 11: 3.

Diep, F. 2017. How do you make a living, paleoartist? *Pacific Standard,* updated June 14, 2017. https://psmag.com/economics/she-supplements-it-with-taxidermy.

Dilkes, D., Reisz, R. 1989. University of Toronto. *SVP News Bulletin* 147: 8–9.

Dineley, D.L. 1977. Shirley Coryndon obituary. *SVP News Bulletin* 111: 39–40.

Dodson, P. 1998. *The Horned Dinosaurs: A Natural History.* Princeton University Press, New Jersey.

Dolgopol de Saez, M. 1927. *Liornis minor*, una especie nueva de ave fósil. *Physis* 31: 584–585.

Dolgopol de Sáez, M. 1940. News about Argentine fossil fishes Celacántidos titonienses of Plaza Huincul. *Notas del Museo de La Plata* 5 (Paleontología 25): 295–298.

Dolgopol de Sáez, M. 1957. Crocodiloideos fósiles argentines: un nuevo crocodiloideo del Mesozoico argentino. *Ameghiniana* 1(1–2): 48–501.

Donner, K. 2014. Between a hardrock and an open place: constraints and freedom of Finnish paleontology. *Annals Zoologici Fennici* 51: 2–16.

Downs, T. 1980. Appreciations. In: Campbell, K.E., Jr. (ed.), *Papers in Avian Paleontology Honoring Hildegarde Howard*. Contributions in Science 330. Los Angeles County Museum of Natural History, vii.

Duffin, C. 1978. The importance of certain vertebrate fossils collected by Charles Moore: a scientific perspective. *Geological Curators Group Newsletter* 2(2): 59–67.

Dunkle, D. 1951. Washington, D.C. *SVP News Bulletin* 31: 12–13.

Dutt, K. 2019. Race and racism in the geosciences. *Nature Geoscience*, published online ahead of print, December 16, 2019. https://doi.org/10.1038/s41561-019-0519 -z.

Dutt, K., Pfaff, D.L., Bernstein, A.F., Dillard, J.S., Block, C.J. 2016. Gender differences in recommendation letters for postdoctoral fellowships in geoscience. *Nature Geoscience* 9: 805–809.

Eberle, J., McKenna, M.C. 2007. The indefatigable Mary R. Dawson: Arctic pioneer. *Bulletin of the Carnegie Museum of Natural History* 39: 7–16.

Edinger, T. 1921. Über *Nothosaurus* [On *Nothosaurus*]. Doctoral dissertation, Universität Frankfurt am Main.

Edinger, T. 1929. Die Fossilien Gerhirne [Fossil Brains]. *Ergebnisse der Anatomie und Entwicklungsgeschichte* 28: 1–249.

Edinger, T. 1951. Report from Germany. *SVP News Bulletin* 31: 25–26.

Edinger, T. 1961. Anthropocentric misconceptions in paleoneurology. In: Bejach, H.E. (ed.), *Medical Society in the City of New York*, Volume 19, *Proceedings*. Karger, Basel, 56–107. https://doi.org/10.1159/000405771.

Edinger, T. 1964. Characters in paleontology: Dollo. *SVP News Bulletin* 70: 40–44.

Edinger, T. 1975. Paleoneurology 1804–1966. An annotated bibliography. *Advances in Anatomy, Embryology, and Cell Biology* 49: 1–258.

Edmonds, J.M. 1978. The fossil collection of the Misses Philpot of Lyme Regis. *Proceedings of the Dorset Natural History and Archaeological Society* 98 (for 1976): 43–48.

Edmonds, W. 1979. *The Iguanodon Mystery*. Kestrel Books, Penguin, Harmondsworth, UK.

Elliott, D.K., Maisey, J.G., Yu X.-B., Miao, D. (eds.). 2010. *Morphology, Phylogeny and Paleobiogeography: Honoring Meemann Chang*. Verlag F. Pfeil, Munich.

Emling, S. 2009. *The Fossil Hunter: Dinosaurs, Evolution, and the Woman Whose Discoveries Changed the World*. St. Martin's Press, Macmillan, London.

Emry, R. 2002. *Good Times in the Badlands*. iUniverse Press, Bloomington, Indiana.

Erdmann, M.V. 1999. An account of the first living coelacanth known to scientists from Indonesian waters. *Environmental Biology of Fishes* 54(4): 439–443.

Esensten, J.H. 2003. Death to intelligent design: cavorting beasties. *Harvard Crimson*, March 31, 2003. https://www.thecrimson.com/article/2003/3/31/death-to -intelligent-design-just-months/.

Evans, M. 2010. The roles played by museums, collections and collectors in the early history of reptile palaeontology. In: Moody, R.T.J., Buffetaut, J.E., Naish, D., Martill, D.M. (eds.), *Dinosaurs and Other Extinct Saurians: A Historical Perspective.* Special Publication 343. Geological Society, London, 5–29.

Ewan, E., Innes, S., Reynolds, S., Pipes, R. (eds.). 2006. *The Biographical Dictionary of Scottish Women: From the Earliest Times to 2004.* Edinburgh University Press, Edinburgh.

Ewing, S. 2017. *Resurrecting the Shark: A Scientific Obsession and the Mavericks Who Solved the Mystery of a 270-Million-Year-Old Fossil.* Pegasus Books, New York.

Fagan, S.R. 1960. Osteology of *Mylagaulus laevis*, a fossorial fossorial rodent from the Upper Miocene of Colorado. *University of Kansas Paleontological Contributions, Vertebrata* 9: 1–32.

Falk, D. 2000. Careers in science offer women an unusual bonus: immortality. *Nature* 407: 833.

Falkingham, P., Marty, D., Richter, A. 2016. *Dinosaur Tracks: The Next Steps.* Indiana University Press, Bloomington.

Fallon, P. 2003. So hard the conquering: A life of Irene Longman. MPhil thesis, Griffith University, Brisbane. https://research-repository.griffith.edu.au/handle /10072/367919.

Farlow, J., Brett-Surman, M. (eds.). 1997. *The Complete Dinosaur.* Indiana University Press, Bloomington.

Fernández, M.H., Gómez Cano, A.R. 2015. Nieves López Martínez. Pioneering palaeontologist and inspirational educator. http://trowelblazers.com/nieves-lopez -martinez/.

Find a Grave. N.d. Carrie A. Barbour. Accessed March 3, 2020. https://www .findagrave.com/memorial/69010964/carrie-a.-barbour.

Fischer, E.V., Adams, A., Barnes, R., Bloodhart, B., Burt, M., Clinton, S., Godfrey, E., Pollack, I., Hernandez, P.R. 2018. Welcoming women into the geosciences. *Eos* 99. https://doi.org/10.1029/2018EO095017.

Fleming, N. 2018. How the gender pay gap permeates science and engineering. *New Scientist* 3167, published March 3, 2018.

Flight from 70 Million Years B.C. Yorkshire Television. Accessed August 29, 2019. https://www.youtube.com/watch?v=EcY0a6VytII.

Forey, P.L. 2005. Slicing through history. *Set in Stone: Natural History Museum Palaeontology Departmental Newsletter* 3(1): 5–7.

Forey, P.L., Gardiner, B.G., Patterson, C. 1991. The lungfish, the coelacanth and the cow revisited. In: Schultze, H.-P., Trueb, L. (eds.), *Origins of the Higher Groups of Tetrapods: Controversy and Consensus.* Cornell University Press, Ithaca, NY, 145–172.

Fox, C.W., Duffey, M.A., Fairbairn, D.J., Meyer, J.A. 2019. Gender diversity of editorial boards and gender differences in the peer review process of six journals of ecology and evolution. *Ecology and Evolution* 9:13636–13649.

Fox, C.W., Paine, C.E. 2019. Gender differences in peer review outcomes and manuscript impact at six journals of ecology and evolution. *Ecology and Evolution* 9: 3599–3619.

Frankforter, W.D. 1953. Stanford Museum (Cherokee, Iowa). *SVP News Bulletin* 39: 16.

Friant, M. 1933. *Contribution a l'etude de la différenciation des dents jugales chez les Mammiferes: Essai d'une théorie de la dentition.* Masson et Cie, Paris.

Furlong, E.L., Hall, E.R., David, L.R., Wallace, R.E., Howard, H. 1946. *Fossil Vertebrates from Western North America and Mexico.* Contributions to Paleontology. Carnegie Institution of Washington Publication 551. Washington, DC.

Gabunia, L., Trofimov, B.A. 1973. Vera Gromova obituary. *SVP News Bulletin* 99: 64–65.

Gabunia, L.K., Shevyrëva, N.S., Gabunia, V.D. 1986. First discovery of fossil marsupials (Marsupialia) in Asia. *Transactions (Doklady) USSR Academy of Sciences, Earth Science Section* 281: 179–180.

Gardiner, B.G. 2000. Beatrix Potter's fossils and her interest in geology. *Linnean* 16: 31–47.

Garrod, D.A.E., Bate, D.M.A. (eds.). 1937. *The Stone Age of Mount Carmel.* Volume 1, *Excavations at the Wady El-Mughara.* Clarendon Press, Oxford.

Garrod, D.A.E., Buxton, L.H.D., Elliot-Smith, G., Bate, D.M.A. 1928. Excavation of a Mousterian rock-shelter at Devil's Tower, Gibraltar. *Journal of the Royal Anthropological Institute* 58: 33–113.

Gaudant, J. 1990. Jeanne Signeux obituary. *SVP News Bulletin* 149: 95–96.

Gayet, M., Babin, C. 2007. *Des Paléontologues de A à Z.* Ellipses, Paris.

Gesner, K. 1565. *De rerum fossilium lapidum et gemmarum.* Jacobus Gesnerus, Tiguri.

Gibb, L. 2005. *Lady Hester: Queen of the East.* Faber, London.

Giles, S., Near, T., Xu, G.-H., Friedman, M. 2017. Early members of "living fossil" lineage imply later origin of modern ray-finned fishes. *Nature* 549(7671): 265–268. https://doi.org/10.1038/nature23654.

Glass, J.B. 2015. We are the 20%: updated statistics on female faculty in the earth sciences in the U.S. In: Holmes, M.A., O'Connell, S., Dutt, K. (eds.), *Women the Geosciences: Practical, Positive Practices toward Parity.* American Geological Institute Special Publication 70. American Geophysical Union, Washington, DC, and John Wiley and Sons, Hoboken, NJ, 17–22.

Goodhue, T. W. 2002. *Curious Bones: Mary Anning and the Birth of Paleontology.* Morgan Reynolds, Greensboro, NC.

Goodhue, T.W. 2004. *Fossil Hunter: The Life and Times of Mary Anning (1799–1847).* Academica Press, Bethesda, MD.

Gordon, E.O. 1894. *The Life and Correspondence of William Buckland, D.D., F.R.S.* John Murray, London.

Gorizdro-Kulczycka, Z. 1934. Sur les Ptyctodontidae du Dévonien supérieur du Massif de S-te Croix. *Travaux du Service Géologique de Pologne* 3(1): 19–38.

Gould, S.J. 1980. Edinger, T. Nov. 13, 1897–May 27, 1967: vertebrate paleontologist, paleoneurologist. In: Sicherman, B., Green, C.H., Kantrov, I., Walker, H. (eds.), *Notable American Women,* Volume 4, Harvard University Press, Cambridge, MA, 218–219.

Gould, S.J. 1993. The invisible woman. *Natural History* 102(6): 14, 16, 18, 20–23.

Gould, S.J. 2001. Foreword: life through our ages. In: *Life Through the Ages,* by Charles Robert Knight. Indiana University Press, 2001 Commemorative Edition, vii–x.

Gould, S.J., Vrba, E. 1982. Exaptation—a missing term in the science of form. *Paleobiology* 8: 4–15.

Grady, W. 1993. *The Dinosaur Project: The Story of the Greatest Dinosaur Expedition Ever Mounted.* Macfarlane Walter & Ross, Toronto.

Greider, C.W., Sheltzer, J.M., Cantalupo, N.C., et al. 2019. Increasing gender diversity in the STEM research workforce. *Science* 366: 692–695.

Grigelis, A., Turner, S. 2006. Discoveries: evolution of the earliest Palaeozoic vertebrates. *Geologija* 54: 69–75.

Grigorescu, D. 2004. Vertebrate palaeontology in Romania. *Noesis* (Editura Academiei Romania): 97–116. http://noesis.crifst.ro/wp-content/uploads/revista/2003/2003_2_02.pdf

Grogan, K. 2019. How the entire scientific community can confront gender bias in the workplace. *Nature Ecology and Evolution* 3: 3–6.

Gromova, V.I. 1949. Istoriya loshadey (roda *Equus*) v Starom Svete [The History of Horse (Genus *Equus*) in the Old World]. *Trudy Paleontologeski Instituta* 17: 1–539.

Gross, R. 1998. *Dinosaur Country: Unearthing the Alberta Badlands.* Badlands Books, New York.

Gunther, A.C.L.G.. 1904. *The History of the Collections Contained in the Natural History Departments of the British Museum,* Volume 1. Trustees of the British Museum (Natural History), London.

Halstead, B., Halstead, J. 1981. *Dinosaurs.* Blandford Color, Poole, Dorset, UK.

Halstead, B., Halstead, J. 1982. *A Brontosaur: The Life Story Unearthed.* William Collins, London.

Halstead, B., Halstead, J. 1983. *Terrible Claw: The Story of a Carnivorous Dinosaur.* William Collins, London.

Halstead, B., Halstead, J. 1984. *A Sea Serpent: The Story of a Nothosaur.* William Collins, London.

Halstead, B., Halstead, J. 1985. *A Pterodactyl: The Story of a Flying Reptile.* William Collins, London.

Halstead, J. 1993. Art in science: illustrating the world of Beverly Halstead. Modern Geology 18(1): 141–155.

Halstead, L.B., Middleton, J. 1972. *Bare Bones: An Exploration in Art and Science.* Oliver and Boyd, Edinburgh.

Ham, B. 2019. Societies take a stand against harassment with a new initiative. *Science* 363: 1408.

Hand, S.J., Archer, M. (eds.). 1987. *The Antipodean Ark.* Angus and Robertson, North Ryde, New South Wales.

Häntzschel, W., El-Baz, F., Amstutz, G.C. 1968. *Coprolites: An Annotated Bibliography.* Memoir 108. Geological Society of America, Boulder, CO.

Hastings, B. 1848. On the freshwater Eocene beds of the Hordle Cliff, Hants. *Reports of the British Association for the Advancement of Science* (Oxford, 1947), pt. 2: 63–64.

Hastings, B. 1852. Description géologique des falaises d'Hordle, sur la côté du Hampshire, en Angleterre, composées de couches tertiaries. *Bulletin de la Société Geologique du France* 9: 191–203.

Hastings, B. 1853. On the Tertiary beds of Hordwell, Hampshire. *London, Edinburgh, and Dublin Philosophical Magazine and Journal of Science*, 4th ser., 6(36): 1–11.

Hastings, B. Undated a. Undated letter 1 to R. Owen. Natural History Museum, London. OC62/ 14: 337–338.

Hastings, B. Undated b. Undated letter 2 to R. Owen. Natural History Museum, London. OC62/ 14: 465–466.

Hastings, M.G., Wiedinmyer, K.R. 2015. Facilitating career advancement for women in the geosciences through the Earth Science Women's Network (ESWN). In: Holmes, M.A., O'Connell, S., Dutt, K. (eds.), *Women in the Geosciences: Practical, Positive Practices toward Parity*. American Geological Institute Special Publication 70. American Geophysical Union, Washington, DC, and John Wiley and Sons, Hoboken, NJ, 149–160.

Hayden, F.V. 1875. *Report of the United States Geological Survey of the Territories.* Volume 2, *Department of the Interior.* Government Printing Office, Washington, DC.

Haywood, H. 1970. *My Book of Prehistoric Creatures.* Purnell, London.

Healey, E. 2001. *Emma Darwin: The Inspirational Wife of a Genius.* Headline Review, London.

Heintz, N. 1962. Nyere iakttagelser over fuglelivet på Bouvetøya. In: *1961 Arbok Pal.* Norsk Polarinstitutt, Oslo: 163.

Hemming, J. 1998. *The Golden Age of Discovery: In Celebration of Land-Rover 50th Anniversary.* Pavilion Books, London.

Herbst, R., Anzótegui, L.M. 2016. Las mujeres en la paleontologia argentina. *Revista del Museo de La Plata* 1(3): 130–137.

Hernandez, P.R., Bloodhart, B., Adams, A.S., Barnes, R.T., Burt, M., Clinton, S.M., Du, W., Godfrey, E., Henderson, H., Pollack, I.B., Fischer, E.V. 2019. Role modeling is a viable retention strategy for undergraduate women in the geosciences. *Geosphere* 14(6): 2585–2593.

Herridge, T. 2019. Behind a lesbian furor over a famous palaeontologist lies a deeper truth. *Guardian*, March 20, 2019. https://www.theguardian.com/commentisfree /2019/mar/20/mary-anning-lesbian-palaeontologist-women-film.

Herries Davies, G.L. 2007. *Whatever Is under the Earth: The Geological Society of London 1807 to 2007.* Geological Society, London.

Hill, C., Corbett, C., St. Rose, A. 2010. *Why So Few? Women in Science, Technology, Engineering, and Mathematics.* American Association of University Women, Washington, DC.

Hill, D., Playford, G., Woods, J.T. 1970. *Cainozoic Fossils of Queensland.* Queensland Palaeontographical Society, Brisbane.

Hitchcock, F.R.M. 1887. On the homologies of *Edestus. Proceedings of the American Association for the Advancement of Science* 2: 260.

Hobson, E.C. 1849. On the jaw of *Diprotodon australis* and its dental formula. *Tasmanian Journal of Natural Science, Agriculture, Statistics &c.* 3: 387–388.

Hoch, E. 1985. On Steno. *Dansk Geologisk Forening*, Årsskrift for 1984: 79–86.

Holman, A. 1996. Emma Lewis Lipps 1919–1996. *SVP News Bulletin* 168: 72–73.

Holmes, M.A. 2011. Does gender bias influence awards given by societies? *Eos* 92(47): 421–422.

Holmes, M.A. 2015a. Dual career, flexible faculty. In: Holmes, M.A., O'Connell, S., Dutt, K. (eds.), *Women in the Geosciences: Practical, Positive Practices toward Parity.* American Geological Institute Special Publication 70. American Geophysical Union, Washington, DC, and John Wiley and Sons, Hoboken, NJ, 67–78.

Holmes, M.A. 2015b. Implicit assumption: what it is, how to reduce its impact. In: Holmes, M.A., O'Connell, S., Dutt, K. (eds.), *Women in the Geosciences: Practical, Positive Practices toward Parity.* American Geological Institute Special Publication 70. American Geophysical Union, Washington, DC, and John Wiley and Sons, Hoboken, NJ, 81–93.

Holmes, M.A., O'Connell, S. 2003. *Where Are the Women Geoscience Professors?* Report on the NSF/AGI Foundation sponsored workshop, September 25–27, 2003, Washington, DC.

Holmes, M.A., O'Connell, S., Dutt, K. (eds.). 2015. *Women in the Geosciences: Practical, Positive Practices toward Parity.* American Geological Institute Special Publication 70. American Geophysical Union, Washington, DC, and John Wiley and Sons, Hoboken, NJ.

Holmes, M.A., O'Connell, S., Frey, C. Ongley L. 2008. Gender imbalance in the US geoscience academia. *Nature Geoscience* 1: 79–82.

Holton, C. 1969. The Whitney geological survey: its conflict with the California legislature. *Journal of the West* 8: 200–208.

Home, E. 1814. Some account of the fossil remains of an animal more nearly allied to fishes than to any other classes of animals. *Philosophical Transactions of the Royal Society of London* 104: 571–577.

Hooijer, D. 1952. Notice of death of Antje Schreuder in Leiden Report. *SVP News Bulletin* 36: 21.

Hooks, B. 2019. Feminism is for everybody. *Lancet* Special Issue 393: 493.

Hopkins, M.L. 1951. *Bison (Gigantobison) latifrons* and *Bison (Simobison) alleni* in southwestern Idaho. *Journal of Mammalogy* 32: 192–197.

Hough, J.R. 1948. The auditory region in some members of the Procyonidae, Canidae and Ursidae. *AMNH Bulletin* 92(2): 67–118.

Howard, H. 1943. Review of "The Fossil Birds of California." *Condor* 45(2): 79–80.

Howard, H. 1948. Southern California. *SVP News Bulletin* 24: 21–22.

Howard, H. 1951. Obituary: Dorothea Mineola Alice Bate. *Auk* 69: 491.

Howard, H. 1971. In memoriam: Loye Holmes Miller. *Auk* 88: 276–285.

Howard, H., Dodson, L.M. 1933. Bird remains form an Indian shellmound near Point Mugu, California. *Condor* 35(6): 235.

Howard, R.W. 1975. *The Dawnseekers: The First History of American Paleontology.* Harcourt Brace Jovanovich, New York.

Huddleston, R., Panofsky, A. 1988. Lore Rose David obituary. *SVP News Bulletin* 144: 37–39.

Huene, E. von. 1933. Zur Kenntnis des württembergischen Rätbonebeds mit Zahnfunden neuer Säuger und säugerähnlicher Reptilien [To the knowledge of the Württemberg Rhaetic bonebeds with tooth finds of new mammals and mammal-like reptiles]. Dissertation Tübingener, *Jahreshefte des Vereins für vaterländische Naturkunde in Württemberg*, Jahrgang 89: 65–128.

Huene, E. von. 1935. Ein Rhynchocephale aus den Rhaet (*Pachystropheus* n. g.) [A rhynchocephalian from the Rhaetic (*Pachystropheus* ng)]. *Neues Jahrbuch für Mineralogie, Geologie und Paläontologie* 74: 441–447.

Husband, R. 1924. Variability in *Bubo virginianus* from Rancho La Brea. *Condor* 26(6): 220–225.

Hutchinson, H.N. 1910. *Extinct Monsters and Creatures of Other Days*. Chapman and Hall, London.

Ilies, R., Hauserman, N., Schwochau, S., Stibal, J. 2003. Reported incidence rates of work-related sexual harassment in the United States: using meta-analysis to explain reported rate disparities. *Personnel Psychology* 56: 607–631.

Irwin-Williams, C. 1990. Women in the field: the role of women in archaeology before 1960. In: Kass-Simon, G., Farnes, P., (eds.), *Women of Science: Righting the Record*. Indiana University Press, Bloomington, 1–41.

Isbell, L., Young, T.P., Harcourt, A.H. 2012. Stag parties linger: continued gender bias in a female -rich scientific discipline. *PLoS ONE* 7(11): e49682 https://doi.org/10.1371/journal.pone.0049682.

Jablonski, D., Shubin, N.H. 2015. The future of the fossil record: paleontology in the 21st century. *Proceedings of the National Academy of Science* 112(16): 4852–4858.

Jacobs, L.L. 1997. Message from the President. *SVP News Bulletin* 171: 2.

Janvier, P. 1996. *Early Vertebrates*. Oxford University Press, Oxford.

Jardine, N., Secord, J.A., Spary, E.C. (eds.) 1996. *Cultures of Natural History*. Cambridge University Press, Cambridge, UK.

Jaussaud, P., Brygoo, E.-R. 2004. *Du Jardins au Muséum en 516 Biographies*. Muséum national d'Histoire naturelle, Paris.

Jenkins, P.D., Hutterer, R. 2017. A tribute to Marjorie Greenwood, née George (1924–2006), a neglected mammalian scientist. *Bonn Zoological Bulletin* 66(1): 11–14.

Jeppsson, L., Anehus, R., Fredholm, D. 1999. The optimal acetate buffered acetic acid technique for extracting phosphatic fossils. *Journal of Paleontology* 73: 964–972.

Jiang, Y-W. 2018. Chinese scientist named 2018 L'Oréal-UNESCO Women in Science laureate. http://en.people.cn/n3/2018/0309/c90000-9434996.html.

Johnson, B.A. (ed.). 2018. *Women and Geology: Who Are We, Where Have We Come from, and Where Are We Going?* Memoir 214. Geological Society of America, Boulder, CO.

Johnson, H., Storer, J.E. 1974 and 2009. *A Guide to Alberta Vertebrate Fossils from the Age of Dinosaurs*. Publication 4. Provincial Museum of Alberta, Edmonton.

Kaiser Family Foundation. 2017. Population distribution by race/ethnicity—timeframe 2016. https://go.nature.com/2ElMaAo.

Kaljo, D. (ed.). 2018. Devonian and its fossil world. *Estonian Journal of Earth Sciences* 67(1): 1–111.

Kaplan, K. 2009. Unmasking the imposter. *Nature* 459: 468–469.

Kass-Simon, G., Farnes, P. (eds.). 1990. *Women of Science: Righting the Record*. Indiana University Press, Bloomington.

Kelly, F. (ed.). 1993. *On the Edge of Discovery: Australian Women in Science.* Text Publishing Company, University of Melbourne, Melbourne.

Kerkhoff, N.C. 1993. Notities over de historie van de Pleistocene zoogdierpaleontologie in Nederland. *Cranium* 10(1): 5–17.

Kermack, D.M., Kermack, K.A. 1971. *Early Mammals.* Supp. 1 to *Zoological Journal of the Linnean Society* 50.

Khain, V.E., Malakhova, I.G. 2009. Scientific institutions and the beginning of geology in Russia. In: Lewis, C.L.E., Knell, S.J. (eds.), *The Making of the Geological Society.* Special Publication 317. Geological Society, London, 203–211.

Kielan-Jaworowska, Z. 1969. Preliminary data on the Upper Cretaceous eutherian mammals from Bayn Dzak, Gobi Desert. *Palaeontologia Polonica* 19: 171–191.

Kielan-Jaworowska, Z. 2005. Autobiografia. Translated by Z. Jaworowska, Edited by H. Dieter-Sues. *Kwartalnik Historii Nauki i Techniki* 50 (1): 7–50.

Kielan-Jaworowska, Z. 2008. Halszka Osmolska obituary 1930–2008. *SVP News Bulletin* 195: 20.

Kielan-Jaworowska, Z., Smith, K. 1965. Polish-Mongolian palaeontological expeditions to the Gobi Desert in 1963 and 1964. *Bulletin de l'Académie Polonaise des Sciences* 13(2): 175–179.

King, G. M. 1990a. *The Dicynodonts. A Study in Palaeobiology.* Chapman & Hall, London.

King, G.M. 1990b. Dicynodonts and the end Permian event. *Palaeontologia Africana* 27:31–39.

King, G.M. 1996. *Reptiles and Herbivory.* Chapman and Hall, London.

King, J.E. 1973. Pleistocene Ross seal (*Ommatophoca rossi*) from New Zealand. *New Zealand Journal of Marine and Freshwater Research* 7: 391–397.

King, J.E. 1983. The Ohope skull—a new species of Pleistocene sea lion from New Zealand. *New Zealand Journal of Marine and Freshwater Research* 17: 105–120.

Kingsmill, S. 1990. Recreating the world of dinosaurs. *Canadian Geographer* 110(2): 16–27.

Kish, E., Russell, D.A. 1977. *A Vanished World: The Dinosaurs of Western Canada.* Natural History 4. National Museums of Canada, National Museum of Natural Sciences, Ottawa.

Knipe, H.R. 1905. *Nebula to Man.* J. M. Dent, London.

Knipe, H.R. 1912. *Evolution in the Past.* Herbert and Daniel, London.

Köhring, R. 1997. Senckenbergische Forscher: Tilly Edinger (1897–1967). *Natur und Museum* 127: 55–66.

Köhring, R., Kreft, G. (eds.). 2003. *Tilly Edinger: Leben und Werk einer jüdischen Wissenschaftlerin.* Senckenberg Buch 76. E. Schweizerbart'sche Verlagsbuchhandlung (Naegele u. Obermiller), Stuttgart.

Kölbl-Ebert, M. 1997a. Charlotte Murchison (née Hugonin) 1788–1869. *Earth Sciences History* 16(1): 39–43.

Kölbl-Ebert, M. 1997b. Mary Buckland (née Morland) 1797–1857. *Earth Sciences History* 16(1): 33–38.

Kölbl-Ebert, M. 2002. British geology in the early nineteenth century: a conglomerate with a female matrix. *Earth Science History* 21(1): 3–25.

Kölbl-Ebert, M. 2004. Barbara Marchioness of Hastings (1810–1858)—fossil collector and "lady geologist." *Earth Sciences History* 23(1): 75–87.

Kölbl-Ebert, M. 2012. Sketching rocks and landscape: drawing as a female accomplishment in the service of geology. *Earth Sciences History* 31(2): 270–286.

Kölbl-Ebert, M. 2017. Geology in Germany 1933–1945: people, politics and organization. *Earth Sciences History* 36(1): 63–100.

Kölbl-Ebert, M. 2020. Ladies with hammers—exploring a social paradox in the early 19th century of Britain. In: Burek, C.V., Higgs, B. (eds.), *Uncovering the historical contribution of women in the Geosciences: Celebrations of first female fellows of GSL.* Geological Society, London, Special Publication 506.

Kölbl-Ebert, M., Turner, S. 2017. Towards a history of women in the geosciences. In: Mayer, W., Clary, R.M., Azuela, L.F., Mota, T.S., Wołkowicz, S. (eds.), *History of Geoscience: Celebrating 50 Years of INHIGEO.* Special Publication 442. Geological Society, London, 205–216.

Konnerth, A. 1966. Tilly Bones. *Oceanus* 12: 6–9.

Koppes, S., Allen, S. 2013. University to bestow five honorary degrees at 515th convocation. *University of Chicago News*, May 15.

Korth, W.W., Massare, J.A. (eds.). 2006. Special issue in memory of Elizabeth "Betsy" Nicholls. *Paludicola* 5(4): 1–266.

Krause, D.W., O'Connor, P.M. Rasoamiaramanana, A.H., Buckley, G.A., Burney, D., Carrano, M.T., Chatrath, P.S., Flynn, J.J., Forster, C.A., Godfrey, L.R., Jungers, W.L., Rogers, R.R., Samonds, K.E., Simons, E.L., Wyss, A.R. 2006. Preserving Madagascar's natural heritage: the importance of keeping the island's vertebrate fossils in the public domain. *Madagascar Conservation and Development* 1(1): 43–47.

Kurochkin, E.N., Barsbold, R. 2000. The Russian-Mongolian expeditions and research in vertebrate paleontology. In: Benton, M.J., Shishkin, M.A., Unwin, D.M., Kurochkin, E.N. (eds.), *The Age of Dinosaurs in Russia and Mongolia.* Cambridge University Press, Cambridge, 235–255.

Kuršs, V., Lukševičs, E., Upeniece, I., Zupiņš, I. 1998. Upper Devonian clastics and associated fish remains in Lode clay quarry, Latvia (part I). *Latvijas Ģeoloģijas Vēstis* 5: 7–19.

Kurtén, B., Anderson, E. 1980. *Pleistocene Mammals of North America.* Columbia University Press, New York.

Kylander, M. 2019. Outnumbered and surrounded: women working in male-dominated research fields. *Geoscientist* 29(4): 24–26.

Lambrecht, K., Quenstedt, W., Quenstedt, A. 1938. *Palaeontologii: Catalogus Bio-Bibliographicus.* Fossilium Catalogus, pt. 72. Junk, The Hague.

Lane, R. 2001. *Anna Gurney: Scholar and Philanthropist.* Larks Press, Norfolk.

Lang, H.G., Meath-Lang, B. 1995. Tilly Edinger. In: *Deaf Persons in the Arts and Sciences.* Greenwood Press, Westport, CT, 105–108.

Lang, K. 1960. *So Lange es Tag is: Leben und Wirken von Dr phil. nat Minna Lang (1891–1959).* Self-published, printed by J. Esslinger, Pforzheim.

Lang, W.D. 1955. Mary Anning and Anna Maria Pinney. *Proceedings of the Dorset Natural History and Archaeological Society* 76: 146–152.

Langin, K. 2019. How scientists are fighting against gender bias in conference speaker lineups. Science Careers, February 11, 2019. https://doi.org/10.1126/science.caredit .aaw9742.

Lanham, U. 1973. *The Bone Hunters.* Columbia University Press, New York.

Laporte, L.F. 2000. *George Gaylord Simpson: Palaeontologist and Evolutionist.* Columbia University Press, New York.

Latimer, J., Cerise, S., Ovesiko, P.V., El-Adhami, W. 2019. Australia's strategy to achieve gender equality in STEM. *Lancet* Special Issue 393(10171): 524–526.

Laurin, M., Bardet, N. 2015. A tribute to France de Lapparent de Broin. *Comptes Rendus Palevol* 14: 437–441.

Laursen, S.L., Austin, A.E., Soto, M., Martinez, D. 2015. Strategic institutional change to support advancement of women scientists in the academy: initial lessons for a study of ADVANCE IT projects. In: Holmes, M.A., O'Connell, S., Dutt, K. (eds.), *Women in the Geosciences: Practical, Positive Practices toward Parity.* American Geological Institute Special Publication 70. American Geophysical Union, Washington, DC, and John Wiley and Sons, Hoboken, NJ, 39–50.

Lavas, J.R. 1993. *Dragons from the Dunes: The Search for Dinosaurs in the Gobi Desert.* Self-published, Auckland.

Leakey, M.D. 1979. *Olduvai Gorge: My Search for Early Man.* HarperCollins, New York.

Leakey, M.D. 1984. *Disclosing the Past.* Doubleday, New York.

Lear, L. 2007. *Beatrix Potter: A Life in Nature.* Allen Lane–Penguin, London.

Lebedev, O. (ed.). Forthcoming 2020. *Palaeozoic fossil fish collectors in Russia.* Nauka, Moscow.

Lehtola, K.A. 1973. Ordovician vertebrates from Ontario. *University of Michigan Museum, Paleontological Contributions* 24(4): 23–30.

Lehtola, K.A. 1983. Articulated Ordovician fish from Canon City, Colorado. *Journal of Paleontology* 57: 605–607.

Leibing, A. 2004. Review of Köhring, R., and Kreft, G., *Tilly Edinger: Leben und Werk einer jüdischen Wissenschaftlerin*, Stuttgart: E. Schweizerbart'sche Verlagsbuch-handlung, 2003. Humanities and Social Sciences Online. http://www.h-net.org /reviews/showrev.php?id=9928.

Leiggi, P., May, P. 1994. *Vertebrate Palaeontological Techniques.* Volume 1. Cambridge University Press, Cambridge.

Lerback, J., Hanson, B. 2017. Journals invite too few women to referee. *Nature* 541: 455–457.

Lescaze, Z. 2017. *Paleoart: Visions of the Prehistoric Past.* Taschen, Cologne.

Lewis, C. 2017. *The Enlightened Mr Parkinson: The Pioneering Surgeon Who Identified Parkinson's Disease.* Icon Books, London.

Lewis, C.L.E., Knell, S.J. (eds.). 2009. *The Making of the Geological Society.* Special Publication 317. Geological Society, London.

Lipps, E., Purdy, R., Martin, R. 1988. An annotated bibliography of the Pleistocene vertebrates of Georgia. *Georgia Journal of Science* 46(2): 109–148.

Lipps, J.H. 2004. Success story: the history and development of the Museum of Paleontology at the University of California, Berkeley. *Proceedings of the California Academy of Sciences* 55 (Supplement I, Article 9): 209–243.

Lipps, L., Ray, C.E. 1967. The Pleistocene fossiliferous deposit at Ladds, Bartow County, Georgia. *Bulletin of the Georgia Academy of Science* 25(3): 113–119.

Liston, J. 2013. From obstetrics to oryctology: inside the mind of William Hunter (1718–1783). In: Duffin, C., Moody, R.T.J., Gardner-Thorpe, C. (eds.), *A History of Geology and Medicine.* Special Publication 375. Geological Society, London, 349–373.

Lomax, D. 2019. Mary Anning—palaeontologist extraordinaire. In: Mary Anning's Work and Legacy: Fossil Dealers Conference, York, June 4–5, Society for the History of Natural History and Geological Curators' Group, Programme and Abstracts, 1 p.

Lucas, S.G. 2001. *Chinese Fossil Vertebrates.* Columbia University Press, New York.

Lull, R.S., Wright, N.E. 1942. *Hadrosaurian dinosaurs of North America.* Special Paper 40. Geological Society of America, New York.

Lurie, E. 1960. *Louis Agassiz: A Life in Science.* 1988 edition. Johns Hopkins University Press, Baltimore.

Lukens, R. 2013. Graceanna Lewis. Unpublished material prepared for the Chester Historical Society, Chester, PA. https://www.sierracollege.edu/ejournals/jscnhm /v6n1/lewis.html.

Lyarskaya, L.A., Mark-Kurik, E. 1972. Eine neue Fundstelle oberdevonischer Fishe im Baltikum. *Neues Jahrbuch für Geologie und Paläontologie* Monatshefte 7: 331–343, 407–414.

Lyon, G.M. 1937. Pinnipeds and a sea otter from the Point Mugu Shell Mound of California. *University of California Publications in Biological Sciences* 1(8): 133–168.

Lyon, G.M. 1941. A Miocene sea lion from Lomita, California. *University of California Publications in Zoology* 47: 23–41.

Lyubina, G.I., Bessudnova, Z.A. 2019. *Maria Vasil'yevna Pavlova 1854–1938.* RFFI, Janus-K, Moscow.

MacFadden, B.J., Lundgren, L., Crippen, K., Dunckel, B.A., Ellis, S. 2016. Amateur paleontological societies and fossil clubs, interactions with professional paleon-tologists, and social paleontology in the United States. *Palaeontologia Electronica* 19.2.1E: 1–19.

Maidment, S.C.R., Woodruff, D.C., Horner, J.R. 2018. A new specimen of the ornithis-chian dinosaur *Hesperosaurus mjosi* from the Upper Jurassic Morrison Formation of Montana, U.S.A., and implications for growth and size in Morrison stegosaurs. *Journal of Vertebrate Paleontology* 38: e1406366. http://doi.org/10.1080/02724634 .2017.1406366.

Maier, G. 2003. *African Dinosaur's Unearthed: The Tendaguru Expeditions.* Indiana University Press, Indianapolis.

Maisey, J.G. 1996. *Discovering Fossil Fishes.* Henry Holt, New York.

Maisey, J.G., Wolfram, K.E. 1984. "Notidanus." In: Eldredge, N., Stanley, S. (eds.), *Living Fossils.* Springer Verlag, Berlin, 170–180.

Malkowski, S. 1949. Zinaida Gorizdro-Kulczycka 1884–1949. *Wladomosci Muzeum Ziemi*, Warsaw 4(1948): 412–415.

Märss, T., Ritchie, A. 1998. Articulated thelodonts (Agnatha) of Scotland. *Transac-tions of the Royal Society of Edinburgh: Earth Sciences* 88(3): 143–195.

Märss, T., Turner, S., Karatajute-Talimaa, V.N. 2007. *Agnatha II—Thelodonti.* Volume 1B of *Handbook of Paleoichthyology.* Edited by H.-P. Schultze. Verlag Dr Friedrich Pfeil, Munich.

Maryańska, T., Osmólska, H. 1974. Pachycephalosauria, a new suborder of ornithischian dinosaurs. *Palaeontologica Polonica* 30: 45–102.

Mantell, G. 1822. *Fossils of the South Downs: or Illustrations on the Geology of Sussex.* Lupton Relfe, London.

Mantell, G. 1844. *The Medals of Creation; or, First Lessons in Geology, and in the Study of Organic Remains.* H.G. Bonn, London.

Mantell, G. 1850. *A Pictorial Atlas of Fossil Remains Consisting of Coloured Illustrations Selected from Parkinson's "Organic Remains of a Former World," and Artist's "Antediluvian Phytology."* H.G. Bohn, London.

Mark-Kurik, E., Newman, M.J., Toom, U., Blaauwen, J.L. den. 2018. A new species of the antiarch *Microbrachius* from the Middle Devonian (Givetian) of Belarus. *Estonian Journal of Earth Sciences* 67(1): 1–11. https://doi.org/10.3176/earth.2017.22.

Marsh, L.J., Currano, E. (eds.). 2020. *The Bearded Lady Project: Challenging the Face of Science.* Columbia University Press, New York.

Massare, J.A., Lomax, D.R. 2014. An *Ichthyosaurus breviceps* collected by Mary Anning: new information on the species. *Geological Magazine* 151(1): 21–28.

Mather, P. (ed.) 1986. *A Time for a Museum: History of the Queensland Museum, 1862–1986.* Queensland Museum Memoirs, Brisbane.

Mathews, N.M. 2014. *The Greatest Woman Painter: Cecilia Beaux, Mary Cassatt, and Issues of Female Fame.* Historical Society of Pennsylvania, Philadelphia.

Mayor, A. 2011. *The First Fossil Hunters: Paleontology in Greek and Roman Times,* second edition. Princeton University Press, Princeton, NJ.

McCall, J. 2001. Family in Rift. *Geoscientist* 11(6): 9.

McCallum, C.A. 1966. Hobson, Edmund Charles (1814–1848). In: *Australian Dictionary of Biography.* Melbourne University Press, Melbourne, 544–545.

McCullagh, E.A., Nowak, K., Pogoriler, A., Metcalf, J.L., Zaringhalam, M., Zelikova, T.J. 2019. Request a woman scientist: a database for diversifying the public face of science. *PLoS Biology* 17(4): e3000212. https://doi.org/10.1371/journal.pbio.3000212.

McGowan, C. 2001. *The Dragon Seekers: The Discovery of Dinosaurs during the Prelude to Darwin.* Little, Brown, London.

McKenna, M., Bell, S.K. 1997. *Classification of Mammals above the Species Level.* Columbia University Press, New York.

McKenna, M.C., Holton, C.P. 1967. A new insectivore from the Oligocene of Mongolia and a new subfamily of hedgehogs. *American Museum Novitates* 2311: 1–11.

McKenna, P. 1964. American Museum of Natural History. *SVP News Bulletin* 72: 12.

McNeill, L. 2018. The Woman who shaped the study of fossil brains. *Smithsonian Magazine.* https://www.smithsonianmag.com/science-nature/woman-who-shaped-study-fossil-brains-180968254/#5fjBgFeGOjS7uBAc.99

Medin, D., Lee, C. 2012. Diversity makes better science. *Association for Psychological Science Observer,* May/June 2012. https://www.psychologicalscience.org/observer/diversity-makes-better-science.

Miao, D. 2010. Ode to an unbreakable spirit—Chang Meemann's contributions to paleoichthyology. In: Elliott, D.K., Maisey, J.G., Yu, X.-B., Miao, D. (eds.), *Morphology, Phylogeny and Paleobiogeography: Honoring Meemann Chang.* Verlag F. Pfeil, Munich, 1–23.

Miguel, S., Hidalgo, M., Stubbs, E., Posadas, P., Ortiz Jaureguizar, E. 2013. Estudio bibloimetrico de genero en el paleontologia de vertebrados: El Caso de la revista argentina Ameghiniana (1957–2011). *Investigacion Bibliotechnologica* 27(61): 133–155.

Miles, R.S. 1971. *Palaeozoic Fishes.* Revised by J.A. Moy-Thomas. Chapman and Hall, London.

Miles, R.S. 2012. Book review (of *Fossiljägarna: En berättelse om besatta vetens-kapsmän och fisken some klev upp på land* by Björn Hagberg & Martin Widman. 2011). *Linnean,* 28(2): 48–50.

Miller, H. 1852. *The Old Red Sandstone, or New Walks in an Old Field,* third edition. Johnstone and Hunter, Edinburgh.

Miller, L., DeMay, I. 1942. The fossil birds of California. *University of California Publications in Zoology* 47(4): 47–142.

Milner, A.C. 2016. Lady Smith Woodward's tablecloth. In: Johanson, Z., Barrett, P.M., Richter, M., Smith, M. (eds.), *Arthur Smith Woodward: His Life and Influence on Modern Vertebrate Palaeontology.* Special Publication 430. Geological Society, London, 89–111.

Milner, A.C., Hughes, B. 1995. Pamela Lamplugh Robinson obituary. *SVP News Bulletin* 164: 72–74.

Mohr, B.R. 2010. Wives and daughters of early Berlin geoscientists and their work behind the scenes. *Earth Science History* 29(2): 291–310.

Monaco, P. 2006. Donna Engard 1948–2005 obituary. *SVP News Bulletin* 190: 48–49.

More Than a Dodo, 2016. Tales from the Jurassic Coast. https://morethanadodo.com /2016/10/11/tales-from-the-jurassic-coast/.

Morgan, N. 2014. Distant thunder—ladies' man. *Geoscientist* 24(1): 24.

Morgan, N. 2019. Distant thunder—behind every good man. . . . *Geoscientist* 29(6): 26.

Moss-Racusin, C.A., Doviidio, J.F., Brescoli, V.L., Graham, M.J., Handelsman, J. 2012. Science faculty's subtle gender bias favor. *Proceedings of the National Academy of Sciences* 109(41): 16474–16479.

Müller, H. 1959. In memoriam Dr rer. nat. Minna Lang *20 Marz 1891–30 Juli 1959. *Meininger Kulturspiegel* September: 270–273.

Münster, S. 1544. *Cosmographia.* Heinrich Petri, Basel.

Munthe, K., Hutchison, H. 1976. A wolf-human encounter on Ellesmere Island, Canada. *Journal of Mammalogy* 59(4): 876–877.

Murray, P.F., Vickers-Rich, P. 2004. *Magnificent Mihirungs: The Colossal Flightless Birds of the Australian Dreamtime.* Indiana University Press, Bloomington.

Nalivkin, D.V. 1979. *Nashi Pervye Zhenshchinye-geologi* [Our first women geologists]. Academy of Sciences USSR, Ministry of Geology of USSR, Nauka, Leningrad.

Nash, S.E. 1990. The collections and life history of Etheldred Benett (1776–1845). *Wiltshire Archaeological and Natural History Magazine* 83: 163–169.

National Academies of Sciences, Engineering, and Medicine. 2018. *Sexual Harassment of Women: Climate, Culture, and Consequences in Academic Sciences, Engineering, and Medicine.* National Academies Press, Washington, DC. https://doi.org/10 .17226/24994.

National Center for Education Statistics. 2016. *Digest of Education Statistics 2016: Tables and Figures.* National Center for Education Statistics, Washington, DC.

National Center for Scientific Education. https://ncse.ngo.

National Geographic Kids. Undated. 11-year-old campaigns for Mary Anning statue. Accessed March 3, 2020. https://www.natgeokids.com/au/discover/science /nature/mary-anning-statue/.

National Research Council. 1997. *Adviser, Teacher, Role Model, Friend: On Being a Mentor to Students in Science and Engineering.* The National Academies Press, Washington, DC. http: //www.nap.edu/openbook.php?record_id=5789.

National Science Board. 2018. *Science and Engineering Indicators 2018.* NSB-2018-1. National Science Foundation, Arlington, VA.

Nature. 2018a. Gender bias tilts success rates of grant applications. *Nature* 554: 15.

Nature. 2018b. A code of ethics to get scientists talking. *Nature* 555: 5.

Nature. 2018c. Announcement. Awards for women in science. *Nature* 556: 150.

Nature Ecology and Evolution. 2017. Harassment: a field study. Editorial. *Nature Ecology and Evolution* 1: 1787–1788.

Nelson, R.G., Rutherford, J.N., Hinde, K., Clancy, K.B.H. 2017. Signaling safety: characterizing fieldwork experiences and their implications for career trajectories. *American Anthropologist* 119(4): 710–722.

Newton, A. 2012. Plugging the leaks. *Nature Geoscience* 5(8): 522.

New York Times. 2007. Brief obituary for Charlotte Holton. October 4, 2007.

Nicholas, M. 1982. *The World's Greatest Cranks and Crackpots.* Reed International Books, London.

Nielsen, M.W., Alegrai, S., Borjeson, L., et al. 2017. Gender diversity leads to better science. *Proceedings of the National Academy of Sciences* 114(8): 1740–1742.

Noonan, R. 2017. *Women in STEM: 2017 Update.* US Department of Commerce, Economics and Statistics Administration, Office of Chief Economist, Washington, DC, November 13, 2017.

Normile, D. 2001. Research kicks into high gear after a long, uphill struggle. *Science* 291(5502): 237–238.

Norquay, G. 2012. *The Edinburgh Companion to Scottish Women's Writing.* Edinburgh University Press, Edinburgh.

Nosek, B.A. 2009. National differences in gender-science stereotypes predict national sex differences in science and math achievement. *Proceedings of the National Academy of Sciences* 106: 10593–10597.

Novitskaya, L.I., Afanassieva, O.B. (eds.). 2004. *Fossil Vertebrates of Russia and Adjacent Countries: Agnathans and Early Fishes.* Geos, Moscow.

NSF (National Science Foundation). 2016. *Survey of Earned Doctorates.* National Center for Science and Engineering Statistics, Alexandria, VA.

NSF (National Science Foundation). Undated. Awards (Sedimentary Geology and Paleontology fund code 7459). Accessed August 2018. https://www.nsf.gov /awardsearch/advancedSearch.jsp.

NSF (National Science Foundation), National Center for Science and Engineering. 2017. *Women, Minorities, and Persons with Disabilities in Science and Engineering.* NSF 18-304. https://www.nsf.gov/statistics/wmpd/.

Obroucheva, N.V. 2014. Dmitry Vladimirovich Obruchev: life and destiny (1900–1970). *Paleontological Journal* 48(9): 950–963.

O'Connell, S. 2015. Multiple and sequential mentoring: building your nest. In: Holmes, M.A., O'Connell, S., Dutt, K. (eds.), *Women in the Geosciences: Practical, Positive Practices toward Parity.* American Geological Institute Special Publication 70. American Geophysical Union, Washington, DC, and John Wiley and Sons, NJ, 108–120.

Ogilvie, M.B. 1986. *Women in Science: Antiquity through the Nineteenth Century.* MIT Press, Cambridge, MA.

Orlov, Y.A. 1946. USSR—letter to Ed Colbert. *SVP News Bulletin* 19: 6.

Orlov, Y.A. ed. 1964. Osnovi Paleontologi [*Fundamentals of Paleontology*]. Pt Mammals. Nauka, Moscow.

Orlov, Y.A. ed. 1968. *Fundamentals of Paleontology.* Pt Mammals. English Edtn. Israel Program for Scientific Translation, Jerusalem.

Orr, M. 2020. Collecting women in Geology: Opening the International Case of a Scottish "Cabinétière," Eliza Gordon Cumming (1815–1842). In: Burek, C.V., Higgs, B. (eds.), *Uncovering the historical contribution of women in the Geosciences: Celebrations of first female fellows of GSL.* Geological Society, London, Special Publication 506.

Osborn, H.F. 1915. *Men of the Old Stone Age.* C. Scribner's Sons, New York.

Osborn, H.F. 1936/1942. *Proboscidea: A Monograph of the Discovery, Evolution, Migration and Extinction of the Mastodonts and Elephants of the World.* Volumes 1 and 2. American Museum of Natural History, New York.

Osmólska, H. 1976. New light on the skull anatomy and systematic position of *Oviraptor. Nature* 262: 683–684.

Osmólska, H. 1987. *Borgavia gracilicrus* n. gen. et n. sp., a new troodontid dinosaur from the Late Cretaceous of Mongolia. *Acta Palaeontolgica Polonica* 32: 133–150.

Osmólska, H., Roniewicz, E., Barsbold, R. 1972. A new dinosaur, *Gallimimus bullatus* n. gen., n. sp. (Ornithomimidae) from the Upper Cretaceous of Mongolia. *Palaeontologia Polonica* 27: 103–143.

Ostrom, J.H. 1966. Functional morphology and evolution of the ceratopsian dinosaurs. *Evolution* 20(3): 290–308.

Overton, W.R. 1982. Creationism in schools: The decision in McLean vs. the Arkansas Board of Education. *Science* 215(4535): 934–943.

Owen, R. 1840. Report on the British fossil reptiles: part 1. *Report of the 9th Meeting of the British Association for the Advancement of Science* 1839: 43–126.

Owen, R. 1848. On the fossils obtained by the Marchioness of Hastings from the freshwater Eocene beds of Hordle Cliffs. *Report of the British Association for the Advancement of Science* 1847(2): 65–66.

Owen, R. 1859. On a new genus (*Dimorphodon*) of pterodactyle, with remarks on the geological distribution of flying reptiles. *Report of the British Association for the Advancement of Science* 28(1858): 97–103.

Owen, R., Bell, T. 1849. *The Fossil Reptilia of the London Clay.* Pt. I Chelonia. Palaeontographical Society Monographs, London.

Page, S. 2007. *The Difference: How the Power of Diversity Helps Create Better Groups, Firms, Schools, and Societies.* Princeton University Press, Princeton, NJ.

Palaeontological Association. 1972. List of British palaeontologists. Members' Copy.

Palmer, K.V.W. 1939. *Basilosaurus* in Arkansas. *Bulletin of the American Association of Petroleum Geologists* 23: 1228–1229.

Palmer, K.V.W. 1942. Tales of ancient whales. *Nature* 35(4): 213–214, 221.

Panciroli, E. 2019. International Women's Day: Frances Mussett. Giant Science lady blog by Elsa Panciroli, March 8, 2019. https://giantsciencelady.blogspot.com/2019/03/international-womens-day-frances-mussett.html.

Panofsky, A. 1973. Current techniques for fossil skeleton mounting. Stanford Linear Accelerator Center Technical Note 73-7.

Parker, C. 2008. Cecilia Beaux. Lines and Colours, April 6, 2008. http://linesandcolors.com/2008/04/06/cecilia-beaux/ —accessed March 30, 2019

Parkinson, J. 1811. *Organic Remains of a Former World: An Examination of the Mineralized Remains of the Vegetables and Animals of the Antediluvian World, Generally Termed Extraneous Fossils; the Fossil Starfish, Echini, Shells, Insects, Amphibia, Mammalia, &c.* Volume 3. Sherwood, Neely and Jones, London.

Parrington, F.R., Westoll, T.S. 1974. David Meredith Seares Watson, 1886–1973. *Biographical Memoirs of Fellows of the Royal Society* 20: 482–504.

Parrot, G.F. 1836. Essay sur les ossements fossiles des bords du lac de Burtneck, en Livonie. *Mémoires de l'Académie des Sciences de Saint-Pétersbourg, Sciences Mathematiques, Physiques et Naturelles* 4: 1–95.

Passaglia, K.L., McCaroll, S.M. 1996. Three-dimensional preparation of fossil vertebrates using the Waller method and acid preparation techniques. *Journal of Vertebrate Paleontology* 16: 57A.

Patterson, B. 1961. Margaret Jean Ringier Hough 1903–1961. *SVP News Bulletin* 62: 36.

Patterson, E.C. 1983. *Mary Somerville and the Cultivation of Science, 1815–1840.* Martinus Nijhoff, Boston.

Pavlova, M. 1887–1906. Études sur l'histoire paléontologique des ongulés en Amérique et en Europe. *Bulletin de la Société Imperiale des Naturalistes de Moscou*, Volumes 1–9.

Pavlova, M. 1899. *Iskopaemyie slonyi* [Fossil elephants]. I. N. Skorohodova, printer, Saint Petersburg.

Pavlova, M., Pavlov, A.P. 1910–1914. *Monograph on the Mammalian Fossils of Russia.* [In French.] J.N. Kouchnéreff et Cie, Moscow.

Pearson, H.S. 1923. Some skulls of *Perchoerus* (*Thinohyus*) from the White River and John Day formations. *Annals and Magazine of Natural History* 48: 61–96.

Pearson, H.S. 1924a. A dicynodont reptile reconstructed. *Proceedings of the Zoological Society of London* 94: 827–855.

Pearson, H.S. 1924b. The skull of the dicynodont reptile *Kannemeyeria. Proceedings of the Zoological Society of London* 94: 793–826. https://doi.org/10.1111/j.1096-3642.1924.tb03316.x.

Pearson, H.S. 1928. Chinese fossil Suidae. *Palaeontologica Sinica* (C) 5(5): 1–75.

Petter, G., Senut, B. 1994. *Histoire de nos ancêtres; Lucy retrouvée.* Flammarion, Paris.

Phillips, Z.F. 2001. *German Children's and Youth Literature in Exile, 1933–1950: Biographies and Bibliographies.* K.G. Saur, Munich.

Piveteau, J., Dechaseaux, C. 1952–1966. *Traité de Paléontologie, Vertébrés.* Masson, Paris.

Piveteau, J., Lehman, J.-P., Dechaseaux, C. 1978. *Précis de Paléontologie des Vertébrés.* Masson, Paris.

Place, R. 1969. *First Land Mammals*. Ginn, London.

Plotnick, R.E., Stigall, A.L., Stefanescu, I. 2014. Evolution of paleontology: long-term gender trends in an earth-science discipline. *GSA Today* 24(11): 44–45.

Poplin, C., Blieck, A., Cloutier, R., Derycke, C., Goujet, D., Janvier, P., Lelièvre, H., Véran, M., Wenz, S. 1997. Paléontologie des vertébrés inferièurs—rénouveau et développements [Paleontology of lower vertebrates—renewal and developments]. *Geobios*, Mémoire spécial 20: 437–446.

Poplin, C.M., De Ricqles, A.J. 1970. A technique of serial sectioning for the study of undecalcified fossils. *Curator* 13(1): 7–20.

Popp, A.L., Lutz, S.R., Khatami, S., van Emmerik, T., Knoben, W.J.M. 2019. A global survey on the perceptions and impacts of gender inequality in the earth and space sciences. *Earth and Space Science* 6: 1460–1468. https://doi.org/10.1029/2019EA000706.

Poropat, S. 2017. *Dinosaur Stampede*. Australian Age of Dinosaurs Museum, Winton, Queensland.

Pough, F.H., Janis, C., Heiser, J.B. 2018. *Vertebrate Life*, tenth edition. Prentice Hall, Upper Saddle River, NJ.

Powell, H.P., Edmonds, J.M. 1976. List of type-fossils in the Philpot Collection, Oxford University Museum. *Dorset Natural History and Archaeological Society Proceedings* 98: 48–53.

Powell, K. 2018. She persisted. *Nature* 561: 421–423.

Princeton Alumni Weekly. 1976. The Trustees of Princeton University. Princeton, NJ.

Prothero, D.R. 2009. The early evolution of North American peccaries (Artiodactyla: Tayassuidae). *Museum of Northern Arizona Bulletin* 65: 509–541.

Psihoyos, L., Knoebber, J. 1994. *Hunting Dinosaurs*. Random House, New York.

Pyenson, L., Sheets-Pyenson, S. 1999. *Servants of Nature: History of Scientific Institutions, Enterprises and Sensibilities*. Fontana Press, HarperCollins, London.

Racicot, R. 2017. Fossil secrets revealed: X-ray CT scanning and applications in paleontology. *Paleontological Society Papers* 22: 21–38.

Radulescu, C. 1999. Paleontology of mammals in Romania—a historical perspective. *Acta Palaeontologica Romaniae* 2: 5–6.

Rage, J.-C. 2015. A tribute to France de Lapparent de Broin. *Comptes Rendus Palevol* 14: 437–441.

Rao, C.R. 1973. Prasantha Chandra Mahalanobis 1893–1972. *Biographical Memoirs of Fellows of the Royal Society* 19: 455–492.

Ray, C.E., Lipps, L. 1968. Additional notes on the Pleistocene mammals from Ladds, Georgia. *Bulletin of the Georgia Academy of Science* 26(2): 63.

Rayfield, E.J. 2007. Finite element analysis and understanding the biomechanics and evolution of living and fossil organisms. *Annual Reviews of Earth and Planetary Science* 35: 541–576.

Raymond, J. 2013. Most of us are biased. *Nature* 495: 33–34.

Rayner, D.H. 1948. The structure of certain Jurassic holostean fishes, with special reference to their neurocrania. *Philosophical Transactions of the Royal Society of London. Series B, Biological Sciences* 233 (601): 287–345.

Redfield, A.M. 1857. *A General View of the Animal Kingdom*. Kellogg, New York.

Reed, C. 1986. Priscilla Freudenheim Turnbull obituary. *SVP News Bulletin* 138: 65–66.

Reed, M.D. 1946. A new species of fossil shark from New Jersey. *Notulae Naturae* 172: 1–3.

Reisz, R.R., Sues, H.-D. 2015. The challenges and opportunities for research in paleontology for the next decade. *Frontiers in Earth Science* 3:9. https://doi.org/10.3389/feart.2015.00009.

Rich, P.V., van Tets, G.F., Knight. F. 1985. *Kadimakara. Extinct Vertebrates of Australia.* Pioneer Design Studios, Clayton, Australia.

Rich, P., Zhang, Y.P., Chow, M.C., Wang, B.Y., Komarower, P., Fan, J.H., Sloss, R., Moody, J.K.M., Dawson, J. 1994. *A Chinese-English and English-Chinese Dictionary of Vertebrate Palaeontology.* Monash University, Clayton, Australia.

Rich, T.H. 1999. Australia: vertebrate paleontology. In: Singer, R. (ed.), *Encyclopedia of Paleontology*, Volume 1. Fitzroy Dearborn, Chicago, 140–149.

Rich, T.H., Rich, P. 1985. Chang (Zhang) Yu-Ping obituary. *SVP News Bulletin* 134: 59–60.

Rich, T.H., Vickers-Rich, P. 2003. *A Century of Australian Dinosaur Collecting.* Queen Victoria Museum, Launceston, Tasmania, and Monash Science Centre, Clayton, Australia.

Roberts, G. 1834. *The History and Antiquities of the Borough of Lyme Regis and Charmouth.* Bagster, London.

Robinson, P. L. 1967. The Indian Gondwana formations—a review. In: *Gondwana Stratigraphy. IUGS Symposium, Buenos Aires, 1967.* Reviews. UNESCO, Paris, 201–268.

Robinson, P.L. 1971. A problem of faunal replacement on Permo-Triassic continents. *Palaeontology* 14: 131–153.

Roe, A. 1946. Personality test. *SVP News Bulletin* 17: 18–19.

Roe, A. 1953. *The Making of a Scientist.* Dodd, Mead, New York.

Rolfe, W.D.I., Crowther, P.R., Wimbledon, W.A. (eds.). 1988. The use and conservation of palaeontological sites. *Special Papers in Palaeontology* 40: 1–200.

Romer, A.S. 1944. The pre-natal ontogenesis of the S.V.P. *SVP News Bulletin* 13: 20–22.

Romer, A.S. 1954. *Man and the Vertebrates.* Penguin Books, London.

Romer, A.S. 1958. Report on the Pre-1928 Bibliography. *SVP News Bulletin* 54: 6–7.

Romer, A.S. 1967. Tilly Edinger, 1897–1967. *SVP News Bulletin* 81: 51–53.

Romer, A.S. 1974. The stratigraphy of the Permian Wichita Redbeds of Texas. *Breviora* 427: 1–31.

Romer, A.S., Simpson, G.G. 1940. Section of Vertebrate Paleontology Information Bulletin, September 15, 1940, unpublished [but see Cain, 1990: Appendix for Editorial Notes for Simpson's Manuscript].

Romillo, A., Ryan, R.T., Salisbury, S.W. 2013. Reevaluation of the Lark Quarry dinosaur tracksite (late Albian-Cenomanian Winton Formation, central-western Queensland, Australia): no longer a stampede. *Journal of Vertebrate Paleontology* 3(1): https://doi.org/10.1080/02724634.2012.694591.

Rose, K., Holbrook, L.T., Rana, R.S., Kumar, K., Jones, K.E., Ahrens, H.E., Missiaen, P., Sahni, A., Smith, T. 2014. Early Eocene fossils suggest that the mammalian order Perissodactyla originated in India. *Nature Communications* 5: 5570. https://doi.org/10.1038/ncomms6570.

Rosen, J. 2014. Volunteers get their hands dirty; and like it; at La Brea Tar Pits. *Los Angeles Times*, August 4, 2014. http://www.latimes.com/science/la-sci-c1-tar-pits -volunteers-20140804-story.html.

Rosen, J. 2017. Data illuminate a mountain of molehills facing women scientists. *Eos* 98. https://doi.org/10.1029/2017EO066733.

Rosser, S. 2019. Athena SWAN and ADVANCE: effectiveness and lessons learned. *Lancet* Special Issue 393: 604–608.

Rossiter. M.W. 1982. *Women Scientists in America: Struggles and Strategies to 1940*. Johns Hopkins University Press, Baltimore, MD.

Royal Society. 2014. *A Picture of the UK Scientific Workforce*. The Royal Society, London.

Rozhanov, S.V. 2018. In the memory of academician E.I. Vorobyeva and Professor L.V. Belousov. *Paleontological Journal* 52(14): 1652–1654.

Rozhdestvensky, A.K. 1977. The study of dinosaurs in Asia. *Journal of the Palaeontological Society of India* 20: 102–119.

Rudwick, M.J.S. 1972. *The Meaning of Fossils*. History of Science Library. MacDonald, London.

Rudwick, M.J.S. 1976. The emergence of a visual language for geological science, 1760–1840. *History of Science* 14: 149–195.

Rudwick, M.J.S. 1992. *Scenes from Deep Time: Early Pictorial Representations of the Prehistoric World*. University of Chicago Press, Chicago.

Rudwick, M.J.S. 2005. *Bursting the Limits of Time: The Reconstruction of Geohistory in the Age of Revolution*. University of Chicago Press, Chicago.

Rupke, N.A. 1983. *The Great Chain of History: William Buckland and the English School of Geology (1814–1849)*. Clarendon Press, Oxford.

Salerno, P.E., Páez-Vacas, M., Guayasamin, J.M., Stynoski, J.L. 2019. Male principal investigators (almost) don't publish with women in ecology and zoology. *PLoS ONE* 14(6): e0218598. https://doi.org/10.1371/journal.pone.0218598.

Sánchez, M. 2012. *Embryos and Deep Time: The Rock Record of Biological Development*. University of California Press, Berkeley.

Sanders, M. 1934. Die fossilien Fische der alttertiären Süsswasserablagerungen aus Mittel-Sumatra [The fossil fishes of early Tertiary fresh water deposits from central Sumatra]. *Verhandelingen van het Geologisch-Mijnbouwkundig Genootschap voor Nederland en Koloniën*, Geologische series, 11(1): 1–144. Doctoral thesis, Amsterdam Municipal University. [In German with Dutch summary.]

Sanders, M. 1948. The importance of upwelling water to vertebrate palaeontology and oil geology. *Verhandelingen Koninklijke Nederlandsche Akademie van Wetenschappen*, Afd. Natuurkunde sectn 2, 45(4): 1–112.

Sankey, J.T., Baszio, S. (eds.). 2008. *Vertebrate Microfossil Assemblages: Their Role in Paleoecology and Paleobiogeography*. Indiana University Press, Bloomington.

Sarjeant, W.A.S. (ed.). 1995. *Vertebrate Fossils and the Evolution of Scientific Concepts. Writings in Tribute to Beverly Halstead, by Some of His Friends*. Gordon and Breach/OPA, Amsterdam.

Sato, T., Cheng Y., Wu, X., Zelenitsky, D.K., Hsiao, Y. 2005. A pair of shelled eggs inside mother dinosaur. *Science* 308: 375.

Savage, R.J.G., Coryndon, S.C. 1976. *Fossil Vertebrates of Africa*. 4. New York & London: Academic Press.

Schaffer, F.X. 1924. *Lehrbuch der Geologie*. Volume 2, *Grundzüge der historischen Geologie (Geschichte der Erde, Formationskunde)*. Deuticke, Leipzig.

Schiebinger, L.L., Henderson, A.D., Gilmartin, S.K. 2008. *Dual-Career Academic Couples: What Universities Need to Know*. Stanford University, Stanford, CA.

Schneider, B., Tranel, L.M., Cremans, M. 2018. The Association for Women Geoscientists: forty years of successes, struggles, and sisterhood. In: Johnson, B.A. (ed.), *Women in Geology: Who Are We, Where Have We Come from, and Where Are We Going?* Memoir 214. Geological Society of America, Boulder, CO, 105–120.

Schultz, C.B. 1943. Carrie Adeline Barbour. *SVP News Bulletin* 8: 4.

Schultz, C.B. 1972. F.M. Sunderland obituary. *SVP News Bulletin* 96: 31.

Schultze, H.P. 2007. Tilly Edinger. Leben und Werk Einer Jüdischen Wissenschaftlerin. *Journal of Vertebrate Paleontology* 27(3): 772–773.

Schultze, H.P., Turner, S., Grigelis, A. 2009. Great northern researchers: discoverers of the earliest Paleozoic vertebrates. *Acta Zoologica* 90(Suppl. A1): 3–21.

Scotchmoor, J.G., Breithaupt, B., Springer, D.A., Fiorello, A.R. 2002. *Dinosaurs: The Science behind the Stories*. American Geologic Institute, Alexandria, VA.

Sellers, C.C. 1980. *Mr. Peale's Museum: Charles William Peale and the First Popular Museum of Natural Science and Art*. W.W. Norton, New York.

Servais, T., Antoine P.-O., Danelian, T., Lefebvre, B., Meyer-Berthaud, B. 2012. Paleontology in France: 200 years in the footsteps of Cuvier and Lamarck. *Palaeontologia Electronica* 15.1.2E: 12pp. https://palaeo-electronica.org/content/2012-issue-1-articles-2/79-palaeontology-in-France.

Sewell, L., Barnett, A.G. 2019. The impact of caring for children on women's research output: a retrospective cohort study. *PLoS ONE* 14(3): e0214047. https://doi.org/10.1371/journal.pone.0214047.

Seymour, K.L. 2004. A tribute to Loris Shano Russell, 1904–1998. *Canadian Field-Naturalist* 118(3): 451–464.

Sheets-Pyenson, S. 1988. *Cathedrals of Science: The Development of Colonial Natural History Museums in the Late 19th Century*. McGill-Queens's University Press, Kingston, Ontario.

Shelton, S. 1994. Conservation of vertebrate paleontology collections. In: Leiggi, P., May, P. (eds.), *Vertebrate Paleontological Techniques*, Volume 1. Cambridge University Press, Cambridge, 3–33.

Shelton, S. 1997. The effect of high market prices on the value and valuation of fossil vertebrate sites and specimens. In: Nudds, J.R., Pettitt, C.W. (eds.), *The Value and Valuation of Natural Science Collections*. Geological Society, London, 149–153.

Shelton, S., Chaney, D. 1994. An evaluation of adhesives and consolidants recommended for fossil vertebrates. In: Leiggi, P., May, P. (eds.), *Vertebrate Paleontological Techniques*, Volume 1. Cambridge University Press, Cambridge, 35–45.

Shen, H. 2013. Inequality quantified: mind the gender gap. *Nature* 495(7439): 22–24.

Shen, H. 2016. Lab life: lone-parent scientist. *Nature* 531(March 3): 129–131. https://doi.org/10.1038/nj7592-129a.

Shindler, K. 2005. *Discovering Dorothea: The Pioneering Fossil-Hunter Dorothea Bate*, second edition. The Natural History Museum, London.

Shindler, K. 2007. A knowledge unique: the life of the pioneering explorer and palaeontologist Dorothea Bate. In: Burek, C.V., Higgs, B. (eds.), *The Role of Women in the History of Geology*. Special Publication 281. Geological Society, London, 295–303.

Shindler, K., 2017. *Discovering Dorothea: The Pioneering Fossil-Hunter Dorothea Bate*, third edition. The Natural History Museum, London.

Shipman, P. 1981. *Life History of a Fossil: An Introduction to Taphonomy and Paleoecology*. Harvard University Press, Cambridge, MA.

Shipman, P. 2001. *The Man Who Found the "Missing Link": Eugene Dubois and His Lifelong Quest to Prove Darwin Right*. Simon and Schuster, New York.

Shubin, N.H. 2003. Giving fish a hand. *Science* 301: 766–767.

Shubin, N.H. 2007. *Your Inner Fish*. Penguin Random House, New York.

Sigogneau-Russell, D. 2003. Docodonts from the British Mesozoic. *Acta Palaeontologica Polonica* 48(3): 357–374.

Simpson, G.G. 1935. The first mammals. *Quarterly Review of Biology* 10(2): 154–180.

Simpson, G.G. 1939–1940. Minutes of the Section of Vertebrate Paleontology. *Proceedings of the Geological Society of America* 1939: 268–269; 1940: 268–271.

Simpson, G.G. 1941. History of the society and its predecessors. *SVP News Bulletin* 1: 1–3.

Simpson, G.G. 1942. History of vertebrate paleontology in America. *Proceedings of the American Philosophical Society* 86(1): 130–188.

Simpson, G.G. 1980. *Splendid Isolation: The Curious History of South American Mammals*. Yale University Press, New Haven, CT.

Skibba, R. 2016. Women postdocs less likely than men to get a glowing reference. *Nature*, October 3, 2016. https://doi.org/10.1038/nature.2016.20715.

Smith, E., Smith, R. 1999. *Black Opal Fossils of Lightning Ridge: Treasures from the Rainbow Billabong*. Kangaroo Press, Australia.

Smith, J.L.B. 1956. *Old Four Eyes: The Story of the Coelacanth*. Longmans, Green, London.

Smith, M.M., Coates, M.J. 1998. Evolution origins of the vertebrate dentition: phylogenetic patterns and developmental evolution. *European Journal of Oral Science* 106(Supp. 1): 482–500.

Snyder, D., Burrow, C., Pardo, J., Anderson, J., Turner, S. 2016. Multiple techniques yield complementary data on gyracanth anatomy. SVP 76th Annual Meeting, Salt Lake City, Utah, October 26–29, 2016, Program and Abstracts, 228.

Sollas, W.J., Sollas, I.B.J. 1904. An account of the Devonian fish *Palaeospondylus gunni* Traquair. *Philosophical Transactions of the Royal Society of London*, Ser. B, 196: 267–294.

Sollas, W.J., Sollas, I.B.J. 1912. A study of the skull of a *Dicynodon* by means of serial sections. *Philosophical Transactions of the Royal Society of London*, Ser. B, 204: 201–225.

Sonnert, G., Holton, G. 1996. Career patterns of women and men in the sciences. *American Scientist* 85: 63–71.

Spalding, D.A.E. 1993. *Dinosaur Hunters*. Key Porter Books, Toronto.

Spalding, D.A.E. 1999. *Into the Dinosaurs' Graveyard: Canadian Digs and Discoveries*. Doubleday, Toronto.

Spamer, E.E., Bogan, A.E., Torrens, H.S. 1989. Recovery of the Etheldred Benett collection of fossils mostly from the Jurassic-Cretaceous strata of Wiltshire,

England, analysis of the taxonomic nomenclature of Benett (1831), and notes and figures of type specimens contained in the collection. *Proceedings of the Academy of Natural Sciences of Philadelphia* 141: 115–180.

Spears, T. 2014. Ely Kish: artist of the ancient Earth (1924–2014). *Ottawa Citizen,* October 24, 2014. https://ottawacitizen.com/life/life-story/ely-kish-artist-of-the -ancient-earth-1924-2014.

Stahl, B.J. 1999. *Chondrichthyes III: Holocephali.* Volume 4 of *Handbook of Paleoichthyology.* Edited by H.-P. Schultze. Dr F. Pfeil Verlag, Munich.

Stein, B.R. 2001. *On Her Own Terms: Annie Montague Alexander and the Rise of Science in the American West.* University of California Press, Berkeley.

Stigall, A. 2013a. The Paleontological Society 2013: a snapshot in time. *Priscum* 20(2): 1–4. http://www.alyciastigall.org/wp-content/uploads/2020/02/Priscum _Summer2013.pdf.

Stigall, A. 2013b. Where are the women in paleontology? *Priscum* 20(1): 1–3. https:// www.paleosoc.org/wp-content/uploads/2015/10/Priscum_Winter2013.pdf.

Stigall, A. 2017. Alumni Symposium: women in geology. Stigall Lab, April 12, 2017. https://www.alyciastigall.org/2017/04/alumni-symposium-women-in-geology/.

Stoet, G., Geary, D.C. 2018.The gender-equality paradox in science, technology, engineering and mathematics education. *Psychological Science* 29(4): 581–593.

Sues, H.-D. 1978. A new small theropod dinosaur from the Judith River Formation (Campanian) of Alberta, Canada. *Zoological Journal of the Linnaean Society* 62: 381–400.

Sues, H.-D., Hook, R.W., Olsen, P.E. 2013. Donald Baird and his discoveries of Carboniferous and early Mesozoic vertebrates in Nova Scotia. *Atlantic Geology* 49: 90–103. https://doi.org/10.4138/atlgeol.2013.004.

Sun A.-L., Zhou M.-Z. 1991. Professor Yang Zhong-Jian (C.C. Young)—founder of Chinese vertebrate palaeontology. In: Wang H.-Z., Yang G.-R., Yang J.-Y. (eds.), *Interchange of Geoscience Ideas between the East and the West.* Proceedings of the XVth International Symposium of INHIGEO, Beijing, October 25–31, 1990. China University of Geosciences Press, Wuhan, 113–116.

Sutton, M., Rahman, I., Garwood, R. 2014. *Techniques for Virtual Paleontology.* Wiley Blackwell, Oxford. E-book online.

Svojtka, M. 2011. Adametz, Karoline (Lotte) (1879–1966), Paläontologin, Prähistorikerin und Beamtin. Austrian Biographical Lexicon and Biographical Documentation. Austrian Academy of Sciences. https://biographien.ac.at/ID-0.3020074-1.

SVP (Society of Vertebrate Paleontology). Undated. Code of conduct. Accessed March 13, 2020. http://vertpaleo.org/Annual-Meeting/Code-of-Conduct.aspx.

Swinton, W.E. 1951. Dorothea M.A. Bate obituary. *SVP News Bulletin* 32: 32–33.

Swinton, W.E., Pinner, E. 1948. *The Corridor of Life.* Jonathon Cape, London.

Switek, B. 2010. Beyond Mary Anning—Celebrating the work of women paleontologists. http://scienceblogs.com/laelaps/2010/04/14/beyond-mary-anning-celebrati/.

Symonds, M.E., Gemmell, N.J., Braisher, T.L., Gorringe, K.I., Elgar, M.A. 2006. Gender differences in publication output: towards an unbiased metric of research performance. *PLoS ONE* 1(1): e127. https://doi.org/10.1371/journal.pone.0000127.

Talbot, M. 1911. *Podekosaurus holyokensis,* a new dinosaur from the Triassic of the Connecticut Valley. *American Journal of Science* 31: 469–479.

Tanke, D.H. 2010. Remember me: Irene Vanderloh (1917–2009). *Alberta Palaeontological Society Bulletin* 25(1): 10–18.

Tanke, D.H. 2017. Jane M. Colwell-Danis: Canada's first academically-trained female vertebrate paleontologist. In: 5th Annual Meeting, Canadian Society of Vertebrate Palaeontology, Dinosaur Provincial Park, Alberta, May 15–17, University of Alberta, Abstracts. *Vertebrate Anatomy Morphology Palaeontology* 4: 44–45.

Tanke, D.H. 2019. *Now There Was a Lady! Hope Johnson, LLD 1916–2010.* Alberta Palaeontological Society, Calgary.

Tapanila, L., Pruitt, J., Pradel, A., Wilga, C., Ramsay, J., Schlader, R., Didier, D. 2013. Jaws for a spiral tooth whorl: CT images reveal novel adaptation and phylogeny in fossil *Helicoprion. Biology Letters* 9(2). https://doi.org/10.1098/rsbl.2013.0057.

Tappert, T.L. 1994. *Out of the Background: Cecilia Beaux and the Art of Portraiture.* Original thesis 2009. Traditional Fine Arts Organization, Arizona. http://www.tfaoi.com/aa/9aa/9aa211.htm.

Taquet, P. 1998. *Dinosaur Impressions: Postcards from a Paleontologist.* Translated by Kevin Padian. Cambridge University Press, Cambridge.

Taquet, P. 2006. *Georges Cuvier: Naissance d'un Genie.* Odile Jacob, Paris.

Taylor, M. 1990. Review: vegetarians feast in the Permian—"The Dicynodonts: A Study in Palaeobiology" by Gillian King. *New Scientist*, May 5, 1990. https://www.newscientist.com/article/mg12617155-200-review-vegetarians-feast-in-the-permian-the-dicynodonts-a-study-in-palaeobiology-by-gillian-king/.

Taylor, M.A., Torrens, H.S. 1986. Saleswoman to a new science: Mary Anning and the fossil fish *Squaloraja* from the Lias of Lyme Regis. *Proceedings of the Dorset Natural History and Archaeological Society* 108: 161–168.

Taylor, M.A., Torrens, H.S. 1995. Fossils by the sea. *Natural History* 104(10): 66–71.

Taylor, M.A., Torrens, H.S. 2014a. An account of Mary Anning (1799–1847), fossil collector of Lyme Regis, Dorset, England, published by Henry Rowland Brown (1837–1921) in the second edition (1859) of Beauties of Lyme Regis. *Proceedings of the Dorset Natural History & Archaeological Society* 135: 63–70.

Taylor, M.A., Torrens, H.S. 2014b. An anonymous account of Mary Anning (1799–1847), fossil collector of Lyme Regis, Dorset, England, published in *All the year round* in 1865, and its attribution to Henry Stuart Fagan (1827–1890), schoolmaster, parson and author. *Proceedings of the Dorset Natural History and Archaeological Society* 135: 71–85.

Tedford, R. 1991a. Musings on New Guinea fossil vertebrate discoveries. In: Vickers-Rich, P., Monaghan, J.M., Baird, R.F., Rich, T.H. (eds.), *Vertebrate Palaeontology of Australasia.* Pioneer Design Studios in cooperation with Monash University Publication Committee, Melbourne, 85–110.

Tedford, R. 1991b. Vertebrate palaeontology in Australia: the American contribution. In: Vickers-Rich, P., Monaghan, J.M., Baird, R.F., Rich, T.H. (eds.), *Vertebrate Palaeontology of Australasia.* Pioneer Design Studios in cooperation with Monash University Publication Committee, Melbourne, 45–84.

Tembe, G., Siddiqui, S. 2014. Applications of computed tomography to fossil conservation and education. *Collection Forum* 28(1–2): 47–62.

Thompson, D.W. 1945. *On Growth and Form.* Cambridge University Press and Macmillan, New York.

Thomson, K.S. 1991. *Living Fossil: The Story of the Coelacanth.* Hutchinson Radius, London.

Thomson, K.S. 1992. *Living Fossil: the Story of the Coelacanth.* W.W. Norton and Company, New York.

Thornton, S. 2012. Louise Leakey, "famed" paleontologist and the director of public education and outreach for the Turkana Basin Institute. *National Geographic News,* April 4. https://www.nationalgeographic.org/news/real-world-geography -louise-leakey/.

Thulborn, R.A. 1990. *Dinosaur Tracks.* Chapman and Hall, London.

Thulborn, R.A., Wade, M. 1984. Winton dinosaur footprints. *Memoirs of the Queensland Museum* 21(2): 413–517.

Tonni, E. 2005. El ultimo medio siglo en el studio de los vertebrados fosiles. *Asociación Palentológica Argentina Publicación Especial* 10: 73–85.

Torrens, H.S. 1985. Women in geology: 2. Benett, Etheldred. *Open Earth* 21: 12–13.

Torrens, H.S. 1995. Mary Anning (1799–1847) of Lyme; "the greatest fossilist the world ever knew." *British Journal for the History of Science* 28: 257–284.

Torrens, H.S. 2004. Buckland (née Morland), Mary. *Oxford Dictionary of National Biography.* https://doi.org/10.1093/ref:odnb/46486.

Torrens, H. 2008. A saw for a jaw. *Geoscientist* 12:18–21.

Torrens, H.S., Benamy, E., Daeschler, E.B., Spamer, E.E., Bogan, A.E. 2000. Etheldred Benett of Wiltshire, England, the first lady geologist—her fossil collection in the Academy of Natural Sciences of Philadelphia, and the rediscovery of "lost" specimens of Jurassic Trigoniidae (Mollusca: Bivalvia) with their soft anatomy preserved. *Proceedings of the Academy of Natural Sciences of Philadelphia* 150: 59–123.

Trofimov, V.A. 1961. 70-years Vera Isaakovni Gromova. *Palaeontologocheskoi Zhurnal* 1961: 1, 3–5.

Trofimov, V.A., Reshetov, V. Ju. 1983. Elizaveta Ivanonova Belajeva obituary. *SVP News Bulletin* 129: 46.

TrowelBlazers. N.d. Dorothea Bate. The Star(key) of Bethlehem. Accessed March 4, 2020. https://trowelblazers.com/dorothea-bate/.

Turnbull, W.D., Zangerl, R. 2004. Barbara Jaffe Stahl obituary. *SVP News Bulletin* 187: 23–24.

Turner, M.A. 1975. *Amebelodon floridanus* (Leidy), a shovel-tusked gomphothere from Mixson's Bone Bed, Levy County, Florida. *Proceedings of the Nebraska Academy of Science and Affiliated Societies* 85: 44–45.

Turner, S. 1970. Timing of the Appalachian/Caledonian Orogen contraction. *Nature* 227(5253): 90.

Turner, S. 1973. Siluro-Devonian thelodonts from the Welsh Borderland. *Journal of the Geological Society of London* 129: 557–584.

Turner, S. 1986. Vertebrate palaeontology in Queensland. *Earth Sciences History* 5(1): 50–65.

Turner, S. 1990. Preface and introduction. In: Turner, S., Thulborn, R.A., Molnar, R.E. (eds.), Proceedings of the De Vis Symposium. *Memoirs of the Queensland Museum* 28(1): vi–viii.

Turner, S. 2007. Invincible but mostly invisible: Australian women's contribution to geology and palaeontology. In: Burek, C.V., Higgs, B. (eds.), *The Role of Women in Geology.* Special Publication 281. Geological Society, London, 165–202.

Turner, S. 2008. Dr Susan Turner (1946–): a peripathetic life and Judy Bracefield—scientific helpmate. *WiseNet Newsletter* 77(May): 15–17.

Turner, S. 2009. Reverent and exemplary: "dinosaur man" Friedrich von Huene (1875–1969). In: Kölbl-Ebert, M. (ed.), *Geology and Religion: Historical Views of an Intense Relationship between Harmony and Hostility.* Special Publication 310. Geological Society, London, 223–242.

Turner, S. 2020. Far-flung female (and bone-hunting) fellows. In: Burek, C.V., Higgs, B. (eds.), *Uncovering the historical contribution of women in the Geosciences: Celebrations of first female fellows of GSL.* Geological Society, London, Special Publication 506.

Turner, S., Berta, A. 2019. "Bone Hunters" Project—Australasian women in vertebrate paleontology. Brisbane Education and Outreach Poster Session, 79th Annual SVP Meeting, October 9–12, 2019, October Program and Abstracts, 209.

Turner, S., Berta, A. In press. Illustrating the unknowable: women paleoartists who drew ancient vertebrates. In: Clary, R.M., Rosenberg, G.D., Evans, D. (eds.), *The Evolution of Paleontological Art.* Geological Society of America special volume.

Turner, S., Burek, C.V., Moody, R.T.J. 2010a. Forgotten women in an extinct saurian (man's) world. In Moody, R.T.J., Buffetaut, J.E., Naish, D., Martill, D.M. (eds.), *Dinosaurs and Other Extinct Saurians: A Historical Perspective.* Special Publication 343. Geological Society, London, 111–153.

Turner, S., Burrow, C.J., Schultze, H.-P., Blieck, A. R.M., Reif, W.-E., Bultynck, P.B., Rexroad, C., Nowlan, G.S. 2010b. Conodont-vertebrate phylogenetic relationships revisited. *Geodiversitas* 32(4): 545–594.

Turner, S., Cadée, G.C. 2006. Dr Margaretha Brongersma-Sanders (1905–1996), Dutch scientist: an annotated bibliography of her work to celebrate 100 years since her birth. *Zoologische Mededelingen* 80: 183–204.

Turner, S., Long, J.A. 2016. The Woodward Factor: Arthur Smith Woodward's legacy to geology in Australia and Antarctica. In: Johansson, Z., Barrett, P.M., Richter, M., Smith, M. (eds.), *Arthur Smith Woodward: His Life and Influence on Modern Vertebrate Palaeontology.* Special Publication 430. Geological Society, London, 261–288.

Turner, S., Malakhova, I. 2010. IUGS-50: how the women faired [*sic*]. In: History of Research in Mineral Resources. SEDPGYM, EUPA, Ministry of Science and Innovation, Institute of Geology and Minerology of Spain, Madrid, Abstracts, 60.

Turner, S., Miles, R.S. In press. Erik Stensiö, Sweden's doyen of fossil fish. In: Talent, J.A. (ed.), *Biographies of Eminent Palaeontologists.* Springer, Berlin.

Turner, S., Thulborn, R.A. (eds.). 1987. De Vis Symposium, Problems in vertebrate biology and phylogeny: an Australian perspective, Brisbane, May 12–14, 1987. Queensland Museum, Brisbane.

Turner, S., Vickers-Rich, P. 2007. Sprigg, Glaessner and Wade and the discovery and international recognition of the Ediacaran fauna. In: Vickers-Rich, P., Komarower,

P. (eds.), *IGCP 493: The Rise and Fall of the Ediacaran Biota*. Special Publication 286. Geological Society, London, 443–445.

Turvey, S.T., Bruun, K., Ortiz, A., Hansford, J., Hu S.-M., Ding, Y., Zhang T.-N., Chaterjee, H. 2018. New genus of extinct Holocene gibbon associated with humans in imperial China. *Science* 360(6395): 1346–1349. https://doi.org/10.1126/science.aao4903.

Uhen, M.D., Barnosky, A.D., Bills, B., Blois, J., Carrano, M.T., Carrasco, M.A., Erickson, G.M., Eronen, J.T., Fortelius, M., Graham, R.W., Grimm, E.C., O'Leary, M.A., Mast, A., Piel, W.H., Polly, P.D., Säilä, L.K. 2013. From card catalogs to computers: databases in vertebrate paleontology. *Journal of Vertebrate Paleontology* 33(1): 13–28. http://dx.doi.org/10.1080/02724634.2012.716114.

UNESCO (United Nations Educational, Scientific and Cultural Organization). 2018. Women in science. Accessed August 26, 2018. http://uis.unesco.org/en/topic/women-science.

Valkova, O. 2008. The conquest of science: women and science in Russia, 1860–1940. *Osiris* 23(1): 136–165.

Van der Lee, R., Ellemers, N. 2015. Gender contributes to personal research funding success in the Netherlands. *Proceedings of the National Academy of Sciences* 112(4): 12349–12353.

Vanderloh, I. 1986. Story of Irene Coultis Vanderloh. In: *The Coultis Connection. By the Descendants of the Steveville Pioneers—Coultis, Davies and Shaw*. Colin Bate Books, Lethbridge, 143–167.

Van Miegroet, H., Glass, C., Callister, R.R., Sullivan, K. 2019. Unclogging the pipeline: advancement to full professor in academic STEM. *Equality, Diversity and Inclusion* 38 (20): 246–264.

Varker, J.W. 2004. Obituary: Dorothy Helen Rayner 1912–2002. *Proceedings of the Yorkshire Geological Society* 55(2): 160.

Vickers-Rich, P., Archbold, N. 1991. Squatters, priests and professors: a brief history of vertebrate palaeontology in *Terra Australis*. In: Vickers-Rich, P., Monaghan, J.M., Baird, R.F., Rich, T.H. (eds.), *Vertebrate Palaeontology of Australasia*. Pioneer Design Studios in cooperation with Monash University Publication Committee, Melbourne, 1–43.

Vickers-Rich, P., Rich, T.H. 1993–2019. *Dinosaurs of Darkness*. Vol. 2. Queen Victoria Museum, Launceston, Tasmania, and Monash Science Centre, Australia. Different editions.

Vincent, P., Taquet, P., Fischer, V., Bardet, N., Falconnet, J., Godefroit, P. 2014. Mary Anning's legacy to French vertebrate palaeontology. *Geological Magazine* 151: 7–20.

Voeten, D.F.A.E., Cubo, J., Margerie, E. de, Röper, M., Beyrand, V., Bureš, S., Tafforeau, P., Sanchez, S. 2018. Wing bone geometry reveals active flight in Archaeopteryx. *Nature Communications* 9(1). https://doi.org/10.1038/s41467-018-03296-8.

Vorobyeva, E.I., Obruchev, D.V. 1964. Class Osteichthyes, bony fishes. Sublass arcopterygii. In: Obruchev, D.V. (ed.), *Fundamentals of Paleontology*. XI. Agnatha, Pisces. Nauka, Moscow, 420–509.

Vrba, E.S. 1980. Evolution, species and fossils: how does life evolve? *South African Journal of Science* 76: 61–84.

Vrba, E.S. 1993. Turnover pulses, the Red Queen and related topics. *American Journal of Science* 293: 418–452.

Walker, K.R. (ed.). 1984. *The Evolution–Creation Controversy: Perspectives on Religion, Philosophy, Science and Evolution: A Handbook.* Proceedings of a symposium convened by Gastaldo, R.A., and Tanner, W.F. Paleontological Society (USA) Special Publication No. 1.

Walters, B., Kissinger, T. 2014. *Discovering Dinosaurs.* Applesauce Press, New York.

Warnock, R., Dunne, E., Giles, S., Saupe, E., Soul, L., Lloyd, G. 2019. Special Report: Are we reaching gender parity among Palaeontologists? *Palaeontological Association Newsletter* 103.

Warren, L. 1998. *Joseph Leidy: The Last Man Who Knew Everything.* Yale University Press, New Haven, CT.

Watson, D.M.S. 1946. Notice of death Dr Anna R. Hartmann-Weinberg. *SVP New Bulletin* 18: 20.

Weatherford, D. 1994. *American Women's History: An A to Z of People, Organisations, Issues, and Events.* Prentice Hall General Reference, New York.

Weinburg, S. 1999. *A Fish Caught in Time: The Search for the Coelacanth.* Fourth Estate, London.

Weishampel, D.B., Dodson, P., Osmolska, H. (eds.). 2004. *The Dinosauria*, second edition. University of California Press, Berkeley.

Weishampel, D.B., Reif, W.-E. 1984. The work of Franz Baron Nopsca (1877–1933): dinosaurs, evolution and theoretical tectonics. *Jahrbuch für Geologie* 127(2): 187–203.

Weishampel, D.B., White, N.M. (eds.). 2003. *The Dinosaur Papers, 1676–1906.* Smithsonian Institution, Washington, DC.

Weishampel, D.B., Young, L. 2001. *Dinosaurs of the East Coast.* Johns Hopkins University Press, Baltimore, MD.

Welzenbach, M. 1990. The dinosaur artist, painting prehistory. *Washington Post*, April 12, 1990. https://www.washingtonpost.com/archive/lifestyle/1990/04/12/the-dinosaur-artist-painting-prehistory/279acba7-bed8-4a51-a482-cb5a163c9c21/.

West, J.D., Jacquet, J., King, M.M., Correll, S.J., Bergstrom, C.T. 2013. The role of gender in scholarly authorship. *PLoS ONE* 8(7): e66212. https://doi.org/10.1371/journal.pone.0066212.

Whatley, R.L., Beck, A.L. 2016. A history of paleontology from Anning to Zanno and beyond. From a presentation at PaleoFest 2016, March 12, 2016. See http://vertpaleo.org/Society-News/Blog/Old-Bones-SVP-s-Blog/March-2016/Historic-Conference-Quite-Possibly-the-First-All-W.aspx.

Whitfield, G.R., Bramwell, C.D. 1971. Palaeoengineering. Birth of a new science. *New Scientist* 52: 202–205.

Whitmore, M., Kool, L. 1991/1996. Techniques used in preparation of terrestrial vertebrates. In: Vickers-Rich, P., Monaghan, J.M., Baird, R.F., Rich, T.H. (eds.), *Vertebrate Palaeontology of Australasia.* Pioneer Design Studios in cooperation with Monash University Publication Committee, Melbourne, 173–200.

Whybrow, P. 1985. A history of fossil collecting and preparation techniques. *Curator* 28: 5–26.

Whybrow, P.J. (ed.). 2000. *Travels with the Fossil Hunters.* Natural History Museum and Cambridge University Press, Cambridge.

Williams, E.E. 1953. Notes on a European tour April to June 1953. *SVP News Bulletin* 39: 22–25.

Wikipedia. 2019. Milo Hellman. Last updated December 4, 2019. https://en.wikipedia .org/wiki/Milo_Hellman.

Wikipedia. 2020. History of science and technology in the People's Republic of China. Last updated January 15, 2020. https://en.wikipedia.org/wiki/History_of_science _and_technology_in_the_People%27s_Republic_of_China.

Williams, G. 2004. Marjorie Eileen Doris Courtenay-Latimer Feb 1907—died East London May 2004. *The West, Perth*, June 5, 2004, Obituaries, 158.

Williams, S.L. 1995. SPNHC: the first ten years. *Collection Forum* 11(2): 69–81.

Williams, S.L., McLaren, S.B. 2005. SNPC: the second ten years. *Collection Forum* 19(1–2): 1–14.

Williams, W.M., Ceci, S.J. 2012. When scientists choose motherhood. *American Scientist* 100: 138–145.

Willis, P., Muirhead, J. 1987. The preparation and reconstruction of fossils. In: Hand, S.J., Archer, M. (eds.), *The Antipodean Ark.* Angus and Robertson, North Ryde, New South Wales, 11–14.

Williston, S.W. 1898. *The University Geological Survey of Kansas.* Volume 4, *Paleontology.* J.S. Parks, Topeka, KS.

Williston, S.W., Hay, O.P. 1903. The Society of Vertebrate Paleontologists of America. *Science* 18: 827–828.

Wilson, C. 2016. *Status of the Geoscience Workforce.* American Geosciences Institute, Alexandria, VA.

Wilson, C. 2017. *Status of Recent Geoscience Graduates.* American Geosciences Institute, Alexandria, VA.

Wilson, C. 2019. *Status of the Geoscience Workforce.* American Geosciences Institute, Alexandria, VA.

Wilson, J.A. 1990. The Society of Vertebrate Paleontology 1940–1990, a fifty-year retrospective. *Journal of Vertebrate Paleontology* 10(1): 1–39.

Wilson, M., Mienis, H.K. 2008. Paleontological legacy of Yael Chalifa. *Complete Mesoangler* 12(1): 2–4.

Witton, M.P. 2018. *The Paleoartist's Handbook.* Crowood Press, Wiltshire.

Wolsan, M. 2008. Ewa Barycka (1979–2008). *Acta Palaeontologica Polonica* 53(2): 284. https://doi.org/10.4202/app.2008.0216.

Women in Paleontology Directory. Undated. Accessed March 13, 2020. https://docs .google.com/forms/d/e/1FAIpQLSeGoQDZArml1EHTpnnnKHvmVWfS8Ynb66fl RDUQ8fDb3WZEig/viewform.

Wood, B. 1997. Mary Leakey 1913–96. *Nature* 385: 28.

Wood, H.E. 1945. Late Miocene beaver from southern Montanan. *American Museum Novitates* 1299: 1–6.

Wood, S. 1846. On the discovery of an alligator and of several new Mammalia in the Hordwell Cliff; with observations upon the geological phenomena of that locality. *London Geological Journal, and Record of Discoveries in British and Foreign Paleontology* 1: 1–7.

Woodward, A.S. 1889–1901. *Catalogue of Fossil Fishes*. Trustees of the British Museum of Natural History, London. 4 vols.

Woolston, C. 2016. Faking it. *Nature* 529: 555–557.

Worsley, L. 2017. *Jane Austen at Home: A Biography*. Hodder and Stoughton, London.

Wyse Jackson, P.N., Spencer Jones, M.E. 2007. The quiet workforce: the various roles of women in geological and natural history museum during the early to mid-1900s. In: Burek, C.V., Higgs, B. (eds.), *The Role of Women in the History of Geology*. Special Publication 281. Geological Society, London, 97–113.

Xiao, S., Yang, Q., Luo, Z.-X. 2010. A golden age of paleontology in China? A SWOT analysis. *Palaeontologia Electronica* 13.1.3E: 4pp. https://palaeo-electronica.org/2010_1/commentary/china.htm.

Xie, Y., Fang, M., Shauman, K. 2015. STEM education. *Annual Reviews of Sociology* 41: 331–357.

Zelenitsky, D.K. 2000. Dinosaur eggs of Asia and North America. *Journal of the Paleontological Society of Korea Special Publication* 4: 13–26.

Zhamoida, A.I., Rozanov, A.Yu. 2016. Hundred years of (the) Paleontological Society of Russia: key events in the history and prospects. *Paleontological Journal* 50(2): 109–116.

Zhang, Y.-P. 1989. A review of Middle and late Eocene mammalian faunas from China. *Acta Paleontologica Sinica* 28(5): 663–682.

For the correct understanding of Science, it is necessary for her to know history.

H.A. Brouwer, University of Amsterdam, 1929;
quoted in Kerkoff, 1993

General Works

Aldrich, M.L. 1982. Women in paleontology in the United States 1840–1960. *History of Geology* 1(1): 14–22.

Aldrich, M.L. 1990. Women in geology. In: Kass-Simon, G., Farnes, P. (eds.), *Women of Science: Righting the Record*. Indiana University Press, Bloomington, 42–71.

Bailey, M.J. 1994. *American Women in Science: A Biographical Dictionary*. ABC-CLIO, Denver, CO.

Buffetaut, E. 1987. *A Short History of Vertebrate Palaeontology*. Croom Helm, London.

Burek, C.V., Higgs, B. (eds.). 2007. *The Role of Women in the History of Geology*. Special Publication 281.Geological Society, London.

Burek, C., Higgs, B. eds. *Uncovering the historical contribution of women in the Geosciences: Celebrations of first female fellows of GSL*. Geological Society, London, Special Publication 506.

Haines, C.M.C., Stevens, H.M. 2001. *International Women in Science: A Biographical Dictionary to 1950*. ABC-CLIO, Santa Barbara, CA.

Holmes, M.A., O'Connell, S., Dutt, K. (eds.). 2015. *Women in the Geosciences: Practical, Positive Practices toward Parity*. American Geological Institute Special Publication 70. American Geophysical Union, Washington, DC, and John Wiley and Sons, Hoboken, NJ.

Ignotofsky, R. 2016. *Women in Science—50 Fearless Pioneers Who Changed the World*. Ten Speed Press, Berkeley, CA.

Johnson, B.A. (ed.). 2018. *Women and Geology: Who Are We, Where Have We Come from, and Where Are We Going?* Memoir 214. Geological Society of America, Boulder, CO.

Kass-Simon, G., Farnes, P. (eds.). 1990. *Women of Science: Righting the Record*. Indiana University Press, Bloomington.

Ogilvie, M.B. 1986. *Women in Science: Antiquity through the Nineteenth Century*. MIT Press, Cambridge, MA.

Ogilvie, M.B., Harvey, J.D. 2000. *The Biographical Dictionary of Women in Science: Pioneering Lives from Ancient Times to the Mid-20th Century.* Routledge, Philadelphia, PA, and Taylor and Francis, London.

Ogilvie, M.B., Meek, K.L. 1996/2018. *Women and Science: An Annotated Bibliography.* Routledge, London.

Rossiter, M.W. 1982. *Women Scientists in America: Struggles and Strategies to 1940.* Johns Hopkins University Press, Baltimore, MD.

Rossiter, M.W. 1984. *Women Scientists in America: Before Affirmative Action, 1940–1972.* Johns Hopkins University Press, Baltimore, MD.

Rossiter, M.W. 1995. *Women Scientists in America: Before Affirmative Action, 1940–1972.* Vol. 2. Johns Hopkins University Press, Baltimore, MD.

Rossiter, M.W. 2012. *Women Scientists in America: Forging a New World since 1972.* Johns Hopkins University Press, Baltimore, MD.

Sarjeant, W.A.S. 1978–1987. *Geologists and the History of Geology.* Robt E. Krieger, Malabar, FL. 3 vols. plus supplements.

Biographies, Autobiographies, and Novels

Chevalier, T. 2009. *Remarkable Creatures.* HarperCollins, London.

Elliot, A.B. 2000. *Charming the Bones: A Portrait of Margaret Matthew Colbert.* Kent State University Press, Kent, OH.

Emling, S. 2009. *The Fossil Hunter: Dinosaurs, Evolution, and the Woman Whose Discoveries Changed the World.* St. Martin's Press, New York.

Kielan-Jaworowska, Z. 1969. *Hunting for Dinosaurs.* MIT Press, Boston, MA.

Kielan-Jaworowska, Z. 2005. Autobiografia. Translated by Z. Jaworowska, edited by H.-D. Sues. *Kwartalnik Historii Nauki I Techniki* 50(1): 7–50.

Leakey, M. 1984. *Disclosing the Past: An Autobiography.* Weidenfeld and Nicholson, London.

Morell, V. 1995. *Ancestral Passions: The Leakey Family and the Quest for Humankind's Beginnings.* Simon and Schuster, New York.

Shindler, K. 2004. *Discovering Dorothea: The Life of the Pioneering Fossil Hunter Dorothea Bate, 1825–1892.* HarperCollins, London.

Wiffen, J. 1991. *Valley of the Dragons: The Story of New Zealand's Dinosaur Woman.* Random Century, Glenfield, New Zealand.

Children's Books (see Table 2.2 on Mary Anning literature)

Buzzeo, T. 2019. *When Sue Found Sue: Sue Hendrickson Discovers Her* T. rex. Abrams Books, New York.

Diehn, A. 2019. *Fossil Huntress: Mary Leakey, Paleontologist.* Illustrated by Katie Mazeika. Picture Book Biography. Nomad Press, White River Junction, VT.

Gold, M.E.L., West, A.R. 2017. *She Found Fossils.* CreateSpace Independent, Scotts Valley, CA.

Miller, C.C. 2016. *When I Grow Up: I'll Be a Paleontologist.* Reprint. RiverStream, Mankato, MN.

Porcello, D. 2017. *Dinogirl: Young Paleontologist,* second edition. CreateSpace, Scotts Valley, CA.

Shroff, V. 2019. *The Adventures of Padma and a Blue Dinosaur.* HarperCollins, Noida, Uttar Pradesh, India.

Strickler, B., McGilliss, A. 2017. *Daring to Dig: Women in American Paleontology.* Paleontological Research Institution, Ithaca, NY.

VP Bibliographies

Bacskai, J., Bartlett, S., Brajnikov, B., Munthe, K., Nichols, R. Matsumoto, M. 1968–1969, 1969–1970, 1970–1971. *Bibliography of Vertebrate Paleontology and Related Subjects.* Nos. 24–26. Society of Vertebrate Paleontology VP, Yale Peabody Museum with NSF grant, Chicago.

Camp, C.L., Nichols, R.H., Brajnikov, B., Fulton., E., Bacskai, J.A. (eds.). 1972. *Bibliography of Fossil Vertebrates, 1964–1968.* Memoir 134. Geological Society of America, Boulder, CO. (C. Camp and colleagues edited eight volumes of the *Bibliography of Fossil Vertebrates,* published by the Geological Society of American from 1940 to 1972; several volumes included women coeditors Rachel H. Nichols and Judy A. Bacskai.)

Camp, C.L., Vanderhoof, V.L. (eds.). 1940. *Bibliography of Fossil Vertebrates, 1928–1933.* Special Paper 27. Geological Society of America, Boulder, CO.

Clark, J. 1968. *Bibliography of Vertebrate Paleontology and Related Subjects, 1966–1967.* No. 22. Society of Vertebrate Paleontology, Chicago.

Gregory, J.T., Bacskai, J.A., Brajnikov, B., Munthe K. (eds.). 1973. *Bibliography of Fossil Vertebrates, 1969–1972.* Memoir 141. Geological Society of America, Boulder, CO. (J.T. Gregory and colleagues edited 18 volumes of the *Bibliography of Fossil Vertebrates* from 1969 to 1996. Women coeditors included Judy A. Bacskai, Laurie J. Bryant, Kathleen Munthe, Bonnie Rauscher, and Melissa Winans.)

Gregory, J.T., Bacskai, J.A., Shkurkin, G.V., Rauscher, B.H. (eds.). 1993. *Bibliography of Fossil Vertebrates, 1993.* Society of Vertebrate Paleontology, Chicago.

Gregory, J.T., Bacskai, J.A., Shkurkin, G.V., Rauscher, B.H. (eds.). 1996. *Bibliography of Fossil Vertebrates, 1996.* Society of Vertebrate Paleontology, Chicago.

Hay, O.P. 1902. *Bibliography and Catalogue of Fossil Vertebrata of North America.* Bulletin 179. US Geological Survey, Washington, DC.

Hay, O.P. 1929. *Second Bibliography and Catalogue of Fossil Vertebrata of North America.* Publication 390(1). Carnegie Institute of Washington, Washington, DC.

Holton, C. 1966. *Bibliography of Vertebrate Paleontology and Related Subjects.* No. 20. Society of Vertebrate Paleontology, Chicago.

Matsumoto, M. 1971. *Bibliography of Vertebrate Paleontology and Related Subjects, 1969–1970.* No. 25. Society of Vertebrate Paleontology, Chicago.

Matsumoto, M., Nichols, R. 1970. *Bibliography of Vertebrate Paleontology and Related Subjects, 1968–1969.* No. 24. Society of Vertebrate Paleontology, Chicago.

Nichols, R. 1960. *Bibliography of Vertebrate Paleontology and Related Subjects.* No. 14 [first to be numbered]. Society of Vertebrate Paleontology, Chicago. (Earlier ones in *SVP News Bulletin* back to 1943)

Nichols, R., Lyons, J.M. 1961. *Bibliography of Vertebrate Paleontology and Related Subjects.* No. 15. Society of Vertebrate Paleontology, Chicago.

Romer, A.S., Wright, N.E., Edinger, T., Frank R. van (eds.). 1962. *Bibliography of Fossil Vertebrates Exclusive of North America.* Memoir 87. Geological Society of America, Boulder, CO. (The two Hay volumes indexed North American VP literature, and the complementary Romer volume included non–North American VP literature.)

Zidek, J. 1981. 1980 literature on fossil vertebrates. *Journal of Vertebrate Paleontology* 1(3/4): 407–422.

Page numbers in *italics* denote figures, boxes, and tables.